Springer Proceedings in Mathematics & Statistics

Volume 349

This book series features volumes composed of selected contributions from workshops and conferences in all areas of current research in mathematics and statistics, including operation research and optimization. In addition to an overall evaluation of the interest, scientific quality, and timeliness of each proposal at the hands of the publisher, individual contributions are all refereed to the high quality standards of leading journals in the field. Thus, this series provides the research community with well-edited, authoritative reports on developments in the most exciting areas of mathematical and statistical research today.

More information about this series at http://www.springer.com/series/10533

Tim Hoffmann · Martin Kilian ·
Katrin Leschke · Francisco Martin
Editors

Minimal Surfaces: Integrable Systems and Visualisation

m:iv Workshops, 2016–19

Springer

Editors
Tim Hoffmann
Fakultät für Mathematik
Technical University Munich
Garching, Germany

Martin Kilian
School of Mathematical Sciences
University College Cork
Cork, Ireland

Katrin Leschke
School of Mathematics & Actuarial Science
University of Leicester
Leicester, UK

Francisco Martin
Department of Geometry & Topology
University of Granada
Granada, Spain

ISSN 2194-1009 ISSN 2194-1017 (electronic)
Springer Proceedings in Mathematics & Statistics
ISBN 978-3-030-68540-9 ISBN 978-3-030-68541-6 (eBook)
https://doi.org/10.1007/978-3-030-68541-6

Mathematics Subject Classification: 53-06, 58-06, 52-06, 49-06

This Springer imprint is published by the registered company Springer Nature Switzerland AG
The registered company address is: Gewerbestrasse 11, 6330 Cham, Switzerland

Introduction

Proceedings of the Workshop Series of *Minimal Surfaces: Integrable Systems and Visualisation*

The study of minimal surface in 3-space started with Euler and Lagrange in the eighteenth century. They studied the problem of determining a graph over a domain $\Omega \subset \mathbb{R}^2$ with the least possible area among all surfaces that assume given values on the boundary of Ω. Mathematicians soon realised that this was not only a problem of extraordinary difficulty but also of unlimited possibilities. Meusnier supplied a geometric interpretation of the minimal graph equation: the mean curvature H of the surface vanishes. This is the reason why it has become customary to use the term minimal surface for surfaces with vanishing mean curvature, independent on whether they minimise the area.

During the nineteenth century significant progress was made. From the point of view of model theory of minimal surfaces, Weierstrass' and Enneper's contributions are particularly important. They introduced the so-called Enneper–Weierstrass representation for minimal surfaces which establishes a close link between this theory and Complex Analysis. In the middle of the nineteenth century, the physicist Plateau observed that minimal surfaces can be physically realised as soap films. Thus, the problem of determining a minimal surface with fixed topology and with prescribed Jordan curve as boundary is now called Plateau's problem.

In the twentieth century, the theory of minimal surfaces has greatly advanced through contributions from other mathematical areas such as PDE theory, Complex Analysis, Algebraic Geometry, Geometric Measure Theory and Geometric Analysis. Very recently, the classification of minimal planar domains by Meeks, Perez and Ros has highlighted the value of the link between minimal surfaces and Integrable System Theory. The international network *Minimal surfaces: integrable systems and visualisation*, funded by the Leverhulme Trust, brought together researchers working on different aspects on these fields to generate synergies between their work.

The four workshops organised by the network all connected the last achievements in this field with techniques developed in two emerging fields, Geometric Flows and Discrete Geometry. The solution of the Poincare conjecture by Perelman using Ricci flow showed the importance of the concepts and methods of geometric flows to solve a wide variety of problems in mathematics and physics. On the other hand, Discrete Geometry plays an important role in Virtual Reality and Computer Graphics and has important applications in this field.

The workshop in Cork in winter 2017 concentrated around topics in surface theory in homogeneous 3-manifolds, the workshop in Granada in winter 2018 focused mainly on the global theory of minimal surfaces, the workshop in Munich in summer 2018 centred its attention around discrete geometry and visualisation and the final work workshop in Leicester in autumn 2019 combined all main research areas.

The proceedings of the network workshops collect important results and surveys on

1. Integrable systems in surface theory,
2. Discrete Differential Geometry,
3. Ricci Flow and Mean Curvature Flow,
4. Minimal surfaces in homogeneous 3-manifolds and
5. Constant Mean Curvature and Constant Gaussian Curvature surfaces.

The editors of these proceedings would like to express their gratitude to the Leverhulme Trust for supporting our work, and the participants of the workshops for helping to create the vibrant and productive atmosphere at the workshops and opening new research links between the various research topics.

March 2020

Tim Hoffmann
Martin Kilian
Katrin Leschke
Francisco Martin
Minimal Surfaces: Integrable Systems and Visualisation
International Network
Leverhulme Trust, IN-2016-019

Contents

Translating Solutions to Mean Curvature Flow

Theodora Bourni, Mat Langford, and Giuseppe Tinaglia

Abstract We describe our recent construction of a new family of translating solutions to mean curvature flow and discuss some implications for the construction of new convex ancient solutions. The full details appear in Bourni et al. (J Differ Geom, 2017) and Bourni et al. (Anal PDE 13:1051–1072, 2020).

Keywords Mean curvature flow · Translators · Ancient solutions

1 Introduction

A smooth one-parameter family $\{M_t^n\}_{t\in I}$ of smoothly immersed hypersurfaces M_t^n of $\mathbb{R}^{n+1}$ is a (classical) *solution to mean curvature flow* if there exists a smooth one parameter family $X : M^n \times I \to \mathbb{R}^{n+1}$ of immersions $X(\cdot, t) : M^n \to \mathbb{R}^{n+1}$ with $M_t^n = X(M^n, t)$ satisfying

$$\frac{\partial X}{\partial t}(x, t) = \vec{H}(x, t) \quad \text{for all} \quad (x, t) \in M^n \times I\,,$$

where $\vec{H}(\cdot, t)$ is the mean curvature vector field of $X(\cdot, t)$. The mean curvature flow is the L^2-gradient flow of the area functional and hence, roughly speaking, deforms a hypersurface in such a way as to decrease its area most rapidly.

An *ancient* solution to a geometric flow, such as mean curvature flow, is one which is defined on a time interval of the form $I = (-\infty, T)$, where $T \leq \infty$. A special class of ancient solutions are the *translating solutions*. As the name suggests, these are solutions $\{M_t^n\}_{t\in(-\infty,\infty)}$ which evolve by translation: $M_{t+s}^n = M_t^n + se$ for some fixed vector $e \in \mathbb{R}^{n+1}$. The timeslices M_t^n of a translating solution $\{M_t^n\}_{t\in(-\infty,\infty)}$ are all

T. Bourni · M. Langford
University of Tennessee, Knoxville TN 37996, US

G. Tinaglia (✉)
King's College London, The strand London, London, UK
e-mail: giuseppe.tinaglia@kcl.ac.uk

T. Hoffmann et al. (eds.), *Minimal Surfaces: Integrable Systems and Visualisation*, Springer Proceedings in Mathematics & Statistics 349,
https://doi.org/10.1007/978-3-030-68541-6_1

congruent and satisfy the *translator equation*, which asserts that the mean curvature vector of M_t^n is equal to the projection of e onto its normal bundle. Translating solutions arise as blow-up limits of type-II singularities along an essential blow-up sequence [4, 5, 12, 23]. Type-II singularities (and, more generally, translating solutions) are still not very well understood, except in certain special cases [3, 6, 13, 16–18, 26, 28, 29]. More general blow-up sequences yield more general ancient solutions. Understanding ancient and translating solutions is therefore important to many applications of the flow which require a controlled continuation of the flow through singularities.

Further interest in ancient and translating solutions to geometric flows arise from their rigidity properties, which are analogous to those of complete minimal surfaces, harmonic maps and Einstein metrics; for example, when $n \geq 2$, under certain geometric conditions—uniform convexity, bounded eccentricity, type-I curvature decay or bounded isoperimetric ratio, for instance—the only compact,[1] convex (or noncollapsing) ancient solutions to mean curvature flow are shrinking spheres [14, 19]. In fact, the convexity/noncollapsing condition can be weakened to a uniform bound for $|A|^2/H^2$, where A is the second fundamental form of the solution [20].

Compact ancient solutions to mean curvature flow are closely related to translating solutions. Indeed, if $\{M_t\}_{t\in(-\infty,0)}$ is a compact, convex ancient solution and $P(e,t) \in M_t$ satisfies $\nu(P(e,t),t) = e \in S^n$, then, by the differential Harnack inequality [2, 12], the translated flow $M_t^s := M_{t+s} - P(e,s)$ converges (in the smooth topology, uniformly on compact subsets) to a convex translating solution with velocity $-H_\infty e$, where $H_\infty \div \lim_{s\to-\infty} H(P(e,s),s)e$.

2 Translators

As we mentioned in the introduction, a translating solution of mean curvature flow is one which evolves purely by translation and, in that case, the time slices are all congruent and satisfy

$$H(x) = -\langle \nu(x), e\rangle \tag{1}$$

for some $e \in \mathbb{R}^{n+1}$, where ν is a choice of local unit normal field near x and $H = \operatorname{div} \nu$ is the corresponding mean curvature. Since we are interested in the classification problem, it is useful to eliminate the scaling invariance and isotropy of (1) by restricting attention to translating solutions which move with unit speed in the 'upwards' direction. That is, we henceforth assume that $e = e_{n+1}$. We will refer to a hypersurface $M^n \subset \mathbb{R}^{n+1}$ satisfying (1) with $e = e_{n+1}$ as a *translator*.

The most prominent example of a translator is the Grim Reaper curve, $\Gamma^1 \subset \mathbb{R}^2$, defined by

$$\Gamma^1 := \left\{(x, -\log\cos x) : |x| < \tfrac{\pi}{2}\right\}.$$

[1]We refer to a solution $\{M_t^n\}_{t\in I}$ to mean curvature flow as *compact, convex, embedded, etc.* if this is the case for each time slice M_t^n.

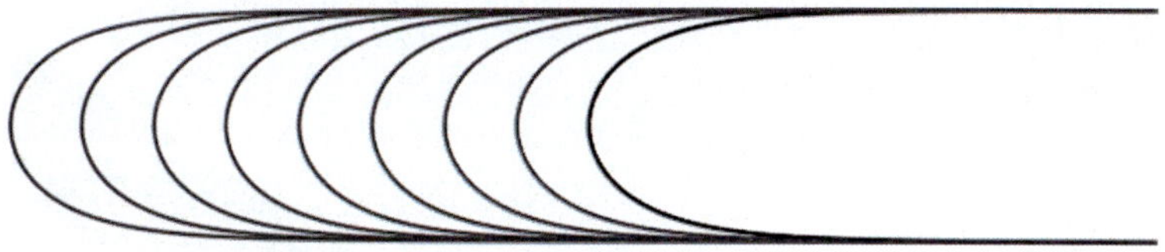

Fig. 1 The Grim Reaper translating to the right under curve shortening flow, killing every compact solution in its way (by the avoidance principle)

Taking products with lines then yields the Grim hyperplanes

$$\Gamma^n := \left\{(x_1, \ldots, x_n, -\log\cos x_1) : |x_1| < \tfrac{\pi}{2}\right\}.$$

The Grim hyperplane Γ^n lies in the slab $\{(x_1, \ldots, x_n) : |x_1| < \frac{\pi}{2}\}$ (and in no smaller slab). More generally, if M^{n-k} is a translator in $\mathbb{R}^{n-k+1}$ then $M^{n-k} \times \mathbb{R}^k$ is a translator in $\mathbb{R}^{n-k+1} \times \mathbb{R}^k \cong \mathbb{R}^{n+1}$ (Fig. 1).

There is also a family of 'oblique' Grim planes $\Gamma^n_{\theta,\phi}$ parametrized by $(\theta, \phi) \in [0, \frac{\pi}{2}) \times S^{n-2}$. These are obtained by rotating the 'standard' Grim plane Γ^n through the angle $\theta \in [0, \frac{\pi}{2})$ in the plane $\mathrm{span}\{\phi, e_{n+1}\}$ for some unit vector $\phi \in \mathrm{span}\{e_2, \ldots e_n\}$ and then scaling by the factor $\cos\theta$. To see that the result is indeed a translator, we need only check that

$$-H_\theta = -\cos\theta H = \cos\theta\,\langle \nu, e_{n+1}\rangle = \langle \cos\theta\nu + \sin\theta\phi, e_{n+1}\rangle = \langle \nu_\theta, e_{n+1}\rangle,$$

where H_θ and ν_θ are the mean curvature and unit normal to $\Gamma^n_{\theta,\phi}$ respectively. We also set $\Gamma^n_\theta := \Gamma^n_{\theta,e_2}$.

The oblique Grim hyperplane $\Gamma^n_{\theta,\phi}$ lies in the slab

$$\Sigma^{n+1}_\theta := \{(x_1, \ldots, x_{n+1}) : |x_1| < \tfrac{\pi}{2}\sec\theta\}$$

(and in no smaller slab). More generally, if M^{n-k} is a translator in $\mathbb{R}^{n-k+1}$ then the hypersurface $M^n_{\theta,\phi}$ obtained by rotating $M^{n-k} \times \mathbb{R}^k$ counterclockwise through angle θ in the plane $\phi \wedge e_{n+1}$ and then scaling by $\sec\theta$ is a translator in $\mathbb{R}^{n+1}$, so long as ϕ is a non-zero vector in $\mathrm{span}\{e_{n-k+1}, \ldots e_n\}$ (Fig. 2).

For each $n \geq 2$, Altschuler and Wu constructed an $O(n)$-invariant, convex, entire translating graph in $\mathbb{R}^{n+1}$ asymptotic to a paraboloid [1] (see also [10]). Wang proved that this solution is the only convex entire translator in $\mathbb{R}^3$ and constructed further convex entire examples in higher dimensions [28]. He also proved the existence of strictly convex translating solutions which lie in slab regions in $\mathbb{R}^{n+1}$ for all $n \geq 2$ and showed that these are the only possibilities:

Theorem 2.1 (Wang's dichotomy for translators [28]) *Every proper, convex translator is either entire or lies in a slab region.*

Fig. 2 The oblique Grim plane Γ_θ^2 with $\theta = \pi/6$. The translation direction is vertical

A major difficulty in the construction of solutions to the translator equation is to obtain curvature estimates for the Dirichlet problem for the graphical translator equation (note that convexity is not guaranteed). Wang sidesteps this problem by exploiting the Legendre transform and the existence of *convex* solutions of certain fully nonlinear equations. Unfortunately, this method loses track of the precise geometry of the domain on which the solution is defined and so it remained unclear exactly which slabs admit translators. On the other hand, there can exist no strictly convex translator in a slab of width less than or equal to π (the Grim hyperplane is a barrier).

Shahriyari [25] and Spruck–Xiao [27] obtained curvature estimates for graphical translators in $\mathbb{R}^3$ by exploiting their stability properties (cf. [24]). Using their curvature estimates, Spruck and Xiao were then able to deduce that every mean convex translator in $\mathbb{R}^3$ is actually weakly convex [27, Theorem 1.1]. Inspired by their work, we were able to obtain a complete resolution to the existence question in all dimensions.

Theorem 2.2 (Existence of convex translators in all admissible slabs [8]) *For every $n \geq 2$ and every $\theta \in (0, \frac{\pi}{2})$ there exists a strictly convex translator W_θ^n which lies in*

$$\Sigma_\theta^{n+1} := \{(x, y, z) \in \mathbb{R} \times \mathbb{R}^{n-1} \times \mathbb{R} : |x| < \tfrac{\pi}{2} \sec \theta\} \subset \mathbb{R}^{n+1}$$

and in no smaller slab.

Around the same time our work was completed, Hoffman, Ilmanen, Martìn and White provided an existence theorem for all slabs of width greater than π in the case $n = 2$ [15, Theorem 1.1]. They were also able to prove uniqueness in this case, thereby completing the classification of translating graphs in $\mathbb{R}^3$. Finally, they gave a different construction of examples of translating graphs in slabs in $\mathbb{R}^{n+1}$, extending an earlier construction of Ilmanen for the case $n = 2$ [15]. These solutions are parametrized by the vector of principal curvatures at the 'tip' (the unique point at which the downward unit normal is $-e_n$).

Let us briefly sketch the proof of Theorem 2.2. The idea is to take a limit of solutions to an appropriate sequence of Dirichlet problems. Since translators automatically satisfy $H \leq 1$, general methods of geometric measure theory can be used to obtain curvature estimates when $n \leq 6$. In order to obtain curvature estimates in higher dimensions, one needs to rule out singular (minimal) tangent cones. This can be achieved using the rotational symmetry hypothesis [9, 21, 22].

Proposition 2.3 (Curvature estimates up to the boundary for translating graphs [8]) *Given any $K > 0$ and $\ell \in \mathbb{N}$, there exists a constant $C_\ell < \infty$ with the following property: Let u be a solution to*

$$\operatorname{div}\left(\frac{Du}{\sqrt{1+|Du|^2}}\right) = \frac{1}{\sqrt{1+|Du|^2}} \quad \text{in } \Omega$$
$$u = \psi \quad \text{on } \partial\Omega,$$

with $\partial\Omega$ and ψ bounded in $C^{\ell_0,\alpha}$ by K for some $\ell_0 \geq 2$ and $\alpha \in (0, 1]$ (and rotationally symmetric with respect to the subspace $\mathbb{E}^{n-1} := \operatorname{span}\{e_2, \ldots, e_n\}$ if $n \geq 7$). Then

$$\sup_{p \in \operatorname{graph} u} |\nabla^\ell A(p)| \leq C_\ell \quad \text{for all } \ell \in \{0, \ldots, \ell_0 - 2\},$$

where A is the second fundamental form of graph u *and $\nabla^0 A := A$.*

Remark 2.4 In case $\ell_0 = 1$ we obtain uniform estimates in $C^{1,\alpha}$.

We emphasize that the estimates of Proposition 2.3 hold all the way to the boundary of Ω. This will be needed later.

We are then able to extend the convexity estimate of Spruck and Xiao to higher dimensions under the rotational symmetry hypothesis.

Proposition 2.5 *Let $M \subset \mathbb{R}^{n+1}$ be a mean convex translator with at most two distinct principal curvatures at each point and bounded norm of the second fundamental form. Then M is convex.*

In order to obtain the solution as a limit of solutions to Dirichlet problems, it then remains to obtain height estimates (to ensure that the limit is complete) and to rule out a 'width-drop' in the limit. We achieve this by constructing appropriate barriers: The function $\underline{u} : \{(x, y) \in \mathbb{R} \times \mathbb{R}^{n-1} : |x| < \frac{\pi}{2} \sec\theta\} \to \mathbb{R}$ defined by

$$\underline{u}(x, y) := -\sec^2\theta \log\cos\left(\tfrac{x}{\sec\theta}\right) + \tan^2\theta \log\cosh\left(\tfrac{|y|}{\tan\theta}\right)$$

is a subsolution to the graphical translator equation. It was discovered by modifying the arrival time of the Angenent oval so that it lies in the correct slab and is asymptotic to the correct oblique Grim hyperplanes.

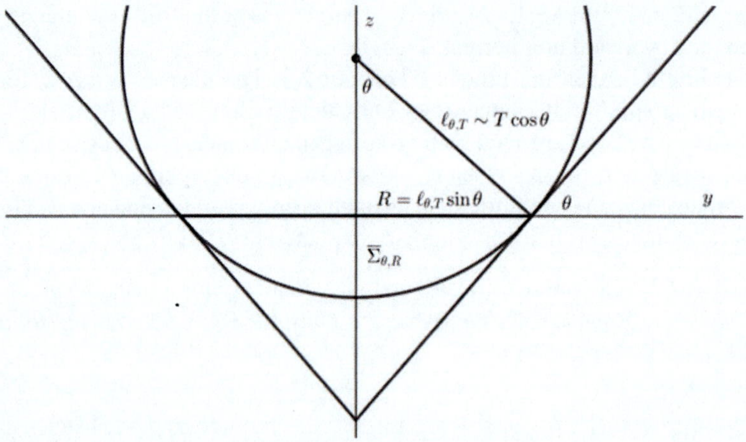

Fig. 3 Given any $\varepsilon \in (0, \varepsilon_0(n, \theta))$, the portion of the rotated time $T = \sec^2\theta\cosh\left(\frac{R}{\tan\theta}\right)$ slice of the Angenent oval of width $\pi\sec\theta$ lying below height $z = -R\frac{\cos(\theta-\varepsilon)}{\sin\theta}$ is a supersolution of the translator equation when $R > R_\varepsilon := \frac{2(n-1)}{\varepsilon}$

A suitable supersolution is obtained by rotating the Angenent oval of width $\pi\sec\theta$ and cutting off at an appropriate height (see Fig. 3).

Given $R > 0$, set

$$\underline{u}_R := \underline{u} - \tan^2\theta \log\cosh\left(\tfrac{R}{\tan\theta}\right)$$

and let u_R be the solution to

$$\begin{cases} \operatorname{div}\left(\dfrac{Du_R}{\sqrt{1+|Du_R|^2}}\right) = \dfrac{1}{\sqrt{1+|Du_R|^2}} & \text{in } \Omega_R \\ u_R = 0 \quad \text{on } \partial\Omega_R\,, \end{cases}$$

where Ω_R is the set of points where $\underline{u}_R < 0$. Since the equation admits upper and lower barriers (0 and $\underline{u}_R$, respectively), existence and uniqueness of a smooth solution follows from well-known methods (see, for example, [11, Chap. 15]). Uniqueness implies that u_R is rotationally symmetric with respect to the subspace $\mathbb{E}^{n-1} = \operatorname{span}\{e_2, \ldots, e_n\}$. Since $\underline{u}_R$ is a subsolution, its graph lies below graph u_R. Since the two surfaces coincide on the boundary $\partial\Omega_R$, the mean curvature H_R of graph u_R satisfies

$$\begin{aligned} H_R = -\langle \nu_R, e_{n+1}\rangle &\geq -\langle \underline{\nu}_R, e_{n+1}\rangle \\ &\geq \cos\theta\cos(x\cos\theta) \\ &\geq \cos\theta\left(1 - \frac{x}{\frac{\pi}{2}\sec\theta}\right) \end{aligned} \tag{2}$$

on $\partial\Omega_R$, where $\underline{\nu}_R$ is the downward pointing unit normal to graph $\underline{u}_R$. On the other hand, using the ancient pancake as an upper barrier yields

$$-u_R(0) \gtrsim \frac{1-\cos\theta}{\sin\theta} R \to \infty \quad \text{as} \quad R \to \infty\,. \tag{3}$$

Let $R_i \to \infty$ be a diverging sequence and consider the translators-with-boundary

$$M_i := \operatorname{graph} u_{R_i} - u_{R_i}(0)e_{n+1}\,.$$

By Proposition 2.3 and the height estimate (3), some subsequence converges locally uniformly in the smooth topology to some limiting translator, M, with bounded second fundamental form. By Proposition 2.5, M is convex.

Certainly M lies in the slab Σ^{n+1}_θ, so it remains only to prove that it lies in no smaller slab (strict convexity will then follow from the splitting theorem and uniqueness of the Grim Reaper). Set

$$v := 1 - \frac{x}{\frac{\pi}{2}\sec\theta}\,,$$

where $x(X) := \langle X, e_1\rangle$. We claim that

$$\inf_{M\cap\{x>0\}} \frac{H}{v} > 0\,. \tag{4}$$

Since $\inf_M H = 0$, we conclude that $\sup_M x = \frac{\pi}{2}\sec\theta$ as desired. To prove (4), first observe that

$$-(\Delta + \nabla_V)v = 0$$

and hence

$$-(\Delta + \nabla_V)\frac{H}{v} = |A|^2\frac{H}{v} + 2\left\langle \nabla\frac{H}{v}, \frac{\nabla v}{v}\right\rangle,$$

where V is the tangential projection of e_{n+1}. The maximum principle then yields

$$\begin{aligned}\min_{M_i\cap\{x>0\}} \frac{H}{v} &\geq \min\left\{\min_{\partial M_i\cap\{x>0\}}\frac{H}{v}, \min_{M_i\cap\{x=0\}}\frac{H}{v}\right\}\\ &= \min\left\{\cos\theta, \min_{M_i\cap\{x=0\}} H\right\}.\end{aligned}$$

If $\liminf_{i\to\infty}\min_{M_i\cap\{x=0\}} H > 0$ then we are done. So suppose that $\liminf_{i\to\infty} H(X_i) = 0$ along some sequence of points $X_i \in M_i \cap \{x = 0\}$. Then, by Proposition 2.3, after passing to a subsequence, the translators-with-boundary

$$\hat{M}_i := M_i - X_i$$

converge locally uniformly in C^∞ to a translator (possibly with boundary) $\hat{M}$ which lies in the slab Σ_θ^{n+1} and satisfies $H \geq 0$ with equality at the origin. By Proposition 2.3 the origin must be an interior point since, recalling (2), $H > \cos\theta$ on $\partial M_i \cap \{x = 0\}$ for all i. The strong maximum principle then implies that $H \equiv 0$ on $\hat{M}$ and we conclude that $\hat{M}$ is either a hyperplane or half-hyperplane. Since, by the reflection symmetry, the limit cannot be parallel to $\{0\} \times \mathbb{R}^{n-1} \times \mathbb{R}$, neither option can be reconciled with the fact that $\hat{M}$ lies in Σ_θ^{n+1}. This proves that the width cannot drop in the limit, and with it Theorem 2.2.

As was the case for ancient pancakes, these solutions necessarily converge to oblique Grim hyperplanes (of width potentially smaller than that of the original slab) after translation parallel to the slab. The following theorem shows that the asymptotic Grim hyperplanes are of full width.

Theorem 2.6 (Unique asymptotics and reflection symmetry [8, 27]) *Given $n \geq 2$ and $\theta \in (0, \frac{\pi}{2})$ let M_θ^n be a convex translator which lies in the slab Σ_θ^{n+1} and in no smaller slab. If $n \geq 3$, assume in addition that M_θ^n is rotationally symmetric with respect to the subspace $\mathbb{E}^{n-1} := \text{span}\{e_2, \ldots, e_n\}$. Given any unit vector $e \in \mathbb{E}^{n-1}$, the curve $\{\sin\omega e - \cos\omega e_{n+1} : \omega \in [0, \theta)\}$ lies in the normal image of M_θ^n and the translators*

$$M_{\theta,\omega}^n := M_\theta^n - P(\sin\omega e - \cos\omega e_{n+1})$$

converge locally uniformly in the smooth topology to the oblique Grim hyperplane $\Gamma_{\theta,e}^n$ as $\omega \to \theta$, where $P : S^n \to M_e^n$ is the inverse of the Gauss map.

Moreover, M_θ^n is reflection symmetric across the hyperplane $\{0\} \times \mathbb{R}^n$.

This result was already obtained by Spruck and Xiao when $n = 2$ using different methods [27]. Note that the translators we construct in Theorem 2.2 satisfy the hypotheses of Theorem 2.6.

It would be useful to have a better understanding of the location of the point $P(\sin\omega e - \cos\omega e_{n+1})$. For example, it remains unclear whether or not the 'flying wing' solution constructed in [8] lies above a (translated) oblique Grim hyperplane $\Gamma_\theta^n - Ce_{n+1}$. This information will be of use in constructing new examples of ancient and translating solutions.

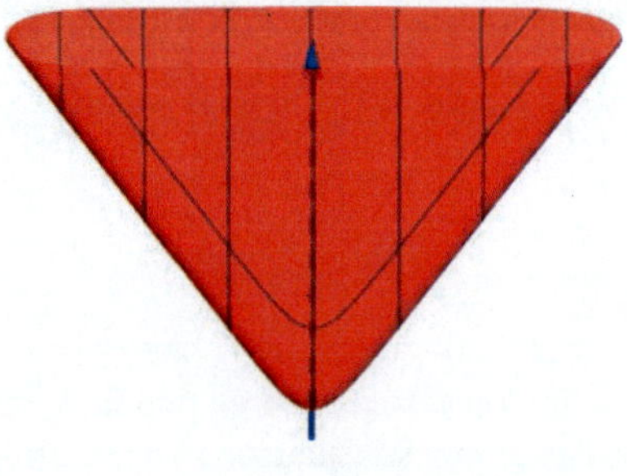

The 'flying wing' W_θ^2 of [8] with $\theta = \frac{\pi}{4}$ (right). The translation direction is vertical.

The rotational symmetry hypothesis—which is not required when $n = 2$—may be necessary in higher dimensions. We note that, in higher dimensions, 'oblique' products of lower dimensional wing families with flat directions provide additional possible asymptotics for higher dimensional translators, so the description of higher dimensional translators is therefore to be far more complex. It is conceivable that there exist convex translators in the slab $\Sigma^4_\theta \subset \mathbb{R}^4$, for example, which are asymptotic to an 'oblique' $M^2_\theta \times \mathbb{R}$, where $M^2_\theta \subset \mathbb{R}^3$ is the translator from Theorem 3.1.

3 Ancient Solutions with Discrete Symmetry Groups

The Angenent oval provides an example of a compact, convex ancient solution to mean curvature flow that lies on a slab.

We shall refer to an ancient solution that satisfies these hypotheses as an *ancient pancake*. In higher dimensions, Xu-Jia Wang has constructed ancient pancakes in $\mathbb{R}^{n+1}$ by taking a limit of solutions to the Dirichlet problem for the level set flow [28].

Recently, we have provided a different construction of an $O(1) \times O(n)$-invariant ancient pancake, including a precise description of its asymptotics using methods that are rather different from Wang's.

Theorem 3.1 (Existence of ancient pancakes [7] (cf. [28])) *There exists a compact, convex, $O(1) \times O(n)$-invariant ancient solution $\{M^n_t\}_{t\in(-\infty,0)}$ to mean curvature flow in $\mathbb{R}^{n+1}$ which lies in the stationary slab $\Sigma := \{x \in \mathbb{R}^{n+1} : |x_1| < \frac{\pi}{2}\}$ and has the following properties.*

(1a) $\{\lambda M_{\lambda^{-2}t}\}_{t\in(-\infty,0)}$ *converges uniformly in the smooth topology to the shrinking sphere* $S^n_{\sqrt{-2nt}}$ *as* $\lambda \to 0$,
(1b) $\{M_{t+s}\}_{t\in(-\infty,-s)}$ *converges locally uniformly in the smooth topology to the stationary solution* $\partial\Sigma$ *as* $s \to -\infty$, *and*
(1c) for any unit vector $e \in \{e_1\}^\perp$, $\{M_{t+s} - P(e,s)\}_{t\in(-\infty,-s)}$ *converges locally uniformly in the smooth topology as* $s \to -\infty$ *to the Grim hyperplane (see Sect. 2) which translates with unit speed in the direction* e, *where, given any* $v \in S^n$, $P(v,t)$ *denotes the unique point of* M^n_t *with outward pointing unit normal* v *(Fig. 4).*

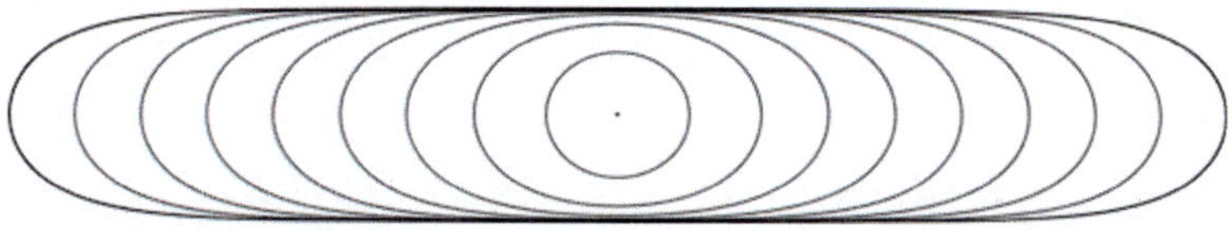

Fig. 4 The Angenent oval solution to curve shortening flow. It shrinks to a round point as $t \to 0$ and sweeps out a strip of width π as $t \to -\infty$

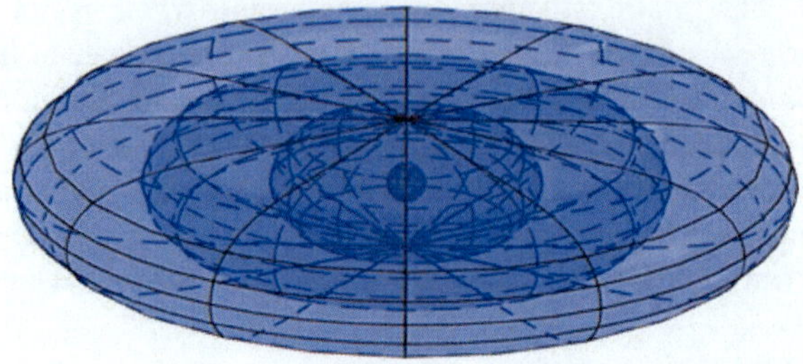

Fig. 5 The rotationally symmetric ancient pancake. It shrinks to a round point as $t \to 0$ and sweeps out a slab of width π as $t \to -\infty$

Moreover, as $t \to -\infty$,

(2a) $\min_{M_t} H = H(P(e_1, t)) \leq o\left(\frac{1}{(-t)^k}\right)$ *for any $k > 0$,*

(2b) $\min_{p \in M_t} |p| = |P(e_1, t)| \geq \frac{\pi}{2} - o\left(\frac{1}{(-t)^k}\right)$ *for any $k > 0$ and*

(3a) $\max_{M_t} H = H(P(\varphi, t)) \geq \left(1 + \frac{n-1}{-t} + o\left(\frac{1}{(-t)^{2-\varepsilon}}\right)\right)$ *for any unit vector $\varphi \in \{e_1\}^{\perp}$ and any $\varepsilon > 0$, and*

(3b) $\max_{p \in M_t} |p| = |P(\varphi, t)| = -t + (n-1)\log(-t) + C + o(1)$ *for any unit vector $\varphi \in \{e_1\}^{\perp}$, where $C \in \mathbb{R}$ is some constant.*

One of our motivations for studying translators lying in slab regions is to study ancient pancake solutions which are not necessarily rotationally symmetric. Based on the description of the translators with $O(1) \times O(n-1)$-symmetry contained in the previous section, the following conjecture appears natural (Fig. 5).

Conjecture 3.2 (*Dihedral pancakes*) Given any $k \geq 3$ there exists an ancient pancake lying in the slab $(-\frac{\pi}{2}\sec\frac{\pi}{k}, \frac{\pi}{2}\sec\frac{\pi}{k}) \times \mathbb{R}^2 \subset \mathbb{R}^3$ (and in no smaller slab) with symmetry group $O(1) \times D_k$, where D_k is the symmetry group of the regular k-sided polygon. Modulo translations and rotations, this is the unique such solution. Let $\{\phi_i\}_{i=1}^k \subset \mathbb{C} \cong \{0\} \times \mathbb{R}^2$ be the kth roots of unity in the $\{x_1 = 0\}$ plane. Up to a rotation, the solution has the following asymptotics: Given a unit vector $\phi \in \{0\} \times \mathbb{R}^2$, the asymptotic translator in the ϕ-direction is the oblique Grim plane $\Gamma^2_{\frac{\pi}{k}}$, except when $\phi \in \{\phi_i\}_{i=1}^k$, in which case the asymptotic translator is the flying wing translator $W^2_{\frac{\pi}{k}}$.

Generalizing these principles leads to the following natural question (Fig. 6).

Question 3.3 *Do there exist translators contained in slab regions of $\mathbb{R}^{n+1}$ with symmetry groups $O(1) \times G^{n-1}$, where G^{n-1} is the symmetry group of a regular $(n-1)$-polytope?*

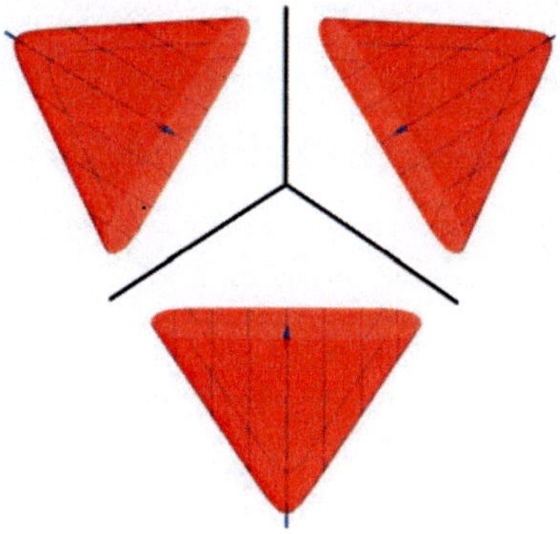

Fig. 6 Gluing three flying wing translators at infinity to form a compact ancient solution

References

1. Altschuler, S.J., Wu, L.F.: Translating surfaces of the non-parametric mean curvature flow with prescribed contact angle. Calc. Var. Part. Differ. Equ. **2**(1), 101–111 (1994)
2. Andrews, B.: Harnack inequalities for evolving hypersurfaces. Math. Z. **217**(2), 179–197 (1994)
3. Andrews, B.: Noncollapsing in mean-convex mean curvature flow. Geom. Topol. **16**(3), 1413–1418 (2012)
4. Angenent, S.B., Velázquez, J.J.L.: Asymptotic shape of cusp singularities in curve shortening. Duke Math. J. **77**(1), 71–110 (1995)
5. Angenent, S.B., Velázquez, J.J.L.: Degenerate neckpinches in mean curvature flow. J. Reine Angew. Math. **482**, 15–66 (1997)
6. Bourni, T., Langford, M.: Type-II singularities of two-convex immersed mean curvature flow. Geom. Flows **2**, 1–17 (2017)
7. Bourni, T., Langford, M., Tinaglia, G.: Collapsing ancient solutions of mean curvature flow. To appear in J. Differ. Geom. (2017). Preprint available at arXiv:1705.06981
8. Bourni, T., Langford, M., Tinaglia, G.: On the existence of translating solutions of mean curvature flow in slab regions. Anal. PDE **13**(4), 1051–1072 (2020)
9. Brito, F., Leite, M.L.: A remark on rotational hypersurfaces of S^n. Bull. Soc. Math. Belg. Sér. B **42**(3), 303–318 (1990)
10. Clutterbuck, J., Schnürer, O.C., Schulze, F.: Stability of translating solutions to mean curvature flow. Calc. Var. Part. Differ. Equ. **29**(3), 281–293 (2007)
11. Gilbarg, D., Trudinger, N.S.: Elliptic partial differential equations of second order. Reprint of the 1998 edition. Springer, Berlin (2001)
12. Hamilton, R.S.: Harnack estimate for the mean curvature flow. J. Differ. Geom. **41**(1), 215–226 (1995)
13. Haslhofer, R.: Uniqueness of the bowl soliton. Geom. Topol. **19**(4), 2393–2406 (2015)
14. Haslhofer, R., Hershkovits, O.: Ancient solutions of the mean curvature flow. Commun. Anal. Geom. **24**(3), 593–604 (2016)
15. Hoffman, D., Ilmanen, T., Martin, F., White, B.: Graphical translators for mean curvature flow. Preprint arXiv:1805.10860
16. Huisken, G., Sinestrari, C.: Convexity estimates for mean curvature flow and singularities of mean convex surfaces. Acta Math. **183**(1), 45–70 (1999)
17. Huisken, G., Sinestrari, C.: Mean curvature flow singularities for mean convex surfaces. Calc. Var. Part. Differ. Equ. **8**(1), 1–14 (1999)
18. Huisken, G., Sinestrari, C.: Mean curvature flow with surgeries of two-convex hypersurfaces. Inven. Math. **175**(1), 137–221 (2009)

19. Huisken, G., Sinestrari, C.: Convex ancient solutions of the mean curvature flow. J. Differ. Geom. **101**(2), 267–287 (2015)
20. Langford, M.: A general pinching principle for mean curvature flow and applications. Calc. Var. Part. Differ. Equ. **56**(4), 107 (2017). Jul
21. Ôtsuki, T.: Minimal hypersurfaces in a Riemannian manifold of constant curvature. Am. J. Math. **92**, 145–173 (1970)
22. Ôtsuki, T.: On integral inequalities related with a certain nonlinear differential equation. Proc. Jpn. Acad. **48**, 9–12 (1972)
23. Savas-Halilaj, A., Smoczyk, K.: Lagrangian mean curvature flow of Whitney spheres. Preprint arXiv:1802.06304
24. Schoen, R.: Estimates for stable minimal surfaces in three-dimensional manifolds. In: Seminar on Minimal Submanifold. Annals of Mathematics Studies, vol. 103, pp. 111–126. Princeton University Press, Princeton (1983)
25. Shahriyari, L.: Translating graphs by mean curvature flow. Geom. Dedicata **175**, 57–64 (2015)
26. Sheng, W., Wang, X.-J.: Singularity profile in the mean curvature flow. Methods Appl. Anal. **16**(2), 139–155 (2009)
27. Spruck, J., Xiao, L.: Complete translating solitons to the mean curvature flow in $\mathbb{R}^3$ with nonnegative mean curvature. Preprint arXiv:1703.01003
28. Wang, X.-J.: Convex solutions to the mean curvature flow. Ann. Math. (2) **173**(3), 1185–1239 (2011)
29. White, B.: The nature of singularities in mean curvature flow of mean-convex sets. J. Am. Math. Soc. **16**(1), 123–138 (electronic) (2003)

Finding Conformal and Isometric Immersions of Surfaces

Check for updates

Albert Chern, Felix Knöppel, Franz Pedit, Ulrich Pinkall, and Peter Schröder

Abstract We introduce a family of variational functionals for spinor fields on a compact Riemann surface M that can be used to find close-to-conformal immersions of M into $\mathbb{R}^3$ in a prescribed regular homotopy class. Numerical experiments indicate that, by taking suitable limits, minimization of these functionals can also yield piecewise smooth isometric immersions of a prescribed Riemannian metric on M.

Keywords Isometric and conformal immersions · Variational problems · Non-linear Dirac equation

1 Introduction

The notion of an abstract Riemannian manifold raises the question of whether every such manifold can be isometrically realized as a submanifold of Euclidean space. This problem has been given an affirmative answer in the smooth category by Nash [21], provided that the codimension of the submanifold is sufficiently large. If one asks the more specific question of whether a given 2-dimensional Riemannian manifold (M, g) can be isometrically immersed into Euclidean 3-space, not too much is known. There are general local existence results for real analytic metrics [28] and for smooth metrics under certain curvature assumptions [19]. Non-existence results are easier to come by: for instance, Hilbert's classical result that the hyperbolic plane does not admit an isometric immersion into $\mathbb{R}^3$, or the fact that a compact

A. Chern · F. Knöppel · U. Pinkall
Institut für Mathematik, Technische Universität Berlin, 10623 Berlin, Germany

F. Pedit (✉)
Department of Mathematics and Statistics, University of Massachusetts Amherst, Amherst, MA 01003, USA
e-mail: pedit@math.umass.edu

P. Schröder
California Institute of Technology, Pasadena, CA 91125, USA

T. Hoffmann et al. (eds.), *Minimal Surfaces: Integrable Systems and Visualisation*, Springer Proceedings in Mathematics & Statistics 349,
https://doi.org/10.1007/978-3-030-68541-6_2

non-positively curved 2-dimensional Riemannian manifold cannot be isometrically immersed into $\mathbb{R}^3$.

Surprisingly though, if one relaxes the smoothness of the immersion, every 2-dimensional Riemannian manifold (M, g) admits a C^1-isometric immersion $f: M \to \mathbb{R}^3$ into Euclidean space [9, 16, 20]. Unfortunately, neither the original existence proofs nor the recent explicit constructions of such isometric immersions [1, 2] reflect much of the underlying geometry of (M, g), as shown in Fig. 3.

On the other hand, there are piecewise linear embeddings of a flat torus which make visible its intrinsic geometry (see Fig. 2). In a more general vein, one could attempt to find isometric immersions in the class of piecewise smooth immersions $f: M \to \mathbb{R}^3$, that is, local topological embeddings whose restrictions to the closed faces of a triangulation of M are smooth. Experiments carried out with a recently developed numerical algorithm [4] provide support of the following

Conjecture *Given a Riemannian surface (M, g), there exists a piecewise smooth isometric immersion $f: M \to \mathbb{R}^3$ in each regular homotopy class.*

The added detail—to realize a given intrinsic geometry within a prescribed regular homotopy class—is advantageous in applications to computer graphics [4] and also for the theoretical approach to the isometric immersion problem. It is the latter which will be discussed in this paper. Our objective is to rephrase the isometric immersion problem of an oriented Riemannian surface (M, g) into Euclidean space $\mathbb{R}^3$ as a variational problem with parameters whose minima, if they were to exist, converge (for limiting parameter values) to isometric immersions $f: M \to \mathbb{R}^3$ in a given regular homotopy class. As was pointed out already, for a generic metric g there will be no smooth isometric immersion into $\mathbb{R}^3$, let alone one within a prescribed regular homotopy class. But experiments with the aforementioned algorithm [4] give some credence to our conjecture that there should be minima in the larger class of piecewise smooth immersions. Adjusting the parameters in our functional, the Willmore energy $\int H^2$, the averaged squared mean curvature of the immersion, is one of its contributors and hence immersions close to a minimizer will avoid excessive creasing. This has the effect that potential minimizers of our functional reflect the intrinsic geometry of (M, g) well, in contrast to the C^1-isometric immersions by Nash and Kuiper [9, 16, 20].

In order to explain our approach in more detail, we first relax the original problem to that of finding a *conformal* immersion $f: M \to \mathbb{R}^3$ of a compact *Riemann surface* M in a given regular homotopy class. It is known that such a conformal immersion always exists in the smooth category [8, 26], and hence our variational problem will have a minimizer if we turn off the contribution from the Willmore energy. But keeping the Willmore energy in the functional has the effect that potential minimizers will minimize the Willmore energy in a given conformal and regular homotopy class, that is, will be constrained Willmore minimizers. There are partial characterizations of constrained Willmore minimizers when M has genus one [10, 18, 22], but hardly anything—besides existence [18] if the Willmore energy is below 8π—is known in higher genus, even though there are some conjectures [11]. One of our future goals

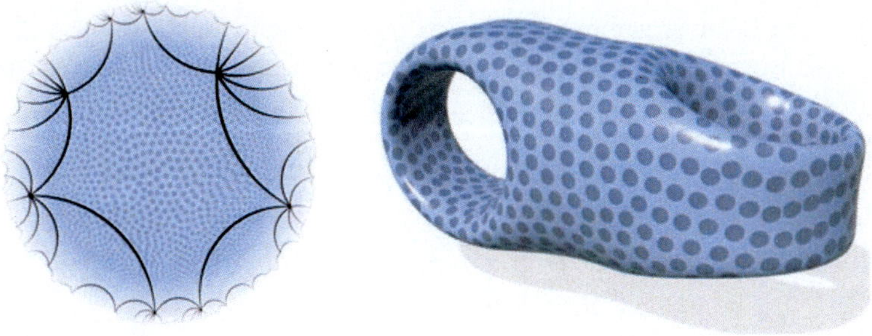

Fig. 1 A smooth, almost isometric immersion of the Riemannian surface of genus 2 with constant curvature shown on the left found by the algorithm in [4]

is to develop a discrete algorithm based on the approach outlined in this paper to find conformal immersions of a compact Riemann surface in a fixed regular homotopy class minimizing the Willmore energy.

Given a (not necessarily conformal) immersion $f\colon M \to \mathbb{R}^3$, we can decompose its derivative $df \in \Omega^1(M, \mathbb{R}^3)$ uniquely into $df = \omega \circ B$ where $\omega \in \Gamma(\mathrm{Conf}(TM, \mathbb{R}^3))$ is a conformal, nowhere vanishing $\mathbb{R}^3$-valued 1-form and $B \in \Gamma(\mathrm{End}(TM))$ is a positive, self-adjoint (with respect to any conformal metric) endomorphism with $\det B = 1$. Obviously $B = \mathrm{id}$ if and only if f is conformal. The space of conformal 1-forms $\mathrm{Conf}(TM, \mathbb{R}^3)$ is a principal bundle with stretch rotations $\mathbb{R}_+\mathbf{SO}(3)$ as a structure group acting from the left. In particular, any two $\omega, \tilde{\omega} \in \Gamma(\mathrm{Conf}(TM, \mathbb{R}^3))$ are related via $\tilde{\omega} = h\,\omega$ for a unique $h\colon M \to \mathbb{R}_+\mathbf{SO}(3)$. Two immersions $f, \tilde{f}\colon M \to \mathbb{R}^3$ are regularly homotopic if and only if their derivatives df and $d\tilde{f}$ are homotopic [13, 27]. The space of positive, self-adjoint bundle maps $B \in \Gamma(\mathrm{End}(TM))$ with $\det B = 1$ is contractible, and we obtain the equivalent reformulation that $f, \tilde{f}\colon M \to \mathbb{R}^3$ are regularly homotopic if and only if their corresponding ω and $\tilde{\omega}$ are homotopic in $\Gamma(\mathrm{Hom}(TM, \mathbb{R}^3))$.

As will be detailed in Sect. 2, any nowhere vanishing $\omega \in \Gamma(\mathrm{Conf}(TM, \mathbb{R}^3))$ induces a spin bundle $L \to M$ and $\omega = (\psi, \psi)$ for a unique (up to sign) nowhere vanishing section $\psi \in \Gamma(L)$ where $(\cdot,\cdot)\colon L \times L \to \mathrm{Hom}(TM, \mathbb{H})$ denotes the spin pairing. Since homotopic $\omega \in \Gamma(\mathrm{Conf}(TM, \mathbb{R}^3))$ give rise to isomorphic spin bundles, we obtain a description of regular homotopy classes of immersions via isomorphism classes of their induced spin bundles $L \to M$. If the genus of M is p, there are 2^{2p} many non-isomorphic spin bundles and hence 2^{2p} many regular homotopy classes of immersions $f\colon M \to \mathbb{R}^3$ (see Fig. 5).

Fixing a regular homotopy class, that is, a spin bundle $L \to M$, our aim is to find a nowhere vanishing section $\psi \in \Gamma(L)$ in such a way that the $\mathbb{R}^3$-valued conformal 1-form $\omega = (\psi, \psi) \in \Gamma(\mathrm{Conf}(TM, \mathbb{R}^3))$ is exact. In this case, the primitive $f\colon M \to \mathbb{R}^3$ of $\omega = df$ is a conformal immersion in the given regular homotopy class. We show (see also [7, 24]) that the closedness of $\omega = (\psi, \psi)$ is equivalent to the nonlinear Dirac equation

$$\bar{\partial}\psi + \tfrac{1}{2} H J \psi(\psi, \psi) = 0\,, \tag{1}$$

where $\bar{\partial}$ is the Dirac structure (see Lemma 1) on the spin bundle L. The function $H: M \to \mathbb{R}$ is the mean curvature, calculated with respect to the induced metric $|df|^2$, of the resulting conformal immersion $f: \tilde{M} \to \mathbb{R}^3$ on the universal cover with translation periods.

As we shall discuss in Sect. 3, the Dirac equation (1) can be given a variational characterization: for non-negative coupling constants $\epsilon = (\epsilon_1, \epsilon_2, \epsilon_3)$, we consider the family of variational problems $E_\epsilon : \Gamma(L^\times) \to \mathbb{R}$ on nowhere vanishing sections of L given by

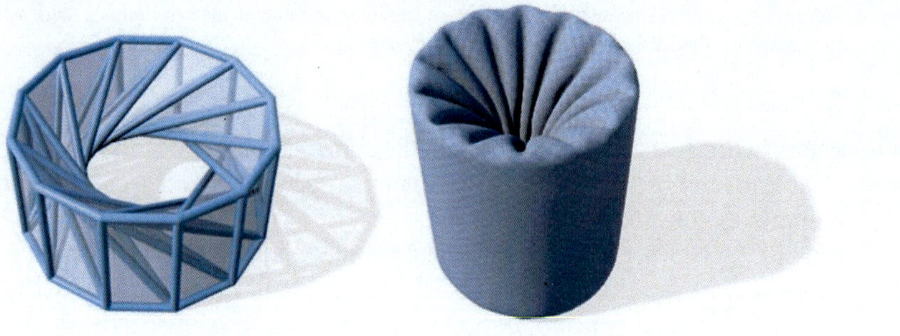

Fig. 2 The angular defects of the embedded, piecewise linear torus shown on the left are the same at all vertices and their sum is zero, implying that the induced metric is flat. The right image shows a smooth, almost isometric immersion of another flat torus found by the algorithm in [4]

$$E_\epsilon(\psi) = \epsilon_1 \int_M \frac{\langle *\bar{\partial}\psi \wedge \bar{\partial}\psi\rangle}{|\psi|^2} + (\epsilon_2 - \epsilon_1) \int_M \frac{\langle *\bar{\partial}\psi \wedge \psi(\psi,\psi)\rangle^2}{|\psi|^4} + (\epsilon_3 - \epsilon_1) \int_M \frac{\langle *\bar{\partial}\psi \wedge J\psi(\psi,\psi)\rangle^2}{|\psi|^4}. \tag{2}$$

Here $|\cdot|^2 : L \to |K|$ denotes the half-density valued quadratic form $|\psi|^2 = |(\psi, \psi)|$ on L and $\langle\cdot,\cdot\rangle : L \times L \to |K|$ is the half-density valued inner product obtained via polarization. The complex structure $*$ on TM^* is the *negative* of the Hodge-star on 1-forms. It is worth noting that the functional E_ϵ is conformally invariant, that is, well-defined on the Riemann surface M, and independent on constant scalings of ψ. In particular, we could normalize ψ by restricting to the L^4-sphere of sections satisfying $\int_M |\psi|^4 = 1$.

The last integral in (2) turns out to be the Willmore functional $\int_M H^2|\psi|^4$ and the first two integrals measure, in L^2, the failure of the non-linear Dirac equation (1) to hold. Thus, for $\epsilon_3 = 0$ and $\epsilon_1, \epsilon_2 > 0$, the functional attains its minimum value $E_\epsilon(\psi) = 0$ at nowhere vanishing sections ψ which correspond to—in general rather singular—conformal immersions. It is therefore conceivable that minimizers of E_ϵ for $\epsilon_3 > 0$, which has the effect of keeping the Willmore energy as a regularizer, will converge as ϵ_3 tends to zero to smooth conformal immersions of M minimizing the Willmore energy, that is, constrained Willmore surfaces. Since the Dirac equation only guarantees the closedness of (ψ, ψ), the resulting conformal immersion given by

$df = (\psi, \psi)$ generally will have translation periods which are controlled by adding the squared lengths of the period integrals $|\int_\gamma (\psi, \psi)|^2$ to the functional (2).

As it turns out, this strategy works surprisingly well [4] when searching for isometric immersions $f: M \to \mathbb{R}^3$ of a compact, oriented Riemannian surface (M, g). Since the conformal class of g gives M the structure of a Riemann surface, we can consider the family of functionals (2) with the additional *isometric constraint*

$$|\psi|^4 = g\,.$$

Then the resulting minimizers under the above described procedure will be isometric immersions $f: M \to \mathbb{R}^3$ whose Willmore energy is "small". The resulting surfaces provide examples of how piecewise smooth, and sometimes even smooth, isometric immersions of compact Riemannian surfaces might look, as can be seen in Figs. 1, 2, and 6 (Figs. 3, 4 and 5).

We should point out that spinorial descriptions of surfaces have been applied to a variety of problems, both in the discrete [5, 6, 14, 30] and smooth settings [7, 12, 15, 17, 23, 24, 29]. The present paper is novel as it focuses on the spinorial construction of conformal and isometric immersions of surfaces in $\mathbb{R}^3$ from a purely intrinsic point of view.

2 Spin Bundles and Regular Homotopy Classes

Given a (not necessarily oriented or compact) 2-dimensional manifold M, we will discuss how to relate a regular homotopy class of immersions $f: M \to \mathbb{R}^3$ and an isomorphism class of spin bundles L over M. The material is somewhat folklore [7, 13, 24, 25, 27], even though there seems to be no single source one could reference. Recall that two smooth immersions $f, \tilde{f}: M \to \mathbb{R}^3$ are *regularly homotopic* if and only if there is a smooth homotopy via immersions $f_t: M \to \mathbb{R}^3$ with $f_0 = f$ and $f_1 = \tilde{f}$. It is well known [13, 27] that two immersions f and $\tilde{f}$ are regularly homotopic if and only if their derivatives df and $d\tilde{f}$ are smoothly homotopic as sections in $\mathrm{Hom}(TM, \mathbb{R}^3)$.

Definition 1 A *spin bundle* over M is a right quaternionic line bundle $L \to M$ together with a non-degenerate quaternionic skew-Hermitian pairing

$$(\cdot,\cdot): L \times L \to \mathrm{Hom}(TM, \mathbb{H})\,, \tag{3}$$

which we refer to as a *spin pairing*.

Two spin bundles $L, \tilde{L} \to M$ are isomorphic if there is a bundle isomorphism $T: L \to \tilde{L}$ intertwining their respective spin pairings.

Later in the paper we use the extension of the spin pairing to the 2-form valued pairing

$$(\cdot,\cdot)\colon \operatorname{Hom}(TM,L)\times L\to \Lambda^2TM^*\otimes\mathbb{H}\,,\qquad (\mu,\psi)_{X,Y}:=(\mu_X,\psi)_Y-(\mu_Y,\psi)_X$$

obtained by inserting an L-valued 1-form μ on the left, where $X, Y\in TM$. Requiring the skew-Hermitian property

$$\overline{(\mu,\psi)}=-(\psi,\mu)$$

to pertain in this scenario, necessitates the analogous definition

$$(\cdot,\cdot)\colon L\times\operatorname{Hom}(TM,L)\to \Lambda^2TM^*\otimes\mathbb{H}\,,\qquad (\psi,\mu)_{X,Y}=(\psi,\mu_X)_Y-(\psi,\mu_Y)_X$$

when inserting the L-valued 1-form μ on the right.

Note that by transversality a quaternionic line bundle $L\to M$ always has a nowhere vanishing smooth section $\psi\in\Gamma(L)$. We denote the (right) $\mathbb{H}^\times$ principal bundle obtained by removing the zero-section of L by $L^\times$, then $\Gamma(L^\times)$ is the space of nowhere vanishing sections. The following are immediate consequences from the definition of a spin bundle:

1. $\overline{(\psi,\psi)}=-(\psi,\psi)$, so that $(\psi,\psi)\in\operatorname{Hom}(TM,\mathbb{R}^3)$ is an $\mathbb{R}^3$-valued 1-form where we identify $\mathbb{R}^3=\operatorname{Im}\mathbb{H}$.

2. Any two sections $\psi,\varphi\in\Gamma(L^\times)$ scale by a nowhere vanishing function $\lambda\in C^\infty(M,\mathbb{H}^\times)$ and hence

$$(\varphi,\varphi)=(\psi\lambda,\psi\lambda)=\bar\lambda(\psi,\psi)\lambda$$

are pointwise related by a stretch rotation in $\mathbb{R}^3$. Therefore, the Riemannian metrics $|(\varphi,\varphi)|^2=|\lambda|^4|(\psi,\psi)|^2$ are conformally equivalent and M inherits a conformal structure, rendering $(\psi,\varphi)\in\Omega^1(M,\mathbb{H})$ conformal.
Whence we observe that an oriented M becomes a Riemann surface in which case we denote its complex structure by $*\colon TM^*\to TM^*$, the negative of the Hodge-star operator on 1-forms. Since $(\psi,\varphi)\in\operatorname{Hom}(TM,\mathbb{H})$ now are conformal 1-forms there is, at each point of M, a unique (quaternionic linear) complex structure $J\in\operatorname{End}(L)$ such that

$$*(\psi,\varphi)=(J\psi,\varphi)=(\psi,J\varphi)\tag{4}$$

for all $\psi,\varphi\in L$. In particular, L becomes a rank 2 complex vector bundle.

Note that if we had started from a Riemann surface in our Definition 1 of a spin bundle, the existence of the complex structure $J\in\Gamma(\operatorname{End}(L))$ and this last compatibility relation (4) would become part of the axioms.

Example 1 (*The induced spin bundle*) Let $f\colon M\to\mathbb{R}^3$ be a (not necessarily conformal) immersion of a conformal surface. We can uniquely decompose the derivative

$$df=\omega\circ B\tag{5}$$

into a nowhere vanishing conformal 1-form $\omega \in \Gamma(\mathrm{Conf}(TM, \mathbb{R}^3))$ and a positive, self-adjoint (with respect to any conformal metric) bundle isomorphism $B \in \Gamma(TM)$ with $\det B = 1$. Note that $B = \mathrm{id}$ if and only if f is conformal. We define the induced spin bundle to be the trivial quaternionic line bundle

$$L_f := M \times \mathbb{H}$$

together with the spin pairing

$$(\psi, \varphi) = \bar{\psi}\,\omega\,\varphi\,.$$

Note that, by construction, $\omega = (1, 1)$ with $1 \in \Gamma(L_f)$ the constant section.

Fig. 3 A square flat torus can be isometrically C^1-embedded into Euclidean 3-space [2]. Pictures by the Hévéa project

In case M is oriented, and thus a Riemann surface, the immersion $f : M \to \mathbb{R}^3$ has an oriented normal $N : M \to S^2$, which, viewed as an imaginary quaternion, satisfies $N^2 = -1$. Then the conformal 1-form $\omega \in \Gamma(\mathrm{Conf}(TM, \mathbb{R}^3))$ satisfies $*\omega = N\omega$ and $J\varphi := -N\varphi$, for $\varphi \in L_f$, is the unique complex structure on L_f which satisfies the compatibility properties (4).

Theorem 1 *The assignment $f \mapsto L_f$ is a bijection between regular homotopy classes of immersions $f : M \to \mathbb{R}^3$ and isomorphism classes of spin bundles $L \to M$.*

Proof Let $f_t : M \to \mathbb{R}^3$ be a regular homotopy between two (not necessarily conformal) immersions $f_0 = f$ and $f_1 = \tilde{f}$. Then their derivatives $df, d\tilde{f}$ are homotopic by [13, 27], and therefore, by uniqueness of (5), we have a homotopy ω_t in $\Gamma(\mathrm{Conf}(TM, \mathbb{R}^3))$ connecting $\omega_0 = \omega$ and $\omega_1 = \tilde{\omega}$. Since $\mathrm{Conf}(TM, \mathbb{R}^3)$ is an $\mathbb{R}_+\mathbf{SO}(3)$ principal bundle, there exists a path $h \in C^\infty([0, 1], \mathbb{R}_+\mathbf{SO}(3))$, starting at the identity $h_0 = 1$, with $\omega_t = h_t\,\omega$. Whence we conclude that there is a lift $\lambda : [0, 1] \to \mathbb{H}^\times$ of h with $\omega_t = \bar{\lambda}_t\omega\lambda_t$, and in particular we have

$$\tilde{\omega} = \bar{\lambda}_1\omega\lambda_1\,.$$

This last implies that the map $\varphi \mapsto T(\varphi) := \lambda_1\varphi$ is a isomorphism between the induced spin bundles L_f and $L_{\tilde{f}}$.

In order to show the converse, let $L \to M$ be a spin bundle and choose a nowhere vanishing section $\psi \in \Gamma(L^\times)$. Then $\omega = (\psi, \psi) \in \Gamma(\mathrm{Conf}(TM, \mathbb{R}^3))$ is a maximal rank 2 conformal bundle map. According to Smale and Hirsch [13, 27], there exists an immersion $f\colon M \to \mathbb{R}^3$ with df homotopic to ω in $\Gamma(\mathrm{Hom}(TM, \mathbb{R}^3))$. From what was said before, we can conclude that $L \cong L_f$. Since all sections of $\Gamma(L^\times)$ are homotopic, the regular homotopy class of the resulting immersion is independent of the nowhere vanishing section chosen. In particular, immersions constructed from isomorphic spin bundles are regularly homotopic. □

So far, we were mainly concerned with the differential topological properties of spin bundles. Understanding how to construct conformal and isometric immersions from spin bundles, we additionally need to investigate their holomorphic aspects. Let $L \to M$ be a spin bundle over a Riemann surface M, in which case the spin pairing (3) is compatible by (4) with the complex structures on M and L. To fix notations and for future use, we list a number of properties of spin bundles over a Riemann surface that follow immediately from their definition.

1. The complex line subbundles

$$E_\pm = \{\varphi \in L\,;\; J\varphi = \pm\varphi i\} \subset L$$

are isomorphic via quaternionic multiplication by j on the right, and thus as a complex rank 2 bundle $L \cong E \oplus E$ is isomorphic to the double of the complex line bundle $E = E_+ \cong E_-$. This isomorphism is also quaternionic linear provided $E \oplus E$ has the right quaternionic structure given by the Pauli matrices.
2. The spin pairing (3) restricts to a non-degenerate complex pairing $E \times E \to K$ with values in the canonical bundle K of M, exhibiting $E \to M$ as a complex spin bundle, that is $E^2 \cong K$. The holomorphic structure $\bar\partial^K$ of the canonical bundle is given by the exterior derivative d on $\Gamma(K) = \Omega^{1,0}(M, \mathbb{C})$. The isomorphism $E^2 \cong K$ induces a unique holomorphic structure $\bar\partial^E$ on E such that $\bar\partial^K = \bar\partial^E \otimes \bar\partial^E$ or, equivalently,

$$d(\psi, \varphi) = (\bar\partial^E\psi, \varphi) + (\psi, \bar\partial^E\varphi)$$

for $\psi, \varphi \in \Gamma(E)$. In particular, if M is compact, $\deg E = p - 1$ is half of the degree of the canonical bundle, where $p = \text{genus}\, M$. Furthermore, E with $\bar\partial^E$ is a holomorphic spin bundle and since there are 2^{2p} many holomorphic square roots of the canonical bundle K—the half lattice points in the Picard torus of isomorphism classes of degree $p - 1$ holomorphic line bundles—there are 2^{2p} many isomorphism classes of holomorphic spin bundles $E \to M$ over a compact Riemann surface.

Lemma 1 *Let $L \to M$ be a spin bundle. Then there exists a unique operator*

$$\bar\partial\colon \Gamma(L) \to \Gamma(\bar K L)$$

called the Dirac structure, *with the following properties.*

1. *$\bar\partial$ is complex linear, that is $[J, \bar\partial] = 0$.*
2. *$\bar\partial$ satisfies the product rule*

$$\bar\partial(\psi\lambda) = (\bar\partial\psi)\lambda + (\psi\, d\lambda)^{0,1}$$

over quaternion valued functions $\lambda \in C^\infty(M, \mathbb{H})$, where $(\cdot)^{0,1}$ denotes the usual type decomposition of complex vector bundle valued 1-forms. In particular, $\bar\partial$ is a (right) quaternionic and (left) complex linear first order elliptic operator.
3. *$\bar\partial$ is compatible with the spin pairing*

$$d(\psi, \varphi) = (\bar\partial\psi, \varphi) + (\psi, \bar\partial\varphi)\,. \tag{6}$$

Proof Since $L \cong E \oplus E$ is the double of a complex holomorphic spin bundle E, the operator $\bar\partial := \bar\partial^E \oplus \bar\partial^E$ can be shown to fulfill the requirements of the lemma. Any other operator satisfying the properties of the lemma has to be of the form $\bar\partial + \alpha$ with $\alpha \in \Gamma(\bar K)$ a 1-form of type (0, 1). But then (6) implies that $\alpha = 0$. □

Corollary 1 *Let M be a compact oriented surface of genus p. Then there are 2^{2p} many isomorphism classes of spin bundles $L \to M$ and therefore, by Theorem 1, also 2^{2p} many regular homotopy classes of immersions $f : M \to \mathbb{R}^3$.*

Proof We know that a spin bundle $L \to M$ induces a unique complex structure on M and L. Since $L \cong E \oplus E$ for a holomorphic spin bundle $E \to M$, we conclude that there are 2^{2p} many isomorphism classes of spin bundles $L \to M$. □

At this point it is helpful to briefly review the notion of a quaternionic holomorphic structure [24] on a quaternionic line bundle $L \to M$ over a Riemann surface. Such a structure is given by an operator

$$D : \Gamma(L) \to \Gamma(\bar K L)$$

satisfying the product rule

$$D(\psi\lambda) = (D\psi)\lambda + (\psi\, d\lambda)^{0,1}$$

over quaternion valued functions $\lambda \in C^\infty(M, \mathbb{H})$. Note that choosing $\lambda \in \mathbb{H}$ constant, the product rule implies that D is quaternionic linear.

If $L \to M$ is a spin bundle, then we can demand the quaternionic holomorphic structure to be compatible with the spin pairing.

Definition 2 Let $L \to M$ be a spin bundle over a Riemann surface. A quaternionic holomorphic structure $D : \Gamma(L) \to \Gamma(\bar K L)$ is called a *quaternionic holomorphic spin structure*, if D is compatible with the spin pairing

$$d(\psi, \varphi) = (D\psi, \varphi) + (\psi, D\varphi) \tag{7}$$

where $\psi, \varphi \in \Gamma(L)$.

Note that by Lemma 1, the Dirac structure $\bar{\partial}$ on a spin bundle $L \to M$ is a quaternionic holomorphic spin structure, in fact the unique one commuting with the complex structure J on L.

The general quaternionic holomorphic spin structure D will not commute with J, and therefore will have a decomposition

$$D = D_+ + D_- \tag{8}$$

into J commuting and J anti-commuting parts. The component $D_+ = \bar{\partial} + \alpha$, a complex holomorphic structure on L, differs from the Dirac structure $\bar{\partial}$ by a $(0, 1)$-form $\alpha \in \Gamma(\bar{K})$, where we identify $\mathrm{End}_+(L) \cong \mathbb{C}$. The component $D_- \in \Gamma(\bar{K}\,\mathrm{End}_-(L))$ is a $(0, 1)$-form with values in the complex antilinear endomorphisms $\mathrm{End}_-(L)$.

In order to characterize quaternionic holomorphic spin structures, it is helpful to identify $\bar{K}\,\mathrm{End}_-(L)$ with *half-densities*. Recall that the bundle of half-densities is the real, oriented line bundle $|K| \to M$, whose fiber over $x \in M$ is given by $|K|_x = \mathbb{R}\sqrt{g_x}$, where g is a Riemannian metric in the conformal class of M. The half-density valued quadratic form

$$|\cdot|^2 : L \to |K|, \qquad |\psi|^2 := |(\psi, \psi)| \tag{9}$$

on the spin bundle L can be polarized to the non-degenerate, symmetric inner product

$$\langle \cdot, \cdot \rangle : L \times L \to |K|. \tag{10}$$

We frequently will identify $|K|^2 \cong \Lambda^2 TM^*$ by assigning a metric g its volume 2-form vol_g. Since $|\psi|^4 \in \Gamma(|K|^2)$ for $\psi \in \Gamma(L)$, a spin bundle carries the conformally invariant L^4-metric $\int_M |\psi|^4$ on $\Gamma(L)$.

Lemma 2 *Let $L \to M$ be a spin bundle over a Riemann surface. Then the complex line bundle $\bar{K}\,\mathrm{End}_-(L)$ is isomorphic to the complexified half-density bundle*

$$|K| \otimes \mathbb{C} \cong \bar{K}\,\mathrm{End}_-(L) \colon U + VJ \mapsto (U + VJ)\eta$$

where $\eta \in \Gamma(\bar{K}\,\mathrm{End}_-(L)|K|^{-1})$ is the nowhere vanishing section

$$\eta := J\psi \frac{(\psi, \cdot)}{|\psi|^2}.$$

Notice that η is well-defined independent of the choice of the nowhere vanishing section $\psi \in \Gamma(L^\times)$.

Proof If $\tilde{\psi} = \psi\lambda$ is another nowhere vanishing section of L, then

$$\tilde{\psi} \frac{(\tilde{\psi}, \cdot)}{|\tilde{\psi}|^2} = \psi\lambda \frac{\bar{\lambda}(\psi, \cdot)}{|\lambda|^2 |\psi|^2} = \psi \frac{(\psi, \cdot)}{|\psi|^2}$$

which shows that η is well-defined. It remains to verify that $*\eta = -J\eta$, that is, $\eta \in \Gamma(\bar{K}\,\mathrm{End}_-(L)|K|^{-1})$. Let $\psi \in \Gamma(L^\times)$ be a nowhere vanishing section so that $J\psi = \psi N$ with $N^2 = -1$. Then, using the compatibility relation (4), we obtain

$$*\eta = J\psi\frac{*(\psi,\cdot)}{|\psi|^2} = J\psi\frac{(J\psi,\cdot)}{|\psi|^2} = J\psi\frac{(\psi N,\cdot)}{|\psi|^2} = J\psi(-N)\frac{(\psi,\cdot)}{|\psi|^2} = -J\eta,$$

which finishes the proof of the lemma. □

With these preparations, we can now give a characterization of quaternionic holomorphic spin structures, which also can be found in [24], albeit from a slightly different perspective.

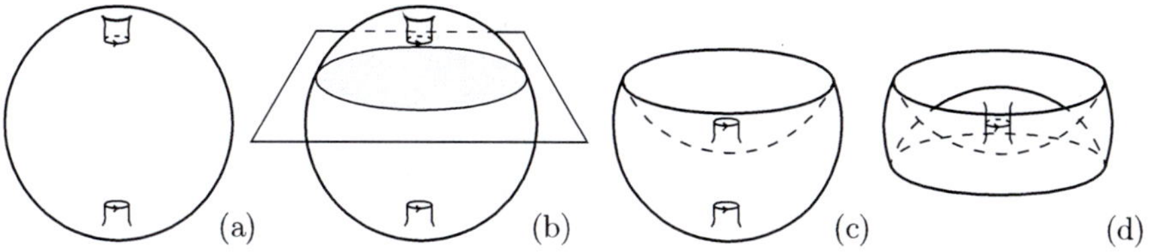

Fig. 4 Consider the Riemannian torus obtained by identifying the two boundary loops of the surface shown on the left (**a**). We believe that this torus does not admit a C^∞-immersion into $\mathbb{R}^3$. However, we can cut the surface by a plane (**b**) and reflect its upper part to obtain the surface (**c**). Applying this construction on the lower part of the surface results in the piecewise smooth isometric immersion of the torus (**d**)

Lemma 3 *Every quaternionic holomorphic spin structure D on a spin bundle $L \to M$ over a Riemann surface is of the form*

$$D = \bar{\partial} + U\eta$$

with $\bar{\partial}$ the Dirac structure and $U \in \Gamma(|K|)$ a real half-density, the Dirac potential.

Proof From (8) and Lemma 2, we know that

$$D = \bar{\partial} + \alpha + (U + JV)\eta$$

with $\bar{\partial}$ the Dirac structure, $\alpha \in \Gamma(\bar{K})$, and $U, V \in \Gamma(|K|)$ half-densities. By Lemma 1 the Dirac structure already fulfills $d(\psi,\varphi) = (\bar{\partial}\psi,\varphi) + (\psi,\bar{\partial}\varphi)$. Thus, D is a quaternionic holomorphic spin structure if and only if the relation

$$((\alpha + (U + JV)\eta)\psi,\varphi) + (\psi,(\alpha + (U + JV)\eta)\varphi) = 0 \tag{11}$$

holds. To evaluate this last, we will use the following easy to verify identities.

1. Every $\alpha \in \Gamma(\bar{K})$ is of the form $\alpha = \beta + *\beta J$ for a unique real 1-from $\beta \in \Omega^1(M, \mathbb{R})$, and then

$$(\alpha\psi, \varphi) + (\psi, \alpha\varphi) = 2(\beta - *\beta) \wedge (\psi, \varphi)\,.$$

2. If $\psi \in \Gamma(L^\times)$ is nowhere vanishing, then

$$|\psi|^2 \left((\eta\psi, \varphi) + (\psi, \eta\varphi)\right) = 0$$

and

$$|\psi|^2 \left((J\eta\psi, \varphi) + (\psi, J\eta\varphi)\right) = 2(\psi, \psi) \wedge (\psi, \varphi)\,.$$

Since the spin pairing is quaternionic Hermitian and η is quaternionic linear, we may put $\varphi = \psi \in \Gamma(L^\times)$ in (11). Together with the above identities, (11) unravels to

$$\begin{aligned} 0 &= ((\alpha + (U + JV)\eta)\psi, \psi) + (\psi, (\alpha + (U + JV)\eta)\psi) \\ &= 2(\beta - *\beta) \wedge (\psi, \psi) + 2\frac{V}{|\psi|^2}(\psi, \psi) \wedge (\psi, \psi)\,. \end{aligned}$$

Letting $J\psi = \psi N$ with $N^2 = -1$, we deduce from the properties of the spin pairing (4) that (ψ, ψ) anti-commutes with N. Hence, the $\mathbb{R}^3$-valued 2-forms $(\beta - *\beta) \wedge (\psi, \psi)$ and $\frac{V}{|\psi|^2}(\psi, \psi) \wedge (\psi, \psi)$ in the last relation take values in complementary subspaces of $\mathbb{R}^3$. Therefore, $D = \bar{\partial} + \alpha + (U + JV)\eta$ is a quaternionic holomorphic spin structure if and only if $V = 0$ and $\beta - *\beta = 0$, which implies $\beta = 0$ and thus $\alpha = 0$. □

In Example 1 we showed how an immersion $f\colon M \to \mathbb{R}^3$ of a surface M induces a spin bundle $L_f \to M$. In case f is a conformal immersion of a Riemann surface, the induced spin bundle L_f additionally carries an induced quaternionic holomorphic spin structure.

Example 2 (*Induced quaternionic holomorphic structure*) Let M be a Riemann surface and $f\colon M \to \mathbb{R}^3$ a conformal immersion. Since $B = \mathrm{id}$ in the decomposition (5), the spin pairing of the induced spin bundle $L_f = M \times \mathbb{H}$ is given by

$$(\psi, \varphi) = \bar{\psi}\, df\, \varphi$$

for $\psi, \varphi \in \Gamma(L_f)$. In particular, $df = (1, 1)$ for the constant section $1 \in \Gamma(L_f^\times)$. If $N\colon M \to S^2$ with $N^2 = -1$ denotes the Gauss normal map of f, the conformality condition reads

$$*df = N\, df = -df\, N\,.$$

The complex structure $J \in \Gamma(\mathrm{End}(L_f))$ on L_f is given by the quaternionic linear endomorphism

$$J\varphi := -N\varphi$$

for $\varphi \in \Gamma(L_f)$ and the compatibility relations (4) hold. There is a natural quaternionic holomorphic structure on L_f given by the (0, 1)-part of the trivial connection

$$D = d^{0,1}\colon \Gamma(L_f) \to \Gamma(\bar{K}L_f)\colon \varphi \mapsto \frac{1}{2}(d\varphi + J * d\varphi)\,.$$

In order to verify that D is in fact a quaternionic holomorphic spin structure, we need to assert the compatibility (7) with the spin pairing

$$\begin{aligned} d(\psi,\varphi) &= d(\bar{\psi}\,df\,\varphi) = d\bar{\psi}\wedge df\,\varphi - \bar{\psi}\,df\wedge d\varphi = \overline{d^{0,1}\psi}\wedge df\,\varphi - \bar{\psi}\,df\wedge d^{0,1}\varphi \\ &= (D\psi,\varphi) + (\psi, D\varphi)\,. \end{aligned}$$

Here we have used $\overline{d^{1,0}\psi}\wedge df = df\wedge d^{1,0}\varphi = 0$ by type considerations. Therefore, $D = d^{0,1}$ is a quaternionic holomorphic spin structure on L_f, and as such decomposes by Lemma 3 into

$$D = \bar{\partial} + U\eta$$

with $\bar{\partial}$ the Dirac structure and $U \in \Gamma(|K|)$ the Dirac potential. One can easily compute [3, 7, 24] that the Dirac potential is given by the mean curvature half-density $U = \frac{1}{2}H|df|$, where $H\colon M \to \mathbb{R}$ is the mean curvature of f calculated with respect to its induced metric $|df|^2$.

The constant section $1 \in \Gamma(L_f^\times)$ lies in the kernel of $D = d^{0,1}$, which is expressed by the non-linear Dirac equation

$$\bar{\partial}1 + \tfrac{1}{2}HJ1(1,1) = 0\,.$$

Here we used $|df| = |(1,1)| = |1|^2$ and the definition of η in Lemma 2 with $\psi = 1 \in \Gamma(L_f^\times)$. The non-linear Dirac equation will be the starting point for our construction of conformal and isometric immersions in a given regular homotopy class.

3 Conformal and Isometric Immersions

Given a Riemann surface M, we want to construct a conformal immersion $f\colon M \to \mathbb{R}^3$ with small Willmore energy in a given regular homotopy class. By Theorem 1, a regular homotopy class is given by a choice of spin bundle $L \to M$ that comes equipped with the Dirac structure $\bar{\partial}$ from Lemma 1. Any nowhere vanishing section $\psi \in \Gamma(L^\times)$ gives rise to a putative derivative $(\psi,\psi) \in \Gamma(\mathrm{Conf}(TM,\mathbb{R}^3))$ of a conformal immersion in the regular homotopy class defined by L. The problem is that, in general, (ψ,ψ) will not be closed, which is necessary for the existence of a conformal immersion $f\colon M \to \mathbb{R}^3$ satisfying $df = (\psi,\psi)$.

Lemma 4 *Let $L \to M$ be a spin bundle over the Riemann surface M and $\psi \in \Gamma(L^\times)$ a nowhere vanishing section of L. Then the conformal 1-form $(\psi,\psi) \in$*

$\Gamma(\mathrm{Conf}(TM, \mathbb{R}^3))$ is closed if and only if ψ solves the non-linear Dirac equation

$$\bar{\partial}\psi + \frac{1}{2}HJ\psi(\psi, \psi) = 0 \tag{12}$$

for some real valued function $H\colon M \to \mathbb{R}$.

The resulting conformal immersion $f\colon \widetilde{M} \to \mathbb{R}^3$ on the universal cover with translation periods satisfying $df = (\psi, \psi)$ has Gauss normal map $N\colon M \to S^2$ given by $J\psi =: -\psi N$. The induced spin bundle $L_f \cong L$ and the induced quaternionic holomorphic spin structure $d^{0,1}$ on L_f corresponds, under this isomorphism, to the quaternionic holomorphic spin structure $D = \bar{\partial} + \frac{1}{2}HJ\psi(\psi, \cdot)$. In particular, H is the mean curvature of f calculated with respect to the induced conformal metric $|df|^2 = |\psi|^4$.

Remark 1 Strictly speaking, Examples 1 and 2 are stated for a (conformal) immersion $f\colon M \to \mathbb{R}^3$ without periods, but all constructions only use information about the derivative df. Whence a (conformal) immersion $f\colon \tilde{M} \to \mathbb{R}^3$ on the universal cover with translation periods induces a spin bundle $L_f \to M$ together with the induced quaternionic holomorphic structure $d^{0,1}$ over M.

Proof From Lemma 3 we know that

$$D = \bar{\partial} + \frac{1}{2}HJ\psi(\psi, \cdot) = \bar{\partial} + U\eta\,,$$

where $U = \frac{1}{2}H|\psi|^2$, is a quaternionic holomorphic spin structure. In particular, D is compatible (7) with the spin pairing

$$d(\psi, \psi) = (D\psi, \psi) + (\psi, D\psi)\,.$$

Therefore, if ψ solves the non-linear Dirac equation, which, expressed in terms of D, reads $D\psi = 0$, the conformal 1-form (ψ, ψ) is closed. The converse follows from computations similar to the proof of Lemma 3. Trivializing $L \cong M \times \mathbb{H}$ via the nowhere vanishing section $\psi \in \Gamma(L^\times)$ provides the isomorphism $L \cong L_f$. The remaining statements follow from Example 2. □

Given a spin bundle $L \to M$, our goal is to set up a variational problem with parameters

$$E_\epsilon\colon \Gamma(L^\times) \to \mathbb{R}$$

on the space of nowhere vanishing sections of L, whose minima will give rise to conformal immersions $f\colon M \to \mathbb{R}^3$. From the previous Lemma 4 we know that a nowhere vanishing section $\psi \in \Gamma(L^\times)$ gives rise to a conformal immersion (with translation periods) whose derivative satisfies $df = (\psi, \psi)$, provided that ψ solves the non-linear Dirac equation (12). In other words, ψ has to satisfy

$$\bar{\partial}\psi = U\eta\psi$$

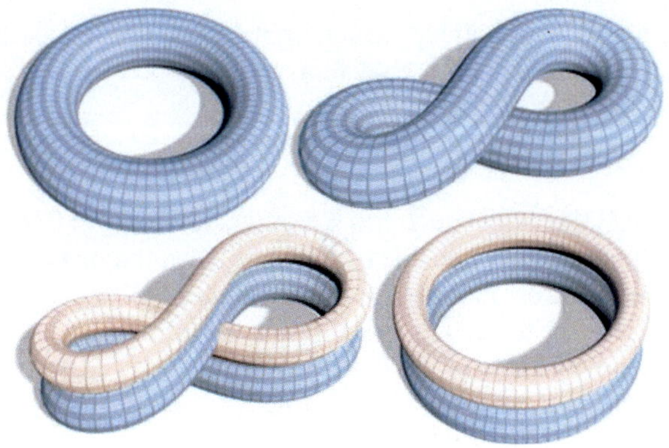

Fig. 5 The four regular homotopy classes of a torus

for some real half-density $U \in \Gamma(|K|)$. In general, since ψ is nowhere vanishing, $\bar{\partial}\psi = Q\psi$ for a (0, 1)-form $Q \in \Gamma(\bar{K}\,\mathrm{End}(L))$ with values in the endomorphims of L. Decomposing Q into the sum $Q = Q_+ + Q_-$ of the J commuting part $Q_+ = \alpha \in \Gamma(\bar{K})$ and the J anti-commuting part $Q_- = (U + VJ)\eta$ with $U, V \in \Gamma(|K|)$ real half-densities, we obtain

$$\bar{\partial}\psi = \alpha\psi + (U + VJ)\eta\,\psi\,. \tag{13}$$

Thus, ψ solves the non-linear Dirac equation, if and only if $\alpha = 0$ and $V = 0$. Before continuing, it is worthwhile to discuss the geometric implications of these conditions.

Remark 2 The nowhere vanishing section $\psi \in \Gamma(L^{\times})$ gives rise to the conformal 1-form $(\psi, \psi) \in \Gamma(\mathrm{Conf}(TM, \mathbb{R}^3))$ which is the putative derivative df of a conformal immersion $f : M \to \mathbb{R}^3$. From Lemma 4 the candidate for the Gauss normal map $N : M \to S^2$ of f is given by $J\psi =: -\psi N$. We can decompose the rank 2 complex bundle $L \to M$ into the sum of complex line bundles, the $\mp N$ eigenspaces

$$L_{\pm} = \{\varphi \in L\,;\; J\varphi = \mp\varphi N\}$$

of the complex structure $J \in \mathrm{End}(L)$. Then $L_+ \subset L$ is a trivial line bundle via the nowhere vanishing section $\psi \in \Gamma(L_+)$. Since

$$*(\psi, \psi) = N(\psi, \psi) = -(\psi, \psi)N$$

due to (4), we have the well-defined complex line bundle isomorphism

$$L_+^2 \to KN^*TS^2 : \psi^2 \mapsto (\psi, \psi)\,. \tag{14}$$

The Dirac structure induces complex holomorphic structures $\bar{\partial}_\pm$ on the summands $L = L_+ \oplus L_-$. Since $*\eta = -J\eta$, the decomposition (13)

$$\bar{\partial}\psi = \alpha\psi + (U + VJ)\eta\,\psi$$

is adapted to the splitting $\bar{K}L = \bar{K}L_+ \oplus \bar{K}L_-$. Therefore, $\alpha = 0$ if and only if the isomorphism (14) is holomorphic, that is, $\bar{\partial}_+$ is the trivial holomorphic structure. In other words, α measures the failure of (14) to be holomorphic.

Since $U \in \Gamma(|K|)$ is the putative mean curvature half-density, it remains to uncover the geometric meaning of the half density $V \in \Gamma(|K|)$ in (13). The derivative $dN \in \Omega^1(M, N^*TS^2)$ of the candidate Gauss map $N\colon M \to S^2$ can be decomposed into conformal and anti-conformal $\mathbb{R}^3$-valued 1-forms

$$dN = dN_+ + dN_- = \frac{2}{|\psi|^2}(U(\psi, \psi) + V * (\psi, \psi)) + q\,.$$

If $(\psi, \psi) = df$ were closed, than the latter would be the decomposition of the shape operator dN into the trace part $H\,df$ and the trace-free part q, the Hopf differential. Therefore, $V = 0$ is exactly the condition that the shape operator dN is self-adjoint for one (and hence any) conformal metric on M.

Incidentally, the above discussion of the geometric content of the decomposition (13) also gives an algorithmic answer to the question "when is a map $N\colon M \to S^2$ from a *compact* Riemann surface M the Gauss normal map of a conformal immersion?" We first choose a spin bundle $L \to M$ which comes with a complex structure $J \in \Gamma(\text{End}(L))$ compatible (4) with the Riemann surface structure of M. According to Theorem 1, the spin bundle L encodes one of the 2^{2p} regular homotopy classes of the resulting conformal immersion, where $p \in \mathbb{N}$ denotes the genus of M. The eigenspace decomposition

$$L_\pm = \{\varphi \in L\,;\; J\varphi = \mp\varphi N\}$$

defines the two complex line subbundles $L_\pm \subset L$ and we need $L_+ \to M$ to admit a global nowhere vanishing section $\psi \in \Gamma(L_+^\times)$. In other words, L_+ has to be trivializable which is equivalent to $\deg L_+ = 0$. Due to (14) this last is guaranteed if and only if $\deg N = 1 - g$, that is, N has the correct degree required by the Gauss-Bonnet Theorem. Moreover, we have seen that L_+ needs to be holomorphically trivial, which puts $2p = \dim_{\mathbb{R}} \text{Jac}(M)$ real conditions on N. Having chosen an N satisfying those conditions, it remains to check whether the half-density $V \in \Gamma(|K|)$ in the decomposition (13) vanishes. Note that globally the only remaining freedom is to rescale ψ by a non-vanishing complex number $\lambda \in \mathbb{C}^\times$, which has the effect of a real scaling and a rotation of the complex half density $U + VJ$. Provided that such a constant rotation renders this complex half density real, there will be a conformal immersion (with translation periods) $f\colon \tilde{M} \to \mathbb{R}^3$ whose Gauss normal map is given by N.

After this brief interlude describing the geometric ramifications of the requirements $\alpha = 0$ and $V = 0$ in the decomposition (13), which guarantee that the conformal $\mathbb{R}^3$-valued 1-form $(\psi, \psi) \in \Gamma(\mathrm{Conf}(TM, \mathbb{R}^3))$ is closed, we discuss the variational aspects of those conditions. On a compact Riemann surface M the requirements $\alpha = 0$ and $V = 0$ are equivalent to the vanishing of the sum of their L^2-norms $\int_M *\bar{\alpha} \wedge \alpha + \int_M V^2 = 0$. Put differently, our variational problem should be designed to measure, in L^2, the failure of $\psi \in \Gamma(L^\times)$ to solve the non-linear Dirac equation (12). In the following lemma we calculate the possible contributions to our functional.

Lemma 5 *Let $L \to M$ be a spin bundle, $\bar{\partial}$ the Dirac structure on L, and $\psi \in \Gamma(L^\times)$ a nowhere vanishing section. Then we have the following expressions for the components of $\bar{\partial}\psi$ in the decomposition* (13)*:*

1. $\langle *\bar{\partial}\psi \wedge \bar{\partial}\psi \rangle = |\psi|^2(*\bar{\alpha} \wedge \alpha + |U|^2 + |V|^2)$,
2. $\langle *\bar{\partial}\psi \wedge \eta\psi \rangle = |\psi|^2 U$,
3. $\langle *\bar{\partial}\psi \wedge J\eta\psi \rangle = |\psi|^2 V$.

Proof The real half-density valued inner product (10) on the quaternionic line bundle L can always be seen as the real part of a quaternionic Hermitian symmetric inner product. Thus, we have

$$\langle \psi\lambda, \psi\mu \rangle = \mathrm{Re}(\bar{\lambda}\mu)|\psi|^2$$

for $\lambda, \mu \in \mathbb{H}$ and $\psi, \varphi \in L$. Moreover, the compatibility (4) of the complex structure $J \in \Gamma(\mathrm{End}(L))$ with the spin pairing implies

$$\langle \psi, \psi \rangle = \langle J\psi, J\psi \rangle \quad \text{and thus} \quad \langle \psi, J\psi \rangle = 0\,.$$

We therefore also have

$$\langle a\psi, b\psi \rangle = \mathrm{Re}(\bar{a}b)|\psi|^2$$

for $a, b \in \mathbb{C}$. Recall that $\eta\psi = J\psi\frac{\omega}{|\psi|^2}$ with $\omega = (\psi, \psi) \in \Omega^1(M, \mathbb{R}^3)$ so that $\bar{\omega} = -\omega$. The (0, 1)-form $\alpha \in \Gamma(\bar{K})$ can be written as $\alpha = \beta + *\beta J$ for a real 1-form $\beta \in \Omega^1(M, \mathbb{R})$. Applying the above identities, after some calculations we obtain the following results.

$$\begin{aligned}
\langle *\alpha\psi \wedge \alpha\psi \rangle &= \mathrm{Re}(*\bar{\alpha} \wedge \alpha)|\psi|^2 \\
|\psi|^2 \langle *\alpha\psi \wedge \eta\psi \rangle &= 2\,\mathrm{Re}(*\beta \wedge \omega)|\psi|^2 = 0 \\
|\psi|^4 \langle \eta\psi \wedge \eta\psi \rangle &= -\mathrm{Re}(\omega \wedge \omega)|\psi|^2 = 0 \\
|\psi|^4 \langle J\eta\psi \wedge J\eta\psi \rangle &= -\mathrm{Re}(\omega \wedge \omega)|\psi|^2 = 0 \\
\langle *\eta\psi \wedge \eta\psi \rangle &= \frac{1}{|\psi|^4}\mathrm{Re}(*\bar{\omega} \wedge \omega)|\psi|^2 = |\psi|^2
\end{aligned}$$

In the third and fourth relation we used the fact that the 2-form $\omega \wedge \omega$ takes values in the orthogonal complement in $\mathbb{R}^3$ of the image of the conformal 1-form ω. In the last relation, we also used the identification $\Lambda^2 TM^* \cong |K|^2$ of 2-forms with conformal metrics on M. Applying those formulas, we deduce

$$\begin{aligned}\langle *\bar{\partial}\psi \wedge \bar{\partial}\psi\rangle &= \mathrm{Re}(*\bar{\alpha} \wedge \alpha)|\psi|^2 + U^2\langle *\eta\psi \wedge \eta\psi\rangle + V^2\langle J * \eta\psi \wedge J\eta\psi\rangle \\ &= |\psi|^2(|\alpha|^2 + |U|^2 + |V|^2),\end{aligned}$$

which is the first identity of the lemma. The second identity follows from

$$\langle *\bar{\partial}\psi \wedge \eta\psi\rangle = \langle U * \eta\psi \wedge \eta\psi\rangle = U|\psi|^2$$

and likewise does the third. □

As we have discussed, a nowhere vanishing section $\psi \in \Gamma(L^\times)$ gives rise to the closed, $\mathbb{R}^3$ valued 1-form (ψ, ψ)—and thus to a conformal immersion with translation periods—if and only if $\int_M *\bar{\alpha} \wedge \alpha + \int_M V^2 = 0$. Here the (0, 1)-form $\alpha \in \Gamma(\bar{K})$ and the real half-density $V \in \Gamma(|K|)$ are the components in the decomposition (13) of $\bar{\partial}\psi$ which, due to the previous lemma, we can express in terms of the section ψ.

Theorem 2 *Let $L \to M$ be a spin bundle over a compact Riemann surface and denote by $\bar{\partial}$ the Dirac structure. For non-negative $\epsilon = (\epsilon_1, \epsilon_2, \epsilon_3)$ the family of functionals*

$$E_\epsilon \colon \Gamma(L^\times) \to \mathbb{R}$$

on nowhere vanishing sections of L, given by

$$E_\epsilon(\psi) = \epsilon_1 \int_M \tfrac{\langle *\bar{\partial}\psi \wedge \bar{\partial}\psi\rangle}{|\psi|^2} + (\epsilon_2 - \epsilon_1)\int_M \tfrac{\langle *\bar{\partial}\psi \wedge \psi(\psi,\psi)\rangle^2}{|\psi|^4} + (\epsilon_3 - \epsilon_1)\int_M \tfrac{\langle *\bar{\partial}\psi \wedge J\psi(\psi,\psi)\rangle^2}{|\psi|^4}$$

is well-defined on the Riemann surface M and invariant under constant, non-zero scalings of ψ. In particular, one could constrain the functional to the L^4-sphere of sections satisfying $\int_M |\psi|^4 = 1$. For $\epsilon_3 = 0$ and arbitrary $\epsilon_1, \epsilon_2 > 0$ the functional assumes its minimum value zero at a section $\psi \in \Gamma(L^\times)$, which gives rise to a conformal immersion (with translation periods) $f : \tilde{M} \to \mathbb{R}^3$ satisfying $df = (\psi, \psi)$ in the prescribed regular homotopy class given by the spin bundle L.

The proof of the theorem follows immediately from Lemma 5, in which the various terms of the functional E_ϵ are calculated. It should be noted that, in order to guarantee exactness of the closed 1-form (ψ, ψ), the functional E_ϵ needs to be augmented by the sum of the squares of the periods $\sum_\gamma |\int_\gamma(\psi, \psi)|^2$, where γ ranges over a basis of the homology group $H_1(M, \mathbb{Z})$. This being said, in the sequel we will always assume that our resulting immersions are defined on M.

Remark 3 For $\epsilon_3 > 0$ the functional E_ϵ contains as a contribution the Willmore energy $\int_M U^2$ of the resulting immersion. It is therefore tempting to minimize E_ϵ for $\epsilon_3 > 0$ while taking $\epsilon_3 \to 0$. The resulting conformal immersion would then be a constrained Willmore surface, that is, a minimizer for the Willmore energy in a fixed conformal and regular homotopy class. At the moment there is no evidence that this strategy, which involves Γ-convergency of our functionals, might be successful. The development of an algorithm based on Theorem 2 to carry out experiments is a work in progress.

We finish this section with a discussion of how to adapt our variational approach to find isometric immersions $f: M \to \mathbb{R}^3$ of an oriented Riemannian surface (M, g) in a given regular homotopy class described by a spin bundle $L \to M$. Every oriented Riemannian surface (M, g) has a unique Riemann surface structure in which g is a conformal metric. The induced metric of the immersion f, constructed from a nowhere vanishing section $\psi \in \Gamma(L^\times)$ satisfying the non-linear Dirac equation, is given by

$$|df|^2 = |(\psi, \psi)|^2 = |\psi|^4 .$$

Hence, we need to minimize our functional under the constraint $|\psi|^4 = g$ in order to find an isometric immersion.

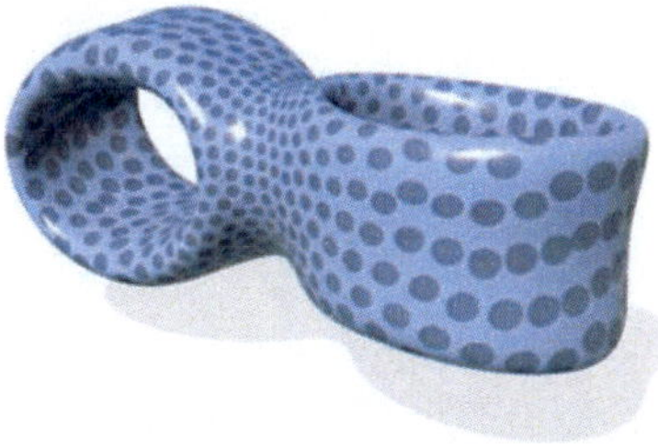

Fig. 6 The Abel–Jacobi map of a compact Riemann surface M induces a Riemannian metric on M. The Gaussian curvature of this metric vanishes at the Weierstrass points. The picture shows an almost isometric smooth realization of the Abel–Jacobi metric on the abstract genus 2 surface in Fig. 1 computed by the algorithm in [4]. The six Weierstrass points lie on the intersection of the surface with its axis of symmetry

Theorem 3 *Let $L \to M$ be a spin bundle over a compact, oriented Riemannian surface and denote by $\bar\partial$ the Dirac structure on L (where we think of M as a Riemann surface). For non-negative $\epsilon = (\epsilon_1, \epsilon_2, \epsilon_3)$ the family of functionals*

$$E_\epsilon \colon \Gamma(L^\times) \to \mathbb{R}$$

on nowhere vanishing sections of L, given by

$$E_\epsilon(\psi) = \epsilon_1 \int_M \frac{\langle *\bar\partial\psi \wedge \bar\partial\psi\rangle}{|\psi|^2} + (\epsilon_2 - \epsilon_1) \int_M \frac{\langle *\bar\partial\psi \wedge \psi(\psi,\psi)\rangle^2}{|\psi|^4} + (\epsilon_3 - \epsilon_1) \int_M \frac{\langle *\bar\partial\psi \wedge J\psi(\psi,\psi)\rangle^2}{|\psi|^4}$$

subject to the constraint $|\psi|^4 = g$, *is well-defined on the Riemannian surface* (M, g). *For* $\epsilon_3 = 0$ *and arbitrary* $\epsilon_1, \epsilon_2 > 0$ *the functional assumes its minimum value, zero, at a section* $\psi \in \Gamma(L^\times)$ *which gives rise to an isometric immersion* $f\colon M \to \mathbb{R}^3$ *satisfying* $df = (\psi, \psi)$ *in the prescribed regular homotopy class given by the spin bundle L.*

For a generic Riemannian surface (M, g) there will not exist a smooth isometric immersion into $\mathbb{R}^3$, even though there always is a C^1 isometric immersion [16, 20]. The methods to construct C^1 isometric immersions result in surfaces in $\mathbb{R}^3$ that do not reflect the intrinsic geometry of (M, g) well (see Fig. 3). On the other hand, minimizing E_ϵ for non-zero $\epsilon_3 > 0$, that is, with the Willmore energy $\int_M |U|^2$ turned on as a contribution to the functional, we expect the limiting isometric immersion as $\epsilon_3 \to 0$ to have small Willmore energy and thus avoid excessive creasing. This has indeed been carried out experimentally with an algorithm based on Theorem 3, which is detailed in [4]. These experiments give some credence to our conjecture, that there should be a piecewise smooth isometric immersion of any Riemannian surface (M, g) in a given regular homotopy class. Again, a theoretical analysis of this conjecture would involve an understanding of the Γ-convergency properties of our family of functionals E_ϵ.

Acknowledgements Authors supported by SFB Transregio 109 "Discretization in Geometry and Dynamics" at Technical University Berlin. Third author partially supported by an RTF grant from the University of Massachusetts Amherst. Fifth author partially supported by the Einstein Foundation. Software support for images provided by SideFX. We thank Stefan Sechelmann for the abstract hyperbolic triangulated surface used in Fig. 1.

References

1. Borrelli, V., Jabrane, S., Lazarus, F., Thibert, B.: Flat tori in three-dimensional space and convex integration. Proc. Nat. Acad. Sci. **109**(19), 7218–7223 (2012)
2. Borrelli, V., Jabrane, S., Lazarus, F., Thibert, B.: Isometric embeddings of the square flat torus in ambient space. Ensaios Matemáticos **24**, 1–91 (2013)
3. Burstall, F.E., Ferus, D., Leschke, K., Pedit, F., Pinkall, U.: Conformal Geometry of Surfaces in S^4 and Quaternions. Lecture Notes in Mathematics. Springer, Berlin (2004)
4. Chern, A., Knöppel, F., Pinkall, U., Schröder, P.: Shape from metric. ACM Trans. Graph. **37**(4), 63 (2018)
5. Crane, K., Pinkall, U., Schröder, P.: Spin transformations of discrete surfaces. ACM Trans. Graph. **30**(4), 104:1–104:10 (2011)
6. Crane, K., Pinkall, U., Schröder, P.: Robust fairing via conformal curvature flow. ACM Trans. Graph. **32**(4), 61:1–61:10 (2013)
7. Ferus, D., Leschke, K., Pedit, F., Pinkall, U.: Quaternionic holomorphic geometry: Plücker formula, Dirac eigenvalue estimates and energy estimates of harmonic 2-tori. Inven. Math. **146**(3), 507–593 (2001)
8. Garsia, A.M.: On the conformal types of algebraic surfaces of Euclidean space. Comment. Math. Helv. **37**(1), 49–60 (1962)
9. Gromov, M.: Partial Differential Relations. Springer, Berlin (1986)
10. Heller, L., Ndiaye, C.B.: First explicit constrained Willmore minimizers of non-rectangular conformal class (2017). arXiv:1710.00533 [math.DG]

11. Heller, L., Pedit, F.: Towards a constrained Willmore conjecture. Willmore Energy and Willmore Conjecture. Monographs and Research Notes in Mathematics, pp. 119–139. Chapman & Hall/CRC, Boca Raton (2017)
12. Heller, S.: A spectral curve approach to Lawson symmetric CMC surfaces of genus 2. Math. Ann. **360**(3–4), 607–652 (2014)
13. Hirsch, M.W.: Immersions of manifolds. Trans. Am. Math. Soc. **93**, 242–276 (1959)
14. Hoffmann, T., Ye, Z.: A discrete extrinsic and intrinsic Dirac operator (2018). arXiv:1802.06278
15. Konopelchenko, B.G.: Weierstrass representations for surfaces in 4D spaces and their integrable deformations via DS hierarchy. Ann. Global Anal. Geom. **18**(1), 61–74 (2000)
16. Kuiper, N.H.: On C^1-isometric imbeddings I & II. Indag. Math. **58**(545–556), 683–689 (1955)
17. Kusner, R., Schmitt, N.: The spinor representation of surfaces in space (1996). arXiv:dg-ga/9610005
18. Kuwert, E., Schätzle, R.: Minimizers of the Willmore functional under fixed conformal class. J. Differ. Geom. **93**(3), 471–530 (2013)
19. Lin, C.S.: The local isometric embedding in R^3 of two-dimensional Riemannian manifolds with Gaussian curvature changing sign cleanly. Commun. Pure App. Math. **39**(6), 867–887 (1986)
20. Nash, J.: C^1 Isometric imbeddings. Ann. Math. **60**(3), 383–396 (1954)
21. Nash, J.: The imbedding problem for Riemannian manifolds. Ann. Math. 20–63 (1956)
22. Ndiaye, C.B., Schätzle, R.: Explicit conformally constrained Willmore minimizers in arbitrary codimension. Calc. Var. Part. Differ. Equ. **51**(1), 291–314 (2014)
23. Ohsawa, T., Kamberov, G., Norman, P., Pinkall, U., Pedit, F.: Quaternions, Spinors, and Surfaces, Contemporary Mathematics, vol. 299. American Mathematical Society, Providence (2002)
24. Pedit, F., Pinkall, U.: Quaternionic analysis on Riemann surfaces and differential geometry. Documenta Mathematica, Extra Volume ICM 1998, pp. 389–400 (1998)
25. Pinkall, U.: Regular homotopy classes of immersed surfaces. Topology **24**(4), 421–434 (1985)
26. Rüedy, R.A.: Embeddings of open Riemann surfaces. Comment. Math. Helv. **46**(1), 214–225 (1971)
27. Smale, S.: A classification of immersions of the 2-sphere. Trans. Am. Math. Soc. **90**(2), 281–290 (1959)
28. Spivak, M.: A Comprehensive Introduction to Differential Geometry, vol. 5. Publish or Perish, Incorporated (1975)
29. Taimanov, I.A.: Surfaces in three-dimensional Lie groups in terms of spinors. RIMS Kokyuroku **1605**, 133–150 (2008)
30. Ye, Z., Diamanti, O., Tang, C., Guibas, L.J., Hoffmann, T.: A unified discrete framework for intrinsic and extrinsic Dirac operators for geometry processing. Comput. Graph. Forum (2018)

Discrete Minimal Nets with Symmetries

Joseph Cho, Wayne Rossman, and Seong-Deog Yang

Abstract In this paper, we extend the notion of Schwarz reflection principle for smooth minimal surfaces to the discrete analogues for minimal surfaces, and use it to create global examples of discrete minimal nets with high degree of symmetry.

Keywords Discrete minimal nets · Reflection principle

1 Introduction

In the case of smooth minimal surfaces in Euclidean 3-space $\mathbb{R}^3$, the Schwarz reflection principle has been used to good effect to extend minimal surfaces and study their global behavior. The Schwarz reflection principle for minimal surfaces comes in two forms. One states that if the minimal surface lies to one side of a plane and has a curvature-line boundary lying in that plane and meeting it perpendicularly, then the surface extends smoothly by reflection to the other side of the plane. The other states that if the minimal surface contains a boundary line segment, then it can be smoothly extended across the line by including the 180° rotation of the surface about that line. When one of these two situations holds on a minimal surface, the other one holds on the conjugate minimal surface.

J. Cho (✉)
Institute of Discrete Mathematics and Geometry, TU Wien, Wiedner Hauptstrasse 8-10/104, 1040 Wien, Austria
e-mail: jcho@geometrie.tuwien.ac.at

W. Rossman
Department of Mathematics, Kobe University, Kobe-shi, Hyogo-ken 657-8501, Japan
e-mail: wayne@math.kobe-u.ac.jp

S.-D. Yang
Department of Mathematics, Korea University, Seoul 02841, Korea
e-mail: sdyang@korea.ac.kr

T. Hoffmann et al. (eds.), *Minimal Surfaces: Integrable Systems and Visualisation*, Springer Proceedings in Mathematics & Statistics 349,
https://doi.org/10.1007/978-3-030-68541-6_3

By the nature of the Schwarz reflection principle, we expect that the surfaces constructed will have relatively high degrees of symmetry. Such symmetry has been seen in numerous works, see, for example, [12, 15, 16, 21, 23, 24, 29, 30, 36].

Such symmetry has also been exploited in the discrete case as well: for discrete S-isothermic minimal nets, see, for example, [4, 9, 10]; for discrete isothermic constant mean curvature nets, see, for example, [17].

In this paper, we investigate how a similar reflection principle will work in the case of discrete isothermic minimal nets and discrete asymptotic minimal nets. The benefit of this is that it provides us a further tool for extending discrete minimal surfaces described locally (which has been well investigated) to surfaces considered at a more global level (which has not received as much attention yet). For example, we will construct the central part of a discrete minimal trinoid, which can then be regarded as existing on a global level, since it is not a simply connected surface, as it is topologically equivalent to the sphere minus three disks. Like in the smooth case, we expect to see relatively high degrees of symmetry in the surfaces we construct in this way.

Our primary results are Proposition 2.8 and Theorem 3.11, which are the two forms of Schwarz reflection in the discrete case. In Corollary 3.10, we also show that the two forms of Schwarz reflection are related by conjugate discrete minimal nets, as defined in [19]. Finally, we use these results to produce examples in Sect. 4.

2 Preliminaries

Let our domain be a $\mathbb{Z}^2$ lattice with $(m, n) \in \mathbb{Z}^2$, and let $(ijkl)$ denote the vertices of an elementary quadrilateral $((m, n), (m + 1, n), (m + 1, n + 1), (m, n + 1))$. For simplicity, we have chosen our domain to be $\mathbb{Z}^2$; however, the theory will hold true for subdomains of $\mathbb{Z}^2$. If F is a discrete net $F : \mathbb{Z}^2 \to \mathbb{R}^3$, then we write $F(m, n) = F_{m,n} = F_i$ over any elementary quadrilateral, and let

$$\mathrm{d}F_{ij} := F_j - F_i.$$

A discrete net F is called a *circular net* if F_i, F_j, F_k, and F_l are concircular, representing a discrete notion of curvature line coordinates [27].

2.1 Discrete Isothermic Nets

First we recall from [6, Definition 4] how the cross ratio of four points in $\mathbb{R}^3$ are defined.

Definition 2.1 Let $x_1, \ldots, x_4 \in \mathbb{R}^3$, and let $\mathbb{R}^3$ be identified with the set of quaternions $\mathbb{H}$ under the usual identification $\mathbb{R}^3 \ni x_i \sim X_i \in \mathbb{H}$. The pair of eigenvalues

$\{q, \bar{q}\}$ of the quaternion

$$(X_1 - X_2)(X_2 - X_3)^{-1}(X_3 - X_4)(X_4 - X_1)^{-1}$$

is called the *cross ratio of* $x_1, \ldots, x_4$. In the case where $x_1, \ldots, x_4$ are concircular, $q = \bar{q} \in \mathbb{R}$, and we write

$$\mathrm{cr}(x_1, x_2, x_3, x_4) = q.$$

Remark 2.2 It was further proved in [6, Lemma 1] that this cross ratio is invariant under Möbius transformations.

Using this definition of cross ratios, discrete isothermic nets are defined as follows in [6, Definition 6]:

Definition 2.3 A circular net F is called a *discrete isothermic net* if on every elementary quadrilateral $(ijkl)$,

$$\mathrm{cr}(F_i, F_j, F_k, F_l) = \frac{a_{ij}}{a_{il}} \in \mathbb{R}_{<0},$$

where a_{ij} (resp. a_{il}) are edge-labeling scalar functions defined on unoriented edges; that is,

$$a_{ij} = a_{lk} \quad \text{and} \quad a_{il} = a_{jk} \tag{1}$$

on every elementary quadrilateral $(ijkl)$. We call a_{ij} and a_{il} the *cross ratio factorizing functions*.

It is shown in [6, Theorem 6] that, for any discrete net F, the discrete isothermicity of F is equivalent to the existence of another discrete net F^* such that

$$\mathrm{d}F^*_{ij} = \frac{a_{ij}}{\|\mathrm{d}F_{ij}\|^2}\mathrm{d}F_{ij}, \quad \mathrm{d}F^*_{il} = \frac{a_{il}}{\|\mathrm{d}F_{il}\|^2}\mathrm{d}F_{il}.$$

If such an F^* exists, F^* is called a *Christoffel transformation* of F, and $(F^*)^* = F$ up to scaling and translation in $\mathbb{R}^3$.

2.2 *Discrete Gaussian and Mean Curvatures*

For any two parallel circular nets F and G, i.e. F and G are both circular nets with parallel corresponding edges, the mixed area of F and G is defined on every elementary quadrilateral as

$$A(F, G)_{ijkl} := \frac{1}{4}(\delta F_{ik} \wedge \delta G_{jl} + \delta G_{ik} \wedge \delta F_{jl})$$

where $\delta F_{ik} := F_k - F_i$ and the exterior algebra $\wedge^2\mathbb{R}^3(\ni u \wedge v)$ is identified with the Lie algebra $\mathfrak{o}(3)$, i.e. for any $u, v, w \in \mathbb{R}^3$,

$$(u \wedge v)\, w = (u \cdot w)v - (v \cdot w)u$$

for the usual inner product of $x, y \in \mathbb{R}^3$ expressed as $x \cdot y$. Note that $A(F)_{ijkl} := A(F, F)_{ijkl}$ gives the area of the quadrilateral spanned by the image of F over an elementary quadrilateral $(ijkl)$.

It is known through [26] that any circular net F has a parallel circular net $N : \mathbb{Z}^2 \to S^2 \subset \mathbb{R}^3$ taking values in the unit sphere. Such an N is called a *discrete Gauss map* of F.

Remark 2.4 If a discrete line bundle $L : \mathbb{Z}^2 \to \{\text{lines in } \mathbb{R}^3\}$ is the normal bundle of F, i.e. $F_i, F_i + N_i \in L_i$, then L constitutes a discrete line congruence in the sense of [13, Definition 2.1], as any two neighboring lines intersect. One can see that after a choice of one normal direction at one vertex of F (an initial condition), the line congruence condition and the parallel mesh condition uniquely determine the normal bundle L over all vertices in the domain, since any two neighboring normal lines must intersect at equal distance from the vertices on the surface.

Furthermore, it is not difficult to see that the parallel net F^t defined as $F^t := F + tN$ for some constant t is also a circular net parallel to F. This allows us to consider the mixed area of F and F^t, and recover the discrete version of the Steiner's formula based on mixed areas (see [28, 35]):

$$\begin{aligned} A(F^t)_{ijkl} &= A(F)_{ijkl} + 2t A(F, N)_{ijkl} + t^2 A(N)_{ijkl} \\ &= (1 - 2t H_{ijkl} + t^2 K_{ijkl}) A(F)_{ijkl} \end{aligned}$$

where H_{ijkl} and K_{ijkl} are defined on each elementary quadrilateral as:

Definition 2.5 We call

$$H_{ijkl} = -\frac{A(F, N)_{ijkl}}{A(F)_{ijkl}}, \quad K_{ijkl} = \frac{A(N)_{ijkl}}{A(F)_{ijkl}}$$

the mean and Gaussian curvatures of a circular net F with Gauss map N.

With the notion of mean curvature on any elementary quadrilateral $(ijkl)$ available, discrete isothermic minimal nets and discrete isothermic constant mean curvature (cmc) nets can be defined as:

Definition 2.6 A circular net F is called a discrete isothermic minimal (resp. cmc) net if $H \equiv 0$ (resp. $H \equiv c \neq 0$ for some non-zero constant c) on every elementary quadrilateral.

2.3 *Planar Reflection Principle for Discrete Isothermic Minimal and cmc Nets*

Since circular nets are a discrete analogue of curvature line coordinates, the following notion is natural.

Definition 2.7 Let $F : \mathbb{Z}^2 \to \mathbb{R}^3$ be a circular net. A discrete space curve F_{m,n_0} (resp. $F_{m_0,n}$) depending on m (resp. n) for each $n_0 \in \mathbb{Z}$ (resp. $m_0 \in \mathbb{Z}$) is called a *discrete curvature line*.

Without loss of generality, let $n_0 \in \mathbb{Z}$, and let F be a discrete isothermic minimal or cmc net, defined on the domain $D := \{(m, n) \in \mathbb{Z}^2 : n \leq n_0\}$ with corresponding Gauss map N. Suppose that the discrete curvature line F_{m,n_0} is contained in a plane $\mathcal{P}$, and further suppose that the unit normal at each vertex (m, n_0) is contained in the plane containing the discrete curvature line, i.e. $F_{m,n_0} + N_{m,n_0} \in \mathcal{P}$.

If we extend F to the domain $\tilde{D} := \{(m, n) \in \mathbb{Z}^2 : n > n_0\}$ by reflecting the vertices across the plane $\mathcal{P}$, then as mentioned in Remark 2.4, the unit normal N also gets uniquely determined on the extended domain. The uniqueness of the unit normal and the symmetry of the discrete net then forces the unit normal to be symmetric with respect to $\mathcal{P}$ as well, giving us the following reflective property of minimal and cmc nets:

Proposition 2.8 *Let $F : D \to \mathbb{R}^3$ be a discrete isothermic minimal (resp. cmc) net with corresponding Gauss map N. Suppose that the discrete curvature line F_{m,n_0} and the normal line congruence L_{m,n_0} along this discrete curve lie in a plane $\mathcal{P}$. Extending F to $\mathbb{Z}^2 = D \cup \tilde{D}$ so that the extension is symmetric with respect to $\mathcal{P}$ results in a discrete minimal (resp. cmc) net on $\mathbb{Z}^2$.*

3 Reflection Properties of Discrete Minimal Nets

In this section, we take a closer look at the reflection properties of discrete minimal nets.

3.1 *Discrete Isothermic Minimal Nets*

Exploiting the relationship between holomorphic functions on the complex plane and conformality, a definition of discrete holomorphic functions was given in [6, Definition 8] as:

Definition 3.1 A map $g : \mathbb{Z}^2 \to \mathbb{R}^2 \cong \mathbb{C}$ is called a *discrete holomorphic function* if

$$\mathrm{cr}(g_i, g_j, g_k, g_l) = \frac{a_{ij}}{a_{il}} \in \mathbb{R}_{<0}$$

for some edge-labeling scalar functions a_{ij} and a_{il}, i.e. satisfying the condition (1).

Using the facts that

- cross ratios are invariant under Möbius transformations,
- a discrete isothermic net on the unit sphere corresponds to a discrete holomorphic function on the complex plane via stereographic projection,
- the Christoffel transform of a discrete minimal net is its own Gauss map, and
- the Christoffel transformation is involutive,

a Weierstrass representation for a discrete minimal net was given in [6, Theorem 9] as follows:

Fact 3.2 *For a discrete holomorphic function g with cross ratio factorizing functions a_{ij} and a_{il}, a discrete isothermic net F defined via*

$$\begin{cases} dF_{ij} = a_{ij} \operatorname{Re}\left((1 - g_i g_j, \sqrt{-1}(1 + g_i g_j), g_i + g_j) \dfrac{1}{dg_{ij}} \right) \\ dF_{il} = a_{il} \operatorname{Re}\left((1 - g_i g_l, \sqrt{-1}(1 + g_i g_l), g_i + g_l) \dfrac{1}{dg_{il}} \right) \end{cases}$$

becomes a discrete isothermic minimal net. Furthermore, any discrete isothermic minimal net can be obtained via some discrete holomorphic function g.

3.2 Discrete Asympotic Minimal Nets

In this section, we make use of shift notations:

$$F = F_{m,n},\ F_1 = F_{m+1,n},\ F_{\bar{1}} = F_{m-1,n},\ F_2 = F_{m,n+1},\ F_{\bar{2}} = F_{m,n-1}.$$

Discrete asymptotic nets were defined as follows in several different contexts (see, for example [8, 32, 33, 37]):

Definition 3.3 A discrete net $\tilde{F} : \mathbb{Z}^2 \to \mathbb{R}^3$ is a *discrete asymptotic net* if each vertex and its neighboring four vertices are coplanar, i.e. $\tilde{F}, \tilde{F}_1, \tilde{F}_{\bar{1}}, \tilde{F}_2, \tilde{F}_{\bar{2}} \in \mathcal{P}_{m,n}$ for some plane $\mathcal{P}_{m,n}$ for each (m, n).

Following [8], we assume that the discrete asymptotic nets here are non-degenerate, i.e. $\tilde{F}_i, \tilde{F}_j, \tilde{F}_k, \tilde{F}_l$ are non-planar.

For a discrete asymptotic net $\tilde{F}$, the Gauss map N is defined as the unit normal to the tangent plane $\mathcal{P}_{m,n}$. Similar to discrete curvature lines, discrete asymptotic lines can be defined as follows:

Definition 3.4 Let $\tilde{F} : \mathbb{Z}^2 \to \mathbb{R}^3$ be a discrete asymptotic net. A discrete space curve $\tilde{F}_{m,n_0}$ (resp. $\tilde{F}_{m_0,n}$) depending on m (resp. n) for each $n_0 \in \mathbb{Z}$ (resp. $m_0 \in \mathbb{Z}$) is called a *discrete asymptotic line*.

Recently, a representation of discrete asymptotic minimal net, where the minimality comes via the edge-constraint condition, was given in [19, Definition 3.1, Theorem 3.14, Lemma 3.17]:

Fact 3.5 *For a discrete holomorphic function g with cross ratio factorizing functions a_{ij} and a_{il}, a discrete asymptotic net $\tilde{F}$ defined via*

$$\begin{cases} d\tilde{F}_{ij} = a_{ij} \operatorname{Re}\left((1 - g_i g_j, \sqrt{-1}(1 + g_i g_j), g_i + g_j)\dfrac{\sqrt{-1}}{dg_{ij}}\right) \\ d\tilde{F}_{il} = a_{il} \operatorname{Re}\left((1 - g_i g_l, \sqrt{-1}(1 + g_i g_l), g_i + g_l)\dfrac{\sqrt{-1}}{dg_{il}}\right) \end{cases}$$

becomes a discrete asymptotic minimal net, in the sense of the discrete minimal edge-constraint nets.

Remark 3.6 It was further shown in [19, Lemma 3.17] that $\tilde{F}$ defined from a discrete holomorphic function g via Fact 3.5 shares the same unit normal as the discrete isothermic minimal net F defined from the same g via Fact 3.2. In such case, $\tilde{F}$ is called the conjugate discrete minimal net of F.

3.3 Reflection Properties of Discrete Minimal Nets

To consider planar discrete space curves, it will be advantageous to use the following notation to denote three consecutive edges:

$$\mathrm{d}F := F_{m+1,n} - F_{m,n}, \quad \mathrm{d}F_1 := F_{m+2,n} - F_{m+1,n}, \quad \mathrm{d}F_{\bar{1}} := F_{m,n} - F_{m-1,n}.$$

We first focus on circular nets: let F be a circular net. Then we have the following lemma, characterizing planar discrete curvature lines in terms of the Gauss map.

Lemma 3.7 *A discrete curvature line on a circular net F is planar if and only if the image of the Gauss map N along the curvature line is contained in a circle.*

Proof Without loss of generality, the planarity of a discrete curvature line is equivalent to the condition

$$\det(\mathrm{d}F_{\bar{1}}, \mathrm{d}F, \mathrm{d}F_1) = 0$$

on any three consecutive edges. However, since F and N are parallel meshes, the above condition is equivalent to

$$\det(\mathrm{d}N_{\bar{1}}, \mathrm{d}N, \mathrm{d}N_1) = 0.$$

Therefore, a discrete curvature line is planar if and only if the image of the Gauss map along the curvature line is planar, i.e. contained in a circle. □

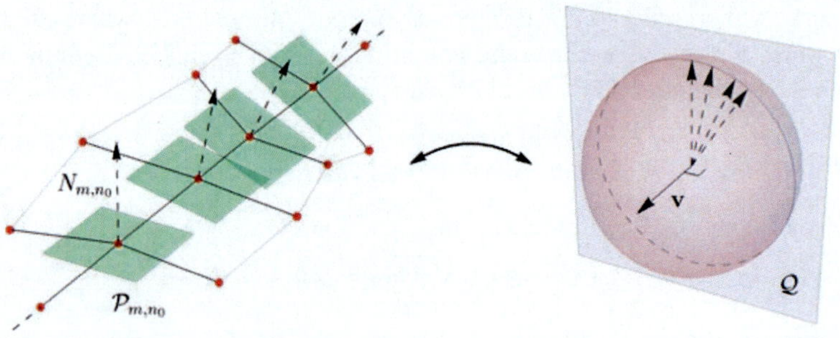

Fig. 1 A discrete asymptotic net containing a straight line and its Gauss map

Hence, by further requiring that the normal line congruence, i.e. the linear span of unit normals placed on the vertices, along the planar curvature line is also included in the same plane, we obtain the following corollary, also mentioned briefly in [9].

Corollary 3.8 *The normal line congruence along a planar discrete curvature line is contained in the same plane if and only if the image of the Gauss map along the curvature line is contained in a great circle.*

Switching our focus to discrete asymptotic nets, now let $\tilde{F}$ be a discrete asymptotic net. Then we can prove the following lemma characterizing a discrete asymptotic line that is a straight line (see also [9]).

Lemma 3.9 *A discrete asymptotic line on a discrete asymptotic net $\tilde{F}$ is a straight line if and only if the image of the Gauss map N along the discrete asymptotic line is contained in a great circle.*

Proof To show one direction, suppose that a discrete asymptotic line $\tilde{F}_{m,n_0}$ is a straight line. Then the tangent planes $\mathcal{P}_{m,n_0}$ at each vertex along $\tilde{F}_{m,n_0}$ must include this straight line. Therefore, N_{m,n_0} must be contained in the plane perpendicular to the straight line, i.e. the image of the Gauss map along the discrete asymptotic line is contained in a great circle.

To show the other direction, now suppose that N_{m,n_0} is contained in a great circle, and let $\mathcal{Q}$ denote the plane containing the great circle with a normal vector $\mathbf{v}$. Then all the tangent planes $\mathcal{P}_{m,n_0}$ must be perpendicular to $\mathcal{Q}$. Hence, from the non-degeneracy condition, any two consecutive tangent planes $\mathcal{P}$ and $\mathcal{P}_1$ must intersect along a line parallel to the normal vector $\mathbf{v}$. However, $\mathcal{P}$ and $\mathcal{P}_1$ intersect along the edge $\mathrm{d}\tilde{F}$, i.e. $\mathrm{d}\tilde{F} \parallel \mathbf{v}$, and it follows that $\tilde{F}_{m,n_0}$ must be a straight line in the direction of $\mathbf{v}$. (See Fig. 1.) □

The fact that a discrete isothermic minimal net F and its conjugate discrete asymptotic minimal net $\tilde{F}$ share the same Gauss map N, as mentioned in Remark 3.6, immediately yields the following corollary.

Corollary 3.10 *The normal line congruence along a planar discrete curvature line on a discrete isothermic minimal net F is contained in the same plane if and only*

if the corresponding discrete asymptotic line on the conjugate discrete asymptotic minimal net $\tilde{F}$ is a straight line.

Now we prove a reflection principle for discrete asymptotic minimal nets. Recall that for some $n_0 \in \mathbb{Z}$, D and $\tilde{D}$ were defined as $D := \{(m, n) \in \mathbb{Z}^2 : n \leq n_0\}$ and $\tilde{D} := \{(m, n) \in \mathbb{Z}^2 : n > n_0\}$, respectively.

Theorem 3.11 *Let $n_0 \in \mathbb{Z}$, and $\tilde{F} : D \subset \mathbb{Z}^2 \to \mathbb{R}^3$ be a discrete asymptotic minimal net with corresponding Gauss map N. Suppose that the discrete asymptotic line $\tilde{F}_{m,n_0}$ is a straight line ℓ. Extending $\tilde{F}$ to the domain $\mathbb{Z}^2 = D \cup \tilde{D}$ so that the extension is symmetric with respect to the line ℓ, the extension is a discrete asymptotic minimal net on $\mathbb{Z}^2$.*

Proof Let $F : D \to \mathbb{R}^3$ be the conjugate discrete isothermic minimal net. Then by Corollary 3.10, we have that the discrete curvature line F_{m,n_0} and the normal line congruence along the curvature line are contained in the same plane $\mathcal{Q}_1$. Therefore, we may invoke Proposition 2.8 to reflect F across $\mathcal{Q}_1$ so that F and N are now defined on $\mathbb{Z}^2$. Now, let $\tilde{F}$ be the conjugate discrete asymptotic minimal net of the extended discrete isothermic minimal net F, where $\tilde{F}\big|_D$ agrees with the original $\tilde{F}$. We now show that $\tilde{F}$ is symmetric with respect to ℓ.

Let $\mathcal{Q}_2$ be the plane such that $N_{m,n_0} \in \mathcal{Q}_2$ for any $m \in \mathbb{Z}$; it follows that ℓ is perpendicular to $\mathcal{Q}_2$. By construction, N is symmetric with respect to the plane $\mathcal{Q}_2$.

Now, let $T \in \mathrm{SO}(3)$ be a rotation around ℓ by $180°$, and consider $\hat{F} := T\tilde{F}$. By the definition of Gauss maps of discrete asymptotic nets, it must follow that one choice of the Gauss map $\hat{N}$ of $\hat{F}$ be $\hat{N} = -TN$. The fact that ℓ is perpendicular to $\mathcal{Q}_2$ implies that $\hat{N}_{m,n}$ is symmetric to $N_{m,n}$ with respect to the plane $\mathcal{Q}_2$. However, because N is symmetric with respect to $\mathcal{Q}_2$, it follows that $N_{m,n_0+k} = \hat{N}_{m,n_0-k}$. Since, $\tilde{F}$ and $\hat{F}$ share the same initial condition along ℓ, we have $\tilde{F}_{m,n_0+k} = \hat{F}_{m,n_0-k}$ by Fact 3.5. □

4 Examples of Discrete Minimal Nets with Symmetry

Let $F : \mathbb{Z}^2 \to \mathbb{R}^3$ be a discrete isothermic minimal surface with Gauss map N, and choose a point $(m_0, n_0) \in \mathbb{Z}^2$. Suppose that the discrete curves F_{m,n_0} and $F_{m_0,n}$, and also the normal line congruences along these curves, are contained in the planes $\mathcal{P}_1$ and $\mathcal{P}_2$, respectively. Since we have that F and N are edge-parallel, N_{m,n_0} and $N_{m_0,n}$ must also be contained in planes $\mathcal{Q}_1$ and $\mathcal{Q}_2$ containing the origin and parallel to $\mathcal{P}_1$ and $\mathcal{P}_2$, respectively (see also Corollary 3.8). Denoting the quadrilateral (m_0, n_0), $(m_0 + 1, n_0)$, $(m_0 + 1, n_0 + 1)$, $(m_0, n_0 + 1)$ by $(ijk\ell)$, we have the following lemma.

Lemma 4.1 *The angle between the planes $\mathcal{P}_1$ and $\mathcal{P}_2$ measured on the side containing the quadrilateral $F_{ijk\ell}$, and the angle between the planes $\mathcal{Q}_1$ and $\mathcal{Q}_2$ measured on the side containing the quadrilateral $N_{ijk\ell}$ are supplementary angles.*

Proof Since F and N are parallel meshes, the angle between $\mathcal{P}_1$ and $\mathcal{P}_2$ equals that between $\mathcal{Q}_1$ and $\mathcal{Q}_2$. However, the Christoffel duality, or the Weierstrass representation, tells us that the orientations of F and N are opposite, giving us the desired conclusion. □

Remark 4.2 Since stereographic projection is a Möbius transformation, it preserves angles. Therefore, to determine the angle between $\mathcal{Q}_1$ and $\mathcal{Q}_2$, one only needs to look at the angle between the circles containing g_{m,n_0} and $g_{m_0,n}$.

Before looking at the examples, we comment on how to change the Weierstrass data of a given smooth minimal surface so that it is parametrized with isothermic coordinates (see, for example, [5, Sect. 2.3]). Let a (smooth) minimal surface $X : \Sigma \subset \mathbb{R}^2 \cong \mathbb{C} \to \mathbb{R}^3$ be represented by

$$X(z) = \mathrm{Re} \int (1 - g(z)^2, \sqrt{-1}(1 + g(z)^2), 2g(z)) f(z) dz$$

over a simply-connected domain Σ on which g is meromorphic, while f and fg^2 are holomorphic. Then the coordinate w satisfying

$$(w_z)^2 = f g_z \quad (\text{resp. } (w_z)^2 = -\sqrt{-1} f g_z), \tag{2}$$

for $w_z = \frac{\partial w}{\partial z}$, becomes an isothermic (resp. conformal asymptotic) coordinate of X, and X can be represented as

$$X(w) = \mathrm{Re} \int (1 - g(w)^2, \sqrt{-1}(1 + g(w)^2), 2g(w)) \frac{1}{g_w(w)} dw$$

$$(\text{resp. } X(w) = \mathrm{Re} \int (1 - g(w)^2, \sqrt{-1}(1 + g(w)^2), 2g(w)) \frac{\sqrt{-1}}{g_w(w)} dw).$$

Example 4.3 Recall that the well-known Enneper surface and higher order Enneper surfaces can be represented via the Weierstrass data $g(z) = z^k$ and $f(z) = 1$ for $k \in \mathbb{N}$. Taking the coordinate change as in (2) (and applying a suitable homothety on the domain depending on k), we obtain new Weierstrass data $g(w) = w^{\frac{2k}{k+1}}$.

Therefore, from the discrete power function z^γ defined in [1] (see also [2, 18]), let g be the discrete power function with $\gamma = \frac{2k}{k+1}$. Then, $g_{m,0} \in \mathbb{R}_{\geq 0}$ while $g_{0,n}$ is on the line $z = re^{\sqrt{-1}\frac{k\pi}{k+1}}$ for $r \in \mathbb{R}_{\geq 0}$. Hence, $F_{m,0}$ and $F_{0,n}$ are on planes meeting at an angle $\frac{\pi}{k+1}$. Reflecting the surface iteratively with respect to these planes give us the discrete isothermic analogue of higher order Enneper surfaces, and by considering its conjugate via Fact 3.5, we obtain a discrete asymptotic net with line symmetries (see Fig. 2).

Example 4.4 Planar Enneper surfaces (see, for example [22]) are examples of minimal surfaces with planar ends. In particular, the planar Enneper surface with 2-fold

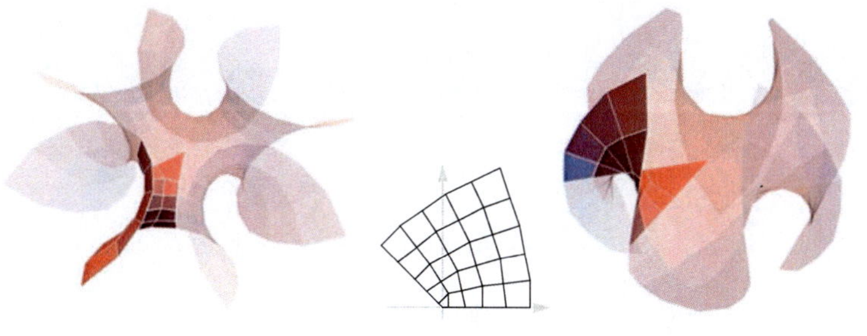

Fig. 2 A discrete higher order Enneper surface from a discrete power function and its conjugate. The left-hand side is a discrete isothermic minimal net having planar symmetry; the right-hand side is a discrete asymptotic minimal net having line symmetry. This figure was drawn with $k = 3$ (See also [31])

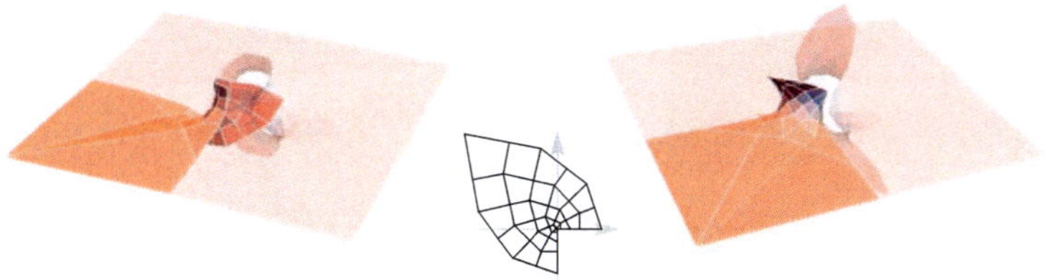

Fig. 3 Discrete planar Enneper surface with 2-fold symmetry from discrete power function z^3 and its conjugate. The left-hand side is a discrete isothermic minimal net having planar symmetry; the right-hand side is a discrete asymptotic minimal net having line symmetry

symmetry is given by the Weierstrass data $g(z) = z^3$ and $f(z) = \frac{1}{g_z(z)}$; hence, z is an isothermic coordinate.

The discrete power function z^3 following [1, 2, 18] becomes immersed on the domain $D := \{(m, n) \in \mathbb{Z}^2 : m \geq 0, n \geq 0\} \setminus \{(0, 0)\}$, and $g_{m,0} \in \mathbb{R}$ while $g_{0,n}$ is on the line $z = -r\sqrt{-1}$ for $r \in \mathbb{R}_{>0}$. Therefore, $F_{m,0}$ and $F_{0,n}$ are on planes meeting at an angle $\frac{\pi}{2}$, and the resulting surface has 2-fold symmetry, and by considering its conjugate via Fact 3.5, we obtain an example of a discrete asymptotic net with line symmetries (see Fig. 3).

Example 4.5 The minimal k-noids (for $k \in \mathbb{N}$, $k \geq 3$) of Jorge-Meeks in [20] are minimal surfaces that are topologically equivalent to the sphere minus k disks with k catenoidal ends, given by the Weierstrass data $g(z) = z^{k-1}$ and $f(z) = \frac{1}{(z^k-1)^2}$. Changing coordinates as in (2) (and applying a suitable homothety on the domain depending on k), we obtain new Weierstrass data $g(w) = (\tanh w)^{\frac{2k-2}{k}}$ with isothermic coordinate w. Under such settings, a fundamental piece of the minimal k-noid can be drawn over the region $w \in [0, \infty] \times \left[0, \frac{\pi}{4}\right] \subset \mathbb{R}^2 \cong \mathbb{C}$ over which $g(w)$ has values

$$g(w) \in D_k := \left\{z = re^{\sqrt{-1}\theta} : 0 \leq r \leq 1, 0 \leq \theta \leq \frac{(k-1)\pi}{k}\right\} \setminus \{1\}.$$

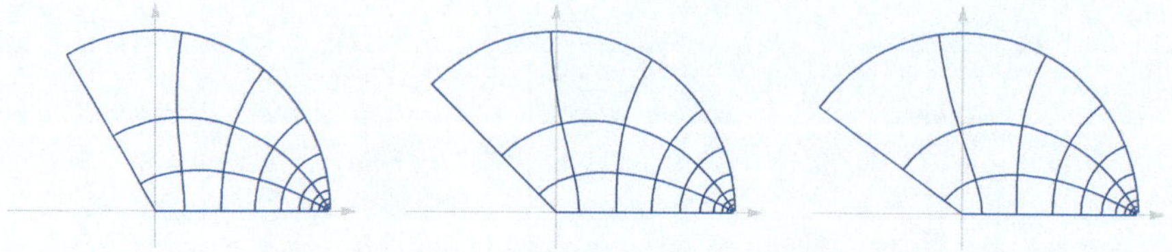

Fig. 4 Images of smooth $g(w)$ giving fundamental pieces of the minimal k-noids, drawn for $k = 3, 4, 5$

In fact, as also demonstrated in Fig. 4,

$$g(w) = \begin{cases} r(w), & \text{if } w \in [0, \infty] \times \{0\} \\ r(w)e^{\sqrt{-1}\frac{(k-1)\pi}{k}}, & \text{if } w \in \{0\} \times \left[0, \frac{\pi}{4}\right] \\ e^{\sqrt{-1}\theta(w)}, & \text{if } w \in [0, \infty] \times \{\frac{\pi}{4}\}. \end{cases}$$

To discretize g (numerically) over the domain $\{(m, n) \in \mathbb{Z}^2 : m \geq 0, 0 \leq n \leq n_{\max}\}$, we require that

- $g_{0,0} = 0$ and $g_{0,n_{\max}} = e^{\sqrt{-1}\frac{(k-1)\pi}{k}}$,
- $g_{m,0} \in [0, 1)$ is a strictly increasing sequence,
- $g_{0,n} = r_n e^{\sqrt{-1}\frac{(k-1)\pi}{k}}$ where $r_n \in [0, 1]$ is a strictly increasing finite sequence,
- $g_{m,n_{\max}} = e^{\sqrt{-1}\theta_m}$ where $\theta_m \in \left(0, \frac{(k-1)\pi}{k}\right]$ is a strictly decreasing sequence,
- the cross ratio of g over any elementary quadrilateral is equal to -1, and
- $g_{m,n} \in D_k$ for all (m, n) in the domain.

By the definition of g, we know that

- the planes containing $F_{m,0}$ and $F_{0,n}$ meet at an angle $\frac{\pi}{k}$, and
- the planes containing $F_{n,0}$ and $F_{m,n_{\max}}$ meet at an angle $\frac{\pi}{2}$,

giving us a discrete analogue of minimal k-noids of Jorge-Meeks (see Figs. 5 and 6).

Example 4.6 By expanding on the idea of using the symmetry of k-noids as boundary conditions for the holomorphic data, we can create other discrete minimal nets with symmetries. In this example, we create discrete minimal nets with symmetry groups of the Platonic solids [38]. As in the k-noids examples, we can ascertain the boundary conditions from the symmetries of the discrete minimal net by calculating the angles at which the great circles meet (see, for example, [3]). Then, by finding discrete holomorphic functions satisfying the given boundary conditions, we can obtain discrete minimal nets with symmetry groups of the Platonic solids. Here, we show two numerical examples of discrete minimal nets with such symmetries in Fig. 7.

Remark 4.7 One may notice that while most of the vertices on the examples have degree 4, i.e. 4 edges meet at the vertex, there are vertices with degree higher than

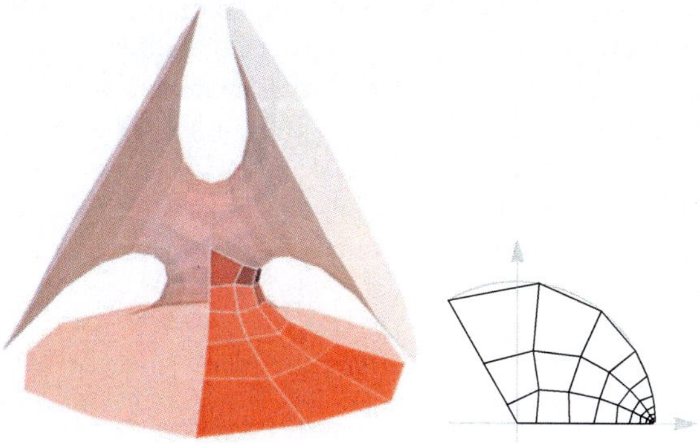

Fig. 5 Numerical solution of discrete trinoid ($k = 3$) given with its discrete holomorphic function satisfying the boundary conditions (with $n_{\max} = 3$)

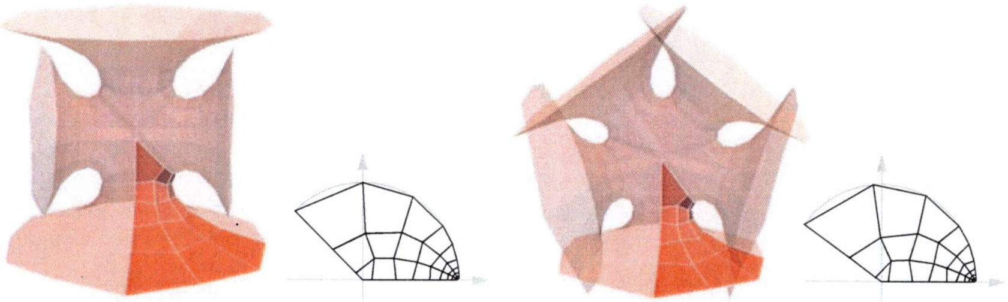

Fig. 6 Numerical solutions of discrete 4-noid and 5-noid given with their discrete holomorphic functions satisfying the boundary conditions (with $n_{\max} = 3$)

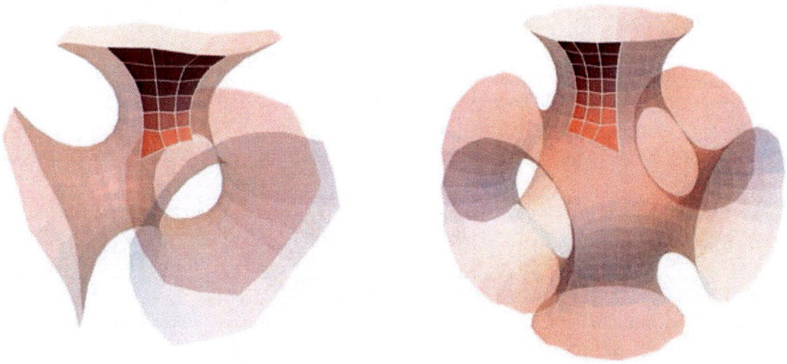

Fig. 7 Numerical solutions of discrete minimal nets with tetrahedral symmetry on the left, and octahedral symmetry on the right

4. While this may indicate the existence of a branch point on the Gauss map, we have avoided this issue by assigning these vertices to be one of the "corner" points of the fundamental piece, and treating the Gauss map as coming from a holomorphic function on a simply-connected domain in the complex plane. In fact, on these vertices, the definition of discrete minimality as in [6, Definition 7] might not be directly applicable; however, the definition via Steiner's formula (as in Definitions 2.5 and 2.6) allows us to consider mean curvatures on the faces around such points, and determine minimality at these points as well.

Acknowledgements The authors would like to express their gratitude to Professor Masashi Yasumoto for fruitful discussions, and the referee for valuable comments. The first author was partially supported by JSPS/FWF Bilateral Joint Project I3809-N32 "Geometric shape generation" and Grant-in-Aid for JSPS Fellows No. 19J10679; the second author was partially supported by two JSPS grants, Grant-in-Aid for Scientific Research (C) 15K04845 and (S) 17H06127 (P.I.: M.-H. Saito); the third author was partially supported by NRF 2017 R1E1A1A 03070929.

References

1. Agafonov, S.I., Bobenko, A.I.: Discrete Z^{γ} and Painlevé equations. Int. Math. Res. Notices **2000**(4), 165–193 (2000). https://doi.org/10.1155/S1073792800000118
2. Ando, H., Hay, M., Kajiwara, K., Masuda, T.: An explicit formula for the discrete power function associated with circle patterns of Schramm type. Funkcial. Ekvac. **57**(1), 1–41 (2014). https://doi.org/10.1619/fesi.57.1
3. Berglund, J., Rossman, W.: Minimal surfaces with catenoid ends. Pac. J. Math. **171**(2), 353–371 (1995). https://doi.org/10.2140/pjm.1995.171.353
4. Bobenko, A.I., Bücking, U., Sechelmann, S.: Discrete minimal surfaces of Koebe type. In: Najman, L., Romon, P. (eds.) Modern Approaches to Discrete Curvature, *Lecture Notes in Mathematics*, vol. 2184, pp. 259–291. Springer, Cham (2017). https://doi.org/10.1007/978-3-319-58002-9_8
5. Bobenko, A.I., Eitner, U.: Painlevé Equations in the Differential Geometry of Surfaces, *Lecture Notes in Mathematics*, vol. 1753. Springer, Berlin (2000). https://doi.org/10.1007/b76883
6. Bobenko, A.I., Pinkall, U.: Discrete isothermic surfaces. J. Reine Angew. Math. **475**, 187–208 (1996). https://doi.org/10.1515/crll.1996.475.187
7. Bobenko, A.I., Pottmann, H., Wallner, J.: A curvature theory for discrete surfaces based on mesh parallelity. Math. Ann. **348**(1), 1–24 (2010). https://doi.org/10.1007/s00208-009-0467-9
8. Bobenko, A.I., Suris, Y.B.: Discrete Differential Geometry. Graduate Studies in Mathematics, vol. 98. American Mathematical Society, Providence, RI (2008)
9. Bücking, U.: Approximation of conformal mappings by circle patterns and discrete minimal surfaces. Ph.D. Thesis, Technische Universität Berlin (2007)
10. Bücking, U.: Minimal surfaces from circle patterns: Boundary value problems, examples. In: Bobenko, A.I., Schröder, P., Sullivan, J.M., Ziegler, G.M. (eds.) Discrete Differential Geometry, *Oberwolfach Semin.*, vol. 38, pp. 37–56. Birkhäuser, Basel (2008). https://doi.org/10.1007/978-3-7643-8621-4_2
11. Burstall, F.E., Hertrich-Jeromin, U., Rossman, W.: Discrete linear Weingarten surfaces. Nagoya Math. J. **231**, 55–88 (2018). https://doi.org/10.1017/nmj.2017.11
12. Costa, C.J.: Example of a complete minimal immersion in R^3 of genus one and three embedded ends. Bol. Soc. Brasil. Mat. **15**(1–2), 47–54 (1984). https://doi.org/10.1007/BF02584707
13. Doliwa, A., Santini, P.M., Mañas, M.: Transformations of quadrilateral lattices. J. Math. Phys. **41**(2), 944–990 (2000). https://doi.org/10.1063/1.533175

14. Ejiri, N., Fujimori, S., Shoda, T.: A remark on limits of triply periodic minimal surfaces of genus 3. Topology Appl. **196**(part B), 880–903 (2015). https://doi.org/10.1016/j.topol.2015.05.014
15. Hoffman, D.A., Meeks III, W.H.: A complete embedded minimal surface in R^3 with genus one and three ends. J. Differ. Geom. **21**(1), 109–127 (1985). https://doi.org/10.4310/jdg/1214439467
16. Hoffman, D.A., Meeks III, W.H.: Embedded minimal surfaces of finite topology. Ann. Math. (2) **131**(1), 1–34 (1990). https://doi.org/10.2307/1971506
17. Hoffmann, T.: Discrete curves and surfaces. Ph.D. Thesis, Technische Universität Berlin (2000)
18. Hoffmann, T., Rossman, W., Sasaki, T., Yoshida, M.: Discrete flat surfaces and linear Weingarten surfaces in hyperbolic 3-space. Trans. Am. Math. Soc. **364**(11), 5605–5644 (2012). https://doi.org/10.1090/S0002-9947-2012-05698-4
19. Hoffmann, T., Sageman-Furnas, A.O., Wardetzky, M.: A discrete parametrized surface theory in $\mathbb{R}^3$. Int. Math. Res. Not. IMRN **2017**(14), 4217–4258 (2017). https://doi.org/10.1093/imrn/rnw015
20. Jorge, L.P., Meeks III, W.H.: The topology of complete minimal surfaces of finite total Gaussian curvature. Topology **22**(2), 203–221 (1983). https://doi.org/10.1016/0040-9383(83)90032-0
21. Karcher, H.: Embedded minimal surfaces derived from Scherk's examples. Manuscripta Math. **62**(1), 83–114 (1988). https://doi.org/10.1007/BF01258269
22. Karcher, H.: Construction of minimal surfaces. Preprint No. 12, SFB 256, Universität Bonn (1989)
23. Karcher, H.: The triply periodic minimal surfaces of Alan Schoen and their constant mean curvature companions. Manuscripta Math. **64**(3), 291–357 (1989). https://doi.org/10.1007/BF01165824
24. Karcher, H., Polthier, K.: Construction of triply periodic minimal surfaces. Philos. Trans. Roy. Soc. Lond. Ser. A **354**(1715), 2077–2104 (1996). https://doi.org/10.1098/rsta.1996.0093
25. Koch, E., Fischer, W.: On 3-periodic minimal surfaces with noncubic symmetry. Z. Krist. **183**(1–4), 129–152 (1988). https://doi.org/10.1524/zkri.1988.183.14.129
26. Konopelchenko, B.G., Schief, W.K.: Three-dimensional integrable lattices in Euclidean spaces: Conjugacy and orthogonality. R. Soc. Lond. Proc. Ser. A Math. Phys. Eng. Sci. **454**(1980), 3075–3104 (1998). https://doi.org/10.1098/rspa.1998.0292
27. Nutbourne, A.W., Martin, R.R.: Differential Geometry Applied to Curve and Surface Design, vol. 1. Ellis Horwood Ltd., Chichester (1988)
28. Pottmann, H., Liu, Y., Wallner, J., Bobenko, A.I., Wang, W.: Geometry of multi-layer freeform structures for architecture. ACM Trans. Graph. (TOG) **26**(3), 65–1–65–11 (2007). https://doi.org/10.1145/1276377.1276458
29. Rossman, W.: Minimal surfaces in $\mathbf{R}^3$ with dihedral symmetry. Tohoku Math. J. (2) **47**(1), 31–54 (1995). https://doi.org/10.2748/tmj/1178225634
30. Rossman, W., Umehara, M., Yamada, K.: Irreducible constant mean curvature 1 surfaces in hyperbolic space with positive genus. Tohoku Math. J. (2) **49**(4), 449–484 (1997). https://doi.org/10.2748/tmj/1178225055
31. Rossman, W., Yasumoto, M.: Discrete linear Weingarten surfaces with singularities in Riemannian and Lorentzian spaceforms. In: Izumiya, S., Ishikawa, G., Yamamoto, M., Saji, K., Yamamoto, T., Takahashi, M. (eds.) Singularities in Generic Geometry, Advanced in Studied Pure Mathematics, vol. 78, pp. 383–410. Math. Soc. Japan, Tokyo (2018). https://doi.org/10.2969/aspm/07810000
32. Sauer, R.: Parallelogrammgitter als Modelle pseudosphärischer Flächen. Math. Z. **52**, 611–622 (1950). https://doi.org/10.1007/BF02230715
33. Sauer, R.: Differenzengeometrie. Springer, Berlin-New York (1970)
34. Sauer, R., Graf, H.: Über Flächenverbiegung in Analogie zur Verknickung offener Facettenflache. Math. Ann. **105**(1), 499–535 (1931). https://doi.org/10.1007/BF01455828
35. Schief, W.K.: On the unification of classical and novel integrable surfaces. II. Differ. Geom. R. Soc. Lond. Proc. Ser. A Math. Phys. Eng. Sci. **459**(2030), 373–391 (2003). https://doi.org/10.1098/rspa.2002.1008

36. Schoen, A.H.: Infinite periodic minimal surfaces without self-intersections. Technical Report NASA-TN-D-5541, C-98, NASA Electronics Research Center, Cambridge, MA (1970)
37. Wunderlich, W.: Zur Differenzengeometrie der Flächen konstanter negativer Krümmung. Österreich. Akad. Wiss. Math.-Nat. Kl. S.-B. IIa. **160**, 39–77 (1951)
38. Xu, Y.: Symmetric minimal surfaces in R^3. Pac. J. Math. **171**(1), 275–296 (1995). https://doi.org/10.2140/pjm.1995.171.275

A Survey on Minimal Isometric Immersions into $\mathbb{R}^3$, $\mathbb{S}^2 \times \mathbb{R}$ and $\mathbb{H}^2 \times \mathbb{R}$

Benoît Daniel

Abstract In this survey, we review some classical and some more recent results related to minimal isometric immersions into Euclidean 3-space and expose some of their extensions when the target manifold is a Riemannian product $\mathbb{S}^2 \times \mathbb{R}$ or $\mathbb{H}^2 \times \mathbb{R}$, where $\mathbb{S}^2$ is the constant curvature 2-sphere and $\mathbb{H}^2$ the hyperbolic plane.

Keywords Isometric immersion · Minimal surface · Homogeneous Riemannian manifold · Associate family · Rigidity

1 Introduction

The two classical examples of minimal surfaces in Euclidean space $\mathbb{R}^3$ are, apart from the plane, the helicoid and the catenoid, which date back from works of Euler and Meusnier in the 18th century. It is well known that these two surfaces are locally intrinsically isometric (up to scaling), and that there even exists a continuous family of locally intrinsically isometric minimal surfaces interpolating the catenoid and the helicoid. This is an illustration of what is called the associate family of a simply connected minimal surface in Euclidean space $\mathbb{R}^3$: any minimal isometric immersion from a simply connected Riemannian surface Σ into $\mathbb{R}^3$ belongs to a continuous one-parameter family of such immersions, and this family is non trivial unless the immersion is totally geodesic.

On the other hand, Calabi proved that two minimal isometric immersions from Σ into $\mathbb{R}^3$ are associate. Hence, for a given Σ, the set of minimal isometric immersions from Σ into $\mathbb{R}^3$, when non empty, is parametrized by one real parameter.

B. Daniel (✉)
Université de Lorraine, Institut Élie Cartan de Lorraine, UMR 7502, CNRS, B.P. 70239, Vandœuvre-lès-Nancy cedex 54506, France
e-mail: benoit.daniel@univ-lorraine.fr
URL: http://www.iecl.univ-lorraine.fr/~Benoit.Daniel/

T. Hoffmann et al. (eds.), *Minimal Surfaces: Integrable Systems and Visualisation*, Springer Proceedings in Mathematics & Statistics 349,
https://doi.org/10.1007/978-3-030-68541-6_4

Given two Riemannian manifolds M and N, we will let $\mathcal{I}(M, N)$ denote the set of equivalence classes of minimal isometric immersions from M into N modulo congruences, i.e., isometries of the target manifold N.

In this survey we will first consider the set $\mathcal{I}(\Sigma, \mathbb{R}^3)$ where Σ is a simply connected Riemannian surface (Sect. 2). We will recall the classical results and outline how they can be derived from the Gauss and Codazzi equations for immersions into $\mathbb{R}^3$. We will also recall a recent result by A. Moroianu and S. Moroianu [20] which gives a necessary and sufficient condition for the existence of a minimal isometric immersion from Σ into $\mathbb{R}^3$, extending the classical Ricci condition. Note that most of these results have analogues for constant mean curvature (CMC) isometric immersions in 3-dimensional space forms.

Next (Sect. 3) we will consider the global problem in $\mathbb{R}^3$, i.e., we will be interested in $\mathcal{I}(\Sigma, \mathbb{R}^3)$ where Σ is a complete Riemannian surface, but not necessarily simply connected. We will state a few rigidity results and indicate a few arguments of their proofs, in particular linked with the notion of flux.

Apart from space forms, the easiest examples of 3-dimensional Riemannian manifolds are probably the product manifolds $\mathbb{S}^2 \times \mathbb{R}$ and $\mathbb{H}^2 \times \mathbb{R}$, where $\mathbb{S}^2$ is the constant curvature 2-sphere and $\mathbb{H}^2$ the hyperbolic plane. The study of minimal surfaces in these manifolds has undergone important progress since the decade 2000. In Sect. 4 we will outline the results obtained by the author [8, 9] regarding the sets $\mathcal{I}(\Sigma, \mathbb{S}^2 \times \mathbb{R})$ and $\mathcal{I}(\Sigma, \mathbb{H}^2 \times \mathbb{R})$, where Σ is a simply connected Riemannian surface. These results are in the same spirit as the classical ones in $\mathbb{R}^3$; they rely on the Gauss and Codazzi equations and other compatibility equations for immersions into $\mathbb{S}^2 \times \mathbb{R}$ and $\mathbb{H}^2 \times \mathbb{R}$.

Finally, in Sect. 5 we propose some possible extensions of these results.

2 The Local Problem in $\mathbb{R}^3$

Given a simply connected Riemannian surface Σ with metric $\mathrm{d}s^2$, the set $\mathcal{I}(\Sigma, \mathbb{R}^3)$ is well understood and most results regarding it can be derived quite elementarily from the Gauss and Codazzi equations in $\mathbb{R}^3$. We recall that, given a field $S : \mathrm{T}\Sigma \to \mathrm{T}\Sigma$ of symmetric operators, then there exists an isometric immersion from Σ into $\mathbb{R}^3$ with S as induced shape operator if and only if

$$K = \det S, \tag{1}$$

$$\forall (X, Y) \in \mathcal{X}(\Sigma)^2, \nabla_X SY - \nabla_Y SX - S[X, Y] = 0, \tag{2}$$

where ∇ and K denote the Levi-Civita connection and the curvature of the metric $\mathrm{d}s^2$. Moreover, the immersion is then unique up to orientation preserving congruences.

2.1 The Associate Family

A first well known fact is that any $x \in \mathcal{I}(\Sigma, \mathbb{R}^3)$ belongs to a one-parameter family $(x_\theta)_{\theta\in\mathbb{R}/2\pi\mathbb{Z}}$ such that $x_\theta \in \mathcal{I}(\Sigma, \mathbb{R}^3)$ for each θ, called the *associate family* of x. This family is non constant in θ unless x is totally geodesic.

Indeed, if we start with $x \in \mathcal{I}(\Sigma, \mathbb{R}^3)$ with shape operator S, then the pair $(\mathrm{d}s^2, S)$ satisfies (1) and (2) on Σ as well as $\operatorname{tr} S = 0$, since the immersion is minimal. If we choose an orientation on Σ, denote by J the $\frac{\pi}{2}$-rotation on $\mathrm{T}\Sigma$ and, for $\theta \in \mathbb{R}$, $e^{\theta J}$ the θ-rotation on $\mathrm{T}\Sigma$, then setting $S_\theta = e^{\theta J} S$ for a fixed θ, one has that S_θ is symmetric, traceless, and that the pair $(\mathrm{d}s^2, S_\theta)$ satisfies (1) and (2) on Σ; hence it gives rise to an isometric immersion $x_\theta : \Sigma \to \mathbb{R}^3$ with S_θ as shape operator, hence minimal. For $\theta = \pm\pi/2$, the immersion $x^* = x_\theta$ is said to be *conjugate* to x.

The most famous example of this is the associate family of the helicoid, which contains the (universal cover of the) catenoid as conjugate immersion.

Using the Weierstrass representation, if x has Weierstrass data (g, ω) where g is a meromorphic function on Σ and ω a holomorphic 1-form on Σ such that

$$\mathrm{d}s^2 = \frac{1}{4}(1 + |g|^2)^2|\omega|^2,$$

then, up to congruences, the associate immersions have Weierstrass data $(g, e^{-i\theta}\omega)$:

$$x = \operatorname{Re}\int \left(\frac{1}{2}(1 - g^2), \frac{i}{2}(1 + g^2), g\right)\omega,$$

$$x_\theta = \operatorname{Re}\int \left(\frac{1}{2}(1 - g^2), \frac{i}{2}(1 + g^2), g\right) e^{-i\theta}\omega.$$

In particular all associate immersions have the same Gauss map g and the coordinate functions of x^* are the harmonic conjugates of the coordinate functions of x.

2.2 Calabi's Rigidity Result

Calabi proved a rigidity result, which again can be easily deduced from the Gauss and Codazzi equations.

Theorem 1 (Calabi [4]) *Two minimal isometric immersions from Σ into $\mathbb{R}^3$ are associate.*

Proof Let $x, \hat{x} \in \mathcal{I}(\Sigma, \mathbb{R}^3)$, S and $\hat{S}$ their respective shape operators. Then $(\mathrm{d}s^2, S)$ and $(\mathrm{d}s^2, \hat{S})$ satisfy the Gauss and Codazzi equations (1) and (2) on Σ. From the Gauss equation we get

$$K = \det S = \det \hat{S} \tag{3}$$

If x is totally geodesic, i.e., if Σ is flat, then $S \equiv 0$ and so we get $\hat{S} \equiv 0$. Hence $\hat{S} = e^{\theta J} S$ for any $\theta \in \mathbb{R}$, and so $\hat{x} = x_\theta$.

If x is not totally geodesic, then on a simply connected open set U on which x has no umbilical points, (3) together with the fact that S and $\hat{S}$ are symmetric and traceless imply that there exists a smooth function $\theta : U \to \mathbb{R}$ such that $\hat{S} = e^{\theta J} S$. Plugging this expression into the Codazzi equation for $\hat{S}$ yields, for all vector fields X and Y on U,

$$\mathrm{d}\theta(X) J e^{\theta J} SY + e^{\theta J} \nabla_X SY - \mathrm{d}\theta(Y) J e^{\theta J} SX - e^{\theta J} \nabla_Y SX - e^{\theta J} S[X, Y] = 0.$$

Now, reporting the Codazzi equation for S in this equation, we get

$$\mathrm{d}\theta(X) SY - \mathrm{d}\theta(Y) SX = 0.$$

Choosing now X and Y as eigenvectors of S, we get $\mathrm{d}\theta = 0$. Hence θ is constant on U, and so x and $\hat{x}$ are associate on U. Finally, since minimal surfaces in $\mathbb{R}^3$ are analytic, we conclude that x and $\hat{x}$ are associate.

In other words, a minimal isometric immersion $x : \Sigma \to \mathbb{R}^3$ is unique up to congruences and its associate family. Hence, if Σ is not flat, then the set $\mathcal{I}(\Sigma, \mathbb{R}^3)$ is either empty or parametrized by $\mathbb{R}/2\pi\mathbb{Z}$.

2.3 The Existence Problem

We now discuss the existence of a minimal isometric immersion $x : \Sigma \to \mathbb{R}^3$, i.e., the non-emptiness of $\mathcal{I}(\Sigma, \mathbb{R}^3)$. First, a necessary condition is, by the Gauss equation (1) and the fact the shape operator of a minimal immersion is traceless, that

$$K \leq 0.$$

Assuming $K < 0$, one can derive quickly from the Gauss and Codazzi equation a necessary and sufficient condition, called the *Ricci condition*.

Theorem 2 (Ricci) *If $K < 0$, then there exists a minimal isometric immersion from Σ into $\mathbb{R}^3$ if and only if the metric $\sqrt{-K}\mathrm{d}s^2$ is flat, or, equivalently, if and only if the metric $-K\mathrm{d}s^2$ has curvature -1.*

Proof The fact that this condition is necessary can be obtained by a somewhat straightforward computation. To prove that this condition is sufficient, we seek a field $S : \mathrm{T}\Sigma \to \mathrm{T}\Sigma$ of traceless symmetric operators such that $(\mathrm{d}s^2, S)$ satisfies the Gauss and Codazzi equations on Σ. The Gauss equation (1) together with the fact that S is symmetric and traceless imply that the only freedom here is the multiplication by a matrix of rotation, whose angle can depend on the point on Σ. The hypothesis that $\sqrt{-K}\mathrm{d}s^2$ is flat is an integrability condition for the existence of such an S.

More specifically, we first notice that the fact that Σ carries the flat metric $\sqrt{-K}\mathrm{d}s^2$ implies, by the Gauss-Bonnet formula, that Σ is not diffeomorphic to a sphere. Hence Σ is diffeomorphic to a plane and in particular there exists an orthonormal frame (e_1, e_2) globally defined on Σ. Since $K < 0$, the function $\lambda = \sqrt{-K}$ is smooth on Σ. We consider a smooth function $\theta : \Sigma \to \mathbb{R}$ and S the field of symmetric operators defined by

$$S = e^{\theta J}\begin{pmatrix} \lambda & 0 \\ 0 & -\lambda \end{pmatrix}$$

in the frame (e_1, e_2). Then $(\mathrm{d}s^2, S)$ satisfies the Gauss equation on Σ and $\operatorname{tr} S = 0$. It now suffices to check the Codazzi equation for $X = e_1$ and $Y = e_2$.

Let α denote the curvature form, i.e.,

$$\nabla_{e_1} e_2 = -\alpha(e_1)e_1, \quad \nabla_{e_2} e_1 = \alpha(e_2)e_2.$$

We have

$$\begin{aligned}\nabla_{e_1} S e_2 &= -\nabla_{e_1}(\lambda e^{\theta J} e_2)\\ &= -\mathrm{d}\lambda(e_1)e^{\theta J} e_2 - \lambda \mathrm{d}\theta(e_1) J e^{\theta J} e_2 + \lambda\alpha(e_1)e^{\theta J} e_1,\end{aligned}$$

$$\begin{aligned}\nabla_{e_2} S e_1 &= \nabla_{e_2}(\lambda e^{\theta J} e_1)\\ &= \mathrm{d}\lambda(e_2)e^{\theta J} e_1 + \lambda \mathrm{d}\theta(e_2) J e^{\theta J} e_1 + \lambda\alpha(e_2)e^{\theta J} e_2,\end{aligned}$$

$$\begin{aligned}S[e_1, e_2] &= S(-\alpha(e_1)e_1 - \alpha(e_2)e_2)\\ &= -\lambda\alpha(e_1)e^{\theta J} e_1 + \lambda\alpha(e_2)e^{\theta J} e_2.\end{aligned}$$

The Codazzi equation is satisfied by S if and only if

$$\lambda^{-1} e^{-\theta J}(\nabla_{e_1} S e_2 - \nabla_{e_2} S e_1 - S[e_1, e_2]) = 0,$$

which, by the previous calculations, is equivalent to

$$\begin{cases} 0 = \mathrm{d}\theta(e_1) + 2\alpha(e_1) - \lambda^{-1}\mathrm{d}\lambda(e_2) \\ 0 = -\lambda^{-1}\mathrm{d}\lambda(e_1) - \mathrm{d}\theta(e_2) - 2\alpha(e_2), \end{cases}$$

i.e., to

$$\mathrm{d}\theta = -2\alpha + \lambda^{-1}\mathrm{d}\lambda \circ J.$$

Hence, such a function θ exists if and only if the 1-form $-2\alpha + \lambda^{-1}\mathrm{d}\lambda \circ J$ is closed.

If (ω_1, ω_2) is the dual coframe of (e_1, e_2), then we have

$$\begin{aligned} \mathrm{d}(-2\alpha + \lambda^{-1}\mathrm{d}\lambda \circ J) &= -2\mathrm{d}\alpha - \lambda^{-2}\mathrm{d}\lambda \wedge (\mathrm{d}\lambda \circ J) + \lambda^{-1}\mathrm{d}(\mathrm{d}\lambda \circ J) \\ &= (2K + \lambda^{-2}||\nabla\lambda||^2 - \lambda^{-1}\Delta\lambda)\omega_1 \wedge \omega_2. \end{aligned}$$

On the other hand, the curvature of the metric $\lambda \mathrm{d}s^2$ is

$$\tilde{K} = \lambda^{-1}\left(K - \frac{1}{2}\Delta \log \lambda\right).$$

From the hypothesis that $\lambda \mathrm{d}s^2$ is flat we have

$$\tilde{K} = 0,$$

from what we deduce that

$$2K = -\lambda^{-2}||\nabla\lambda||^2 + \lambda^{-1}\Delta\lambda$$

and so

$$\mathrm{d}(-2\alpha + \lambda^{-1}\mathrm{d}\lambda \circ J) = 0.$$

In conclusion, there exists $\theta : \Sigma \to \mathbb{R}$ such that S satisfies the Codazzi equation. This completes the proof.

It is only quite recently that a necessary and sufficient condition was obtained in the general case.

Theorem 3 (A. Moroianu and S. Moroianu, [20]) *There exists a minimal isometric immersion from Σ into $\mathbb{R}^3$ if and only if $K \leq 0$ and*

$$K\Delta K - ||\nabla K||^2 - 4K^3 = 0. \tag{4}$$

Note that, on an open set on which $K < 0$, (4) is equivalent to the fact that the metric $\sqrt{-K}\mathrm{d}s^2$ is flat.

The difficult part is to prove that (4) is sufficient. Writing locally the metric as $\mathrm{d}s^2 = e^{2f}|\mathrm{d}z|^2$ where z is a local complex parameter, (4) implies that the function $e^{4f}K$ is log-harmonic, i.e., $\Delta \log |e^{4f}K| = 0$ on open sets where $e^{4f}K \neq 0$. A. Moroianu and S. Moroianu then prove that, in the neighbourhood of any point, there exists a holomorphic function h such that $e^{4f}K = -|h|^2$; the main difficulty is to prove that the zeroes of $e^{4f}K$ are isolated. Then they recover the Weierstrass data (g, ω) of the immersion, using the spinorial version of the Weierstrass representation. This is done locally, but since Σ is simply connected, it is then possible to extend the Weierstrass data to the whole Σ (note that the Gauss map g is unique up to an isometry of $\mathbb{S}^2$, and then ω is unique up to the multiplication by a complex number of modulus 1, which corresponds to the associate family).

2.4 Generalizations

The existence of an associate family can in the same way be proved for constant mean curvature (CMC) immersions into the 3-dimensional space form $\mathbb{M}^3(c)$ (if x is a CMC immersion with shape operator S, $2H = \operatorname{tr} S$ and $K = \det S + c$ by the Gauss equation in $\mathbb{M}^3(c)$, then the shape operators of the associate immersions are $e^{\theta J}(S - H\mathrm{I}) + H\mathrm{I}$ for $\theta \in \mathbb{R}/2\pi\mathbb{Z}$), and Calabi's theorem extends to these classes of surfaces by a similar argument (see [16]). For a generalization of Ricci's theorem for codimension 1 immersions into n-dimensional space forms, see [11].

In the particular case of immersions of surfaces into the 3-dimensional space form $\mathbb{M}^3(c)$, the Gauss equation implies that

$$K \leq c + H^2.$$

Assuming $K < c$, do Carmo and Dajczer [11] proved a necessary and sufficient condition for the existence of a minimal isometric immersion of a given simply connected Riemannian surface in $\mathbb{M}^3(c)$.

Thanks to Lawson's correspondence [16], this can be generalized to constant mean curvature isometric immersions. Indeed, by Lawson's correspondence, CMC H_1 surfaces in $\mathbb{M}^3(c_1)$ are locally isometric to CMC H_2 surfaces in $\mathbb{M}^3(c_2)$ when

$$H_1^2 + c_1 = H_2^2 + c_2.$$

More precisely, if Σ is a simply connected Riemannian surface, then there exist a CMC H_1 isometric immersion $x_1 : \Sigma \to \mathbb{M}^3(c_1)$ if and only if there exist a CMC H_2 isometric immersion $x_2 : \Sigma \to \mathbb{M}^3(c_2)$; moreover these immersions can be chosen so that their shape operators S_1 and S_2 are related by the following relation:

$$S_1 - H_1\mathrm{I} = S_2 - H_2\mathrm{I}.$$

Two such immersions are then called *cousin* immersions.

In particular, choosing $H_2 = 0$, this implies that CMC H surfaces in $\mathbb{M}^3(c)$ are locally isometric to minimal surfaces in $\mathbb{M}^3(c + H^2)$. Combining all these results, we get the following condition.

Theorem 4 (Ricci; Lawson [16]; do Carmo and Dajczer [11]) *Let Σ be a simply connected Riemannian surface with curvature K. Let $H \in \mathbb{R}$. Assume that $K < c + H^2$. Then there exists a CMC H isometric immersion $\Sigma \to \mathbb{M}^3(c)$ if and only if the metric $(-K + c + H^2)\mathrm{d}s^2$ has curvature $1 + \frac{c+H^2}{K-c-H^2}$.*

Remark 1 It is important to notice that, in this condition, the metric $(-K + c + H^2)\mathrm{d}s^2$ does not have constant curvature in general, except when $c + H^2 = 0$. This case corresponds to minimal surfaces in $\mathbb{R}^3$ (Ricci's condition, Theorem 2) and their cousin surfaces in hyperbolic spaces, which share many properties with them, such as a representation in terms of meromorphic data (Bryant's representation [2]).

For results on minimal isometric immersions of Riemannian manifolds into space forms in any dimension and codimension, see for instance [3, 5, 26]. For a general discussion about the existence of the associate family, see [18].

3 The Global Problem in $\mathbb{R}^3$

Here we assume that Σ is complete (but not necessarily simply connected). We say that an immersion $x \in \mathcal{I}(\Sigma, \mathbb{R}^3)$ is *rigid* (among minimal isometric immersions) if $\mathcal{I}(\Sigma, \mathbb{R}^3) = \{x\}$.

Choi, Meeks and White [6] studied the rigidity of minimal isometric immersions. From the local study, if $x \in \mathcal{I}(\Sigma, \mathbb{R}^3)$, then the associate immersions are minimal isometric immersions $x_\theta : \tilde{\Sigma} \to \mathbb{R}^3$ of $\tilde{\Sigma}$, the universal cover of Σ, but they may be not well defined on Σ itself. If γ is a closed curve in Σ, then a lift $\tilde{\gamma}$ of γ in $\tilde{\Sigma}$ may be not closed. The associate immersion x_θ is well defined on Σ if and only if for any loop $\gamma : [0, 1] \to \Sigma$ with lift $\tilde{\gamma} : [0, 1] \to \tilde{\Sigma}$ we have $x_\theta(\tilde{\gamma}(0)) = x_\theta(\tilde{\gamma}(1))$. This is the *period problem*.

Let γ be an oriented closed curve in Σ. We give an orientation to a neighbourhood of γ in Σ. Then the *force* or *flux vector* of γ in $x(\Sigma)$ is

$$\mathbf{F}_{(\gamma,x)} = \int_\gamma \eta$$

where η is the unit conormal vector of $x \circ \gamma$ in $x(\Sigma)$. Using the fact that the coordinates of the conjugate immersion x^* are the harmonic conjugates of the coordinates of x, we get

$$\mathbf{F}_{(\gamma,x)} = \pm \int_{\tilde{\gamma}} \mathrm{d}x^*.$$

Hence the curve $x^* \circ \tilde{\gamma}$ is closed if and only if

$$\mathbf{F}_{(\gamma,x)} = \mathbf{0}.$$

If this is the case, then moreover, for any $\theta \in \mathbb{R}$, $x_\theta \circ \tilde{\gamma}$ is closed, as $x_\theta = (\cos\theta)x + i(\sin\theta)x^*$.

Consequently, the associate immersions x_θ are well defined on Σ if and only if all closed curves in Σ have vanishing force in $x(\Sigma)$. As a corollary, x is rigid if and only if there exists a closed curve γ in Σ with non vanishing force in $x(\Sigma)$.

A similar criterium for constant mean curvature isometric immersions was found by Smyth and Tinaglia [24], in terms of force and torque. We also refer to [13, 14] for discussions on force and torque.

Theorem 5 (Choi, Meeks and White, [6]) *Any properly embedded minimal surface with more than one end admits a unique isometric minimal immersion into* $\mathbb{R}^3$.

In other words, if Σ is the underlying Riemannian surface of such a minimal surface and $\iota : \Sigma \to \mathbb{R}^3$ is the inclusion map, then $\mathcal{I}(\Sigma, \mathbb{R}^3) = \{\iota\}$.

Conjecture 1 (Meeks) Except for the helicoid, the inclusion map of a complete embedded constant mean curvature surface is the unique such isometric immersion with the same constant mean curvature.

Clearly, the inclusion map of the helicoid is not rigid among minimal isometric immersions: as it is simply connected, its associate family is well defined and non trivial (however, the associate immersions are not embeddings; in particular, the conjugate immersion is an immersion of the universal cover of a catenoid). The plane is simply connected but its associate family is trivial, as it is totally geodesic; hence the inclusion map of the plane is rigid among minimal isometric immersions.

The conjecture has been proved in some cases.

Theorem 6 (Meeks and Tinaglia, [19]) *Conjecture 1 is true in the following cases:*

- *if moreover the surface is minimal and has finite genus,*
- *or if moreover the surface is non-minimal, has finite genus and bounded Gaussian curvature.*

The strategy of the proof of Theorem 6 is to study the asymptotic behaviour of these surfaces: it is proved that they have ends asymptotic to helicoids or Delaunay surfaces, which is used to prove that there exist closed curves on the surface with non vanishing force.

Note that extending such results to other 3-dimensional space forms seems very challenging, since the flux and torque argument to determine whether a curve is closed or not on the conjugate immersion does not seem to extend for CMC immersions into other space forms.

In $\mathbb{S}^3$ we mention the following result.

Theorem 7 (Ramanathan, [21]) *If Σ is a compact Riemannian surface, then $\mathcal{I}(\Sigma, \mathbb{S}^3)$ is finite.*

4 The Local Problem in $\mathbb{S}^2 \times \mathbb{R}$ and $\mathbb{H}^2 \times \mathbb{R}$

We fix $c \neq 0$. We now consider a simply connected Riemannian surface Σ and we are interested in the set $\mathcal{I}(\Sigma, \mathbb{M}^2(c) \times \mathbb{R})$ where $\mathbb{M}^2(c)$ denotes the simply connected constant curvature c surface, that is, a sphere $\mathbb{S}^2(c)$ for $c > 0$ and a hyperbolic plane $\mathbb{H}^2(c)$ for $c < 0$. We will let ξ denote a unit vertical vector field in $\mathbb{M}^2(c) \times \mathbb{R}$, that is, tangent to the factor $\mathbb{R}$.

In $\mathbb{M}^2(c) \times \mathbb{R}$, the Gauss–Codazzi equations for an isometric immersion $x : \Sigma \to \mathbb{M}^2(c) \times \mathbb{R}$, with unit normal vector field N, involve not only the metric $\mathrm{d}s^2$ and the induced shape operator S, but also the *angle function* $\nu = \langle N, \xi \rangle$, which is a function on Σ, and the vector field $T \in \mathcal{X}(\Sigma)$ so that $\mathrm{d}x(T)$ is the projection of ξ on $\mathrm{T}x(\Sigma)$. We will call $(\mathrm{d}s^2, S, T, \nu)$ the *Gauss–Codazzi data* of the immersion.

It was proved by the author [8] that a necessary and sufficient condition for the existence of an isometric immersion $x : \Sigma \to \mathbb{M}^2(c) \times \mathbb{R}$ with Gauss–Codazzi data $(\mathrm{d}s^2, S, T, \nu)$ is that $(\mathrm{d}s^2, S, T, \nu)$ satisfy the following set of equations:

$$K = \det S + c\nu^2, \tag{5}$$

$$\nabla_X SY - \nabla_Y SX - S[X, Y] = c\nu(\langle Y, T\rangle X - \langle X, T\rangle Y), \tag{6}$$

$$\nabla_X T = \nu SX, \tag{7}$$

$$\mathrm{d}\nu(X) + \langle SX, T\rangle = 0. \tag{8}$$

$$||T||^2 + \nu^2 = 1. \tag{9}$$

where K denotes the curvature of the metric $\mathrm{d}s^2$. Moreover, the immersion is then unique up to congruences preserving the orientations of $\mathbb{M}^2(c)$ and $\mathbb{R}$. Note that (5) and (6) are respectively the Gauss and Codazzi equations.

From this it is easy to see that an isometric minimal immersion $x : \Sigma \to \mathbb{M}^2(c) \times \mathbb{R}$ admits an associate family as in $\mathbb{R}^3$ [8]. Indeed, if $(\mathrm{d}s^2, S, T, \nu)$ satisfies (5)–(9) with S symmetric and traceless, then $(\mathrm{d}s^2, e^{\theta J}S, e^{\theta J}T, \nu)$ also satisfies these equations and $e^{\theta J}S$ is symmetric and traceless, and so give rise to an isometric minimal immersion $x_\theta \in \mathcal{I}(\Sigma, \mathbb{M}^2(c) \times \mathbb{R})$. This was also proved by Hauswirth, Sá Earp and Toubiana [12] using the fact that the horizontal and vertical projection of a minimal surface are harmonic. For $\theta = \pm\frac{\pi}{2}$, the immersion $x^* = x_\theta$ is called the conjugate immersion.

For instance, in $\mathbb{H}^2 \times \mathbb{R}$ (normalizing $c = -1$), the conjugate surface of a helicoid of pitch smaller than 1 is a catenoid (a rotational minimal surface), but the conjugate surface of a helicoid of pitch 1 or greater than 1 is a surface that is invariant by respectively parabolic or hyperbolic horizontal isometries of the ambient manifold.

However, Calabi's theorem (Theorem 1) does not hold in $\mathbb{H}^2 \times \mathbb{R}$. Indeed, we can check that the intrinsic curvature of a helicoid of pitch 1 is identically -1 [23]. Hence the inclusion map of this helicoid and the inclusion map of a totally geodesic horizontal surface $\mathbb{H}^2 \times \{0\}$ are two minimal isometric immersions from $\mathbb{H}^2$ into $\mathbb{H}^2 \times \mathbb{R}$ that are not associate, since the helicoid is not totally geodesic. Figure 1 shows the conjugate surface of the helicoid of pitch 1; this surface, is the limit of catenoids when we fix a point of the waist circle and make its radius go to $+\infty$.

Regarding the existence of an isometric immersion $x \in \mathcal{I}(\Sigma, \mathbb{M}^2(c) \times \mathbb{R})$, it is not known whether an analogue of Ricci's theorem (Theorem 2) can be proved.

We now state the main results obtained by the author in [9] and we will outline the main ideas of their proofs, which are quite elementary.

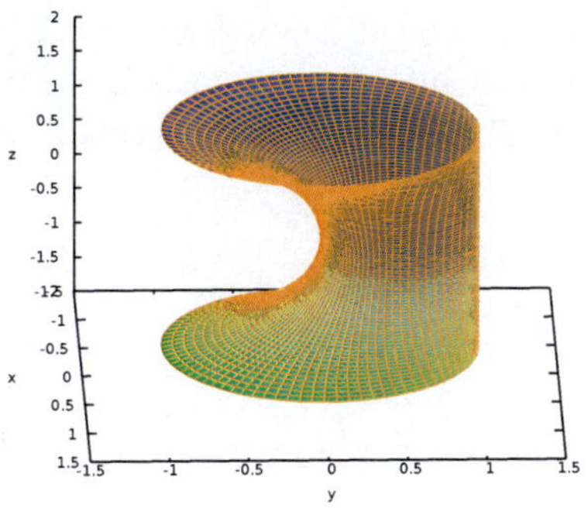

Fig. 1 A minimal surface in $\mathbb{H}^2 \times \mathbb{R}$ $(c = -1)$ foliated by horocycles, with intrinsic curvature -1, and conjugate to a helicoid of pitch 1

Theorem 8 ([9]) *Let Σ be a minimal surface in $\mathbb{M}^2(c) \times \mathbb{R}$ with constant intrinsic curvature K. Then*

- *either Σ is totally geodesic and $K = 0$ or $K = c$,*
- *either $c < 0$, $K = c$ and Σ is part of an associate surface of the helicoid of pitch* 1.

Note that associate surfaces of the helicoid of pitch 1 are simply connected.

Theorem 9 ([9]) *Let Σ be a simply connected Riemannian surface with non constant curvature. Then $\mathcal{I}(\Sigma, \mathbb{M}^2(c) \times \mathbb{R})$ is empty or consists of n families of associate minimal immersions for some integer $n \in [1, 6]$.*

Corollary 1 ([9]) *Let Σ be a simply connected Riemannian surface with non constant curvature. Let $(f_t)_{t \in I}$ be a continuous family of minimal isometric immersions from Σ to $\mathbb{M}^2(c) \times \mathbb{R}$, where I is a real interval. Then all immersions f_t, $t \in I$, are associate.*

The first step is to show that, in the context of minimal isometric immersions, the system (5)–(9) can be reduced to

$$||\nabla \nu||^2 = -(1 - \nu^2)(K - c\nu^2), \tag{10}$$

$$\Delta \nu - 2K\nu + c(1 + \nu^2)\nu = 0. \tag{11}$$

More precisely, assuming that $\nu^2 < 1$, then $(\mathrm{d}s^2, S, T, \nu)$ satisfies (5)–(9) with $\operatorname{tr} S = 0$ if and only if $(\mathrm{d}s^2, \nu)$ satisfy (10)–(11). Note that the case $\nu^2 \equiv 1$ can be easily dealt with in a separate manner (one only gets horizontal totally geodesic surfaces), and, in the remaining case, by analiticity a study on an open set where $\nu^2 < 1$ is sufficient.

The fact that (10)–(11) is necessary is quite straightforward; indeed, (10) is closely related to the Gauss equation and (11) follows from the fact that $\nu = \langle N, \xi \rangle$ is a Jacobi function on Σ, since ξ is a Killing field of the ambient manifold. To prove that (10)–(11) is sufficient, the idea is to show that (11) is an integrability condition for the existence of S and T such that $(\mathrm{d}s^2, S, T, \nu)$ satisfies (5)–(9) and $\operatorname{tr} S = 0$.

The second step is now to study the system (10)–(11). This is a system of one order 1 and one order 2 partial differential equations. Its compatibility condition is given by the Bochner-Weitzenböck formula:

$$\frac{1}{2}\Delta||\nabla\nu||^2 = \langle\nabla\nu, \nabla\Delta\nu\rangle + ||\nabla^2\nu||^2 + K||\nabla\nu||^2,$$

which, in our context, reads as

$$6c\nu\langle\nabla\nu, \nabla K\rangle = ||\nabla K||^2 - (K - c\nu^2)\Delta K + 4(K - c)(K - c\nu^2)(K + 2c\nu^2). \tag{12}$$

If K is constant, then (12) simplifies as $0 = 4(K - c)(K - c\nu^2)(K + 2c\nu^2)$. If $K \neq c$, this implies that ν is constant, and the classification in this case can be easily obtained. If $K = c$, then this equation becomes $0 = 0$; however, (10) implies that $c < 0$ and, in this case, horizontal totally geodesic surfaces and helicoids of pitch one provide solutions ν for all possible initial conditions; this allows to complete the classification in this case.

Note that, when K is constant, (10)–(11) shows that $||\nabla\nu||$ and $\Delta\nu$ are functions of ν, i.e., ν is an isoparametric function.

From now on we assume that K is not constant and, by analiticity, we can restrict ourselves to an open set on which ∇K does not vanish. Then (12) is a new order 1 partial differential equation for ν. A somewhat long calculation using (10), (12), their derivatives and (11) shows that ν satisfies an order-zero equation, that is, an algebraic equation (no differentiation for the unknown function appears) of the form

$$\forall x \in \Sigma, P(x, \nu(x)) = 0,$$

where, for every $x \in \Sigma$, $P(x, \cdot)$ is an even polynomial map of degree at most 12, the map $x \mapsto P(x, \cdot)$ is real analytic and the set $\{x \in \Sigma \mid P(x, \cdot) \not\equiv 0\}$ is dense. Hence there exist at most 12 solutions $\nu : \Sigma \to \mathbb{R}$, and ν is solution if and only if $-\nu$ is solution. Finally, since ν and $-\nu$ give rise to congruent immersions, we obtain at most 6 isometric immersions up to congruences and associate families.

Remark 2 We do not know if the integer 6 in Theorem 9 is optimal.

Sá Earp [22] found an example of a Riemannian surface Σ for which the number of pairwise non associate minimal isometric immersions from Σ into $\mathbb{H}^2 \times \mathbb{R}$ is $n \geq 2$; for this example the author proved in [9] that $n = 2$.

In $\mathbb{S}^2 \times \mathbb{R}$, we do not even know if there exist non associate minimal isometric immersions of a given Riemannian surface.

Remark 3 We remark that if we set $c = 0$ in (12) then we get Moroianu-Moroianu's equation (4) in $\mathbb{R}^3$.

5 Possible Extensions

A natural extension is to investigate, for a given real number $H \neq 0$, the set of constant mean curvature (CMC) H isometric immersions into $\mathbb{S}^2 \times \mathbb{R}$ and $\mathbb{H}^2 \times \mathbb{R}$. Important results in this direction were obtained by Torralbo and Urbano.

Theorem 10 (Torralbo and Urbano, [25]) *Let $H > 0$ and Σ be a simply connected Riemannian surface. Then there exists a one to one correspondence between congruence classes of parallel mean curvature (PMC) isometric immersions $\Sigma \to \mathbb{M}^2(c) \times \mathbb{M}^2(c)$ with mean curvature vector of norm H and pairs of congruence classes of CMC H isometric immersions $\Sigma \to \mathbb{M}^2(c) \times \mathbb{R}$.*

They also characterized the case where the parallel mean curvature isometric immersion takes values in a totally geodesic $\mathbb{S}^2(c) \times \mathbb{S}^1$ or $\mathbb{H}^2 \times \mathbb{R}$: a necessary and sufficient conditions is that the two constant mean curvature isometric immersions are congruent.

The proof of this result is based on the integrability equations for parallel mean curvature isometric immersions into $\mathbb{M}^2(c) \times \mathbb{M}^2(c)$. An important role is played by the Kähler angles of the immersion with respect to the two Kähler structures of $\mathbb{M}^2(c) \times \mathbb{M}^2(c)$; in this correspondence, the Kähler angles of the PMC immersion are mapped to the angle functions of the two CMC immersions.

Torralbo and Urbano also classified pairs of non congruent CMC H isometric immersions of a Riemannian surface having the same angle function.

In a recent joint work of the author with I. Domingos and F. Vitório [10], CMC isometric immersions are investigated from the point of view of [9] explained in Sect. 4. For CMC isometric immersions the system (5)–(9) can be reduced, though the new system is not as simple as the system (10)–(11) of the minimal case. In particular, a classification of surfaces with constant mean curvature H and constant intrinsic curvature is obtained. These surfaces are vertical cylinders, Abresch–Rosenberg–Leite surfaces and some helicoidal surfaces with non constant angle function. The two latter cases only occur in $\mathbb{H}^2 \times \mathbb{R}$. Abresch–Rosenberg–Leite surfaces are constant mean curvature surfaces foliated by horizontal horocycles and with constant angle function. They were introduced by Abresch and Rosenberg [1] and were proved to have constant intrinsic curvature by Leite [17]. Using Torralbo and Urbano's work [25], this also yields a classification of PMC surfaces in $\mathbb{M}^2(c) \times \mathbb{M}^2(c)$ with constant intrinsic curvature.

We also mention that Lawn and Ortega [15] proved that only minimal isometric immersions admit a one parameter family of isometric deformations of a certain type, called a generalized associate family with rotating structure vector field.

Regarding codimension 1 problems, it would also be interesting to understand the set $\mathcal{I}(\Sigma, N)$ where N is a simply connected homogeneous 3-manifold. If N is an $\mathbb{E}(\kappa, \tau)$ manifold, i.e., if N has a 4-dimensional isometry group (for instance, if N is the Heisenberg group Nil_3), then this study can be deduced from the CMC case in $\mathbb{M}^2(c) \times \mathbb{R}$, due to the author's Lawson-type isometric correspondence between

CMC immersions into these manifolds [7]. For other such homogeneous manifolds N (for instance the Lie group Sol_3), the problem remains completely open.

Acknowledgements The author would like to thank the referee for his careful reading and his relevant suggestions.

References

1. Abresch, U., Rosenberg, H.: A Hopf differential for constant mean curvature surfaces in $S^2 \times R$ and $H^2 \times R$. Acta Math. **193**(2), 141–174 (2004)
2. Bryant, R.: Surfaces of mean curvature one in hyperbolic space. Astérisque, (154–155):12, 321–347, 353 (1988), 1987. Théorie des variétés minimales et applications (Palaiseau, 1983–1984)
3. Calabi, E.: Metric Riemann surfaces. Contributions to the theory of Riemann surfaces. Ann. Math. Stud. **30**, 77–85. Princeton University Press, Princeton (1953)
4. Calabi, E.: Quelques applications de l'analyse complexe aux surfaces d'aire minima. In: Topics in Complex Manifolds. Les Presses de l'Université de Montréal (1967)
5. Chern, S.S., Osserman, R.: Remarks on the Riemannian metric of a minimal submanifold. In: Geometry Symposium, Utrecht 1980 (Utrecht, 1980), Lecture Notes in Mathematics, vol. 894, pp. 49–90. Springer, Berlin (1981)
6. Choi, H.I., Meeks III, W.H., White, B.: A rigidity theorem for properly embedded minimal surfaces in R^3. J. Differ. Geom. **32**(1), 65–76 (1990)
7. Daniel, B.: Isometric immersions into 3-dimensional homogeneous manifolds. Comment. Math. Helv. **82**(1), 87–131 (2007)
8. Daniel, B.: Isometric immersions into $\mathbb{S}^n \times \mathbb{R}$ and $\mathbb{H}^n \times \mathbb{R}$ and applications to minimal surfaces. Trans. Am. Math. Soc. **361**(12), 6255–6282 (2009)
9. Daniel, B.: Minimal isometric immersions into $\mathbb{S}^2 \times \mathbb{R}$ and $\mathbb{H}^2 \times \mathbb{R}$. Indiana Univ. Math. J. **64**(5), 1425–1445 (2015)
10. Daniel, B., Domingos, I., Vitório, F.: Constant mean curvature isometric immersions into $\mathbb{S}^2 \times \mathbb{R}$ and $\mathbb{H}^2 \times \mathbb{R}$ and related results (2019). Preprint, arXiv:1911.12630
11. do Carmo, M., Dajczer, M.: Necessary and sufficient conditions for existence of minimal hypersurfaces in spaces of constant curvature. Bol. Soc. Brasil. Mat. **12**(2), 113–121 (1981)
12. Hauswirth, L., Sa Earp, R., Toubiana, E.: Associate and conjugate minimal immersions in $M \times \mathbf{R}$. Tohoku Math. J. **60**(2), 267–286 (2008)
13. Korevaar, N., Kusner, R., Solomon, B.: The structure of complete embedded surfaces with constant mean curvature. J. Differential Geom. **30**(2), 465–503 (1989)
14. Kusner, R.: Global geometry of extremal surfaces in three-space. Ph.D. thesis, University of California, Berkeley (1988)
15. Lawn, M.-A., Ortega, M.: Associated families of surfaces in warped products and homogeneous spaces. Preprint, arXiv:1504.00589, to appear in *Bull. Belg. Math. Soc. Simon Stevin*
16. Lawson Jr., H.B.: Complete minimal surfaces in S^3. Ann. Math. **2**(92), 335–374 (1970)
17. Leite, M.L.: An elementary proof of the Abresch-Rosenberg theorem on constant mean curvature immersed surfaces in $\mathbb{S}^2 \times \mathbb{R}$ and $\mathbb{H}^2 \times \mathbb{R}$. Q. J. Math. **58**(4), 479–487 (2007)
18. Lodovici, S.D.B., Piccione, P.: Associated family of G-structure preserving minimal immersions in semi-Riemannian manifolds. Results Math. **60**(1–4), 453–473 (2011)
19. Meeks III, W.H., Tinaglia, G.: The rigidity of embedded constant mean curvature surfaces. J. Reine Angew. Math. **660**, 181–190 (2011)
20. Moroianu, A., Moroianu, S.: Ricci surfaces. Ann. Sc. Norm. Super. Pisa Cl. Sci. (5) **14**(4), 1093–1118 (2015)
21. Ramanathan, J.: Rigidity of minimal surfaces in S^3. Manuscripta Math. **60**(4), 417–422 (1988)

22. Sa Earp, R.: Parabolic and hyperbolic screw motion surfaces in $\mathbb{H}^2 \times \mathbb{R}$. J. Aust. Math. Soc. **85**(1), 113–143 (2008)
23. Sa Earp, R., Toubiana, E.: Screw motion surfaces in $\mathbb{H}^2 \times \mathbb{R}$ and $\mathbb{S}^2 \times \mathbb{R}$. Illinois J. Math. **49**(4), 1323–1362 (2005)
24. Smyth, B., Tinaglia, G.: The number of constant mean curvature isometric immersions of a surface. Comment. Math. Helv. **88**(1), 163–183 (2013)
25. Torralbo, F., Urbano, F.: Surfaces with parallel mean curvature vector in $\mathbb{S}^2 \times \mathbb{S}^2$ and $\mathbb{H}^2 \times \mathbb{H}^2$. Trans. Am. Math. Soc. **364**(2), 785–813 (2012)
26. Vlachos, T.: Intrinsic obstructions to the existence of isometric minimal immersions. Pacific J. Math. **205**(2), 491–510 (2002)

Families of Minimal Surfaces in $\mathbb{H}^2 \times \mathbb{R}$ Foliated by Arcs and Their Jacobi Fields

Leonor Ferrer, Francisco Martín, Rafe Mazzeo, and Magdalena Rodríguez

Abstract This note provides some new perspectives and calculations regarding an interesting known family of minimal surfaces in $\mathbb{H}^2 \times \mathbb{R}$. The surfaces in this family are the catenoids, parabolic catenoids and tall rectangles. Each is foliated by either circles, horocycles or circular arcs in horizontal copies of $\mathbb{H}^2$. All of these surfaces are well-known, but the emphasis here is on their unifying features and the fact that they lie in a single continuous family. We also initiate a study of the Jacobi operator on the parabolic catenoid, and compute the Jacobi fields associated to deformations to either of the two other types of surfaces in this family.

Keywords Minimal surfaces · Jacobi fields · Moduli spaces

1 Introduction

In these notes we study properties of an interesting family of minimal surfaces in $\mathbb{H}^2 \times \mathbb{R}$. The surfaces in this family are foliated by circles or circular arcs in parallel slices of $\mathbb{H}^2 \times \mathbb{R}$; those surfaces foliated by entire circles are catenoids, those foliated

L. Ferrer, F. Martín and M. M. Rodríguez—are partially supported by MINECO/FEDER grant MTM2014-52368-P and MTM2017-89677-P.
F. Martin—is also partially supported by the Leverhulme Trust grant IN-2016-019.
R. Mazzeo—is supported by the NSF grant DMS-1608223.

L. Ferrer (✉) · F. Martín · M. Rodríguez
Departamento de Geometría y Topología, Universidad de Granada, 18071 Granada, Spain
e-mail: lferrer@ugr.es

F. Martín
e-mail: fmartin@ugr.es

M. Rodríguez
e-mail: magdarp@ugr.es

R. Mazzeo
Department of Mathematics, Stanford University, Stanford, CA 94305, USA
e-mail: mazzeo@math.stanford.edu

T. Hoffmann et al. (eds.), *Minimal Surfaces: Integrable Systems and Visualisation*, Springer Proceedings in Mathematics & Statistics 349,
https://doi.org/10.1007/978-3-030-68541-6_5

by horocycles are parabolic catenoids, and those foliated by arcs equidistant from a geodesic are the so-called tall rectangles. Somewhat surprisingly, using the Poincaré disk model of $\mathbb{H}^2$, these surfaces all appear as intersections of regular surfaces in $\mathbb{R}^3$ with the unit cylinder $\{(x, y, t) \in \mathbb{R}^3 : x^2 + y^2 < 1\}$.

2 Preliminaries and Notation

To set notation, we shall use both the Poincaré disk and upper half-space models of $\mathbb{H}^2$; these are denoted by $\mathbb{D} = \{z \in \mathbb{C} : |z| < 1\}$ and $\mathfrak{H} = \{z = x + \mathrm{i}\, y \in \mathbb{C} : y > 0\}$. We also denote by $o = (0, 0) \in \mathbb{D}$. The isometry between these two models is the Möbius transformation $g : \mathbb{D} \to \mathfrak{H}$,

$$g(z) = \mathrm{i}\,\frac{1 - z}{1 + z}.$$

These have metrics

$$d\rho^2 = 4\,\frac{|dz|^2}{(1 - |z|^2)^2}, \quad \text{and} \quad \frac{1}{y^2}\,|dz|^2,$$

respectively; the corresponding metric for $\mathbb{H}^2 \times \mathbb{R}$ is $d\sigma^2 = d\rho^2 + dt^2$. We denote by $\partial\mathbb{H}^2$ the boundary at infinity of $\mathbb{H}^2$, usually identified with the boundary $S^1 = \{z \in \mathbb{C} : |z| = 1\}$ of $\mathbb{D}$, but sometimes also with the boundary $\mathbb{R}$ of $\mathfrak{H}$ together with the extra point at infinity. Recall from [4] that there are several interesting compactifications of the boundary at infinity of $\mathbb{H}^2 \times \mathbb{R}$, but for the purposes of this paper we consider only the portion $\partial\mathbb{H}^2 \times \mathbb{R}$, which lies in the boundary of any of these useful compactifications.

3 Catenoids of Revolution in $\mathbb{H}^2$ x $\mathbb{R}$

We begin by studying the minimal surfaces of revolution in $\mathbb{H}^2 \times \mathbb{R}$. These analogues of catenoids in $\mathbb{R}^3$ were originally described in the seminal article [6] of Nelli and Rosenberg. We take a slightly different approach to their construction here. Consider a conformal harmonic parametrization of a minimal annulus $\mathbb{A}$:

$$X = (F, h) : \Delta_R = \{R < |z| < 1\} \longrightarrow \mathbb{A} \subset \mathbb{H}^2 \times \mathbb{R};$$

thus

$$F : \Delta_R \to \mathbb{H}^2, \qquad h : \Delta_R \to \mathbb{R}$$

are both harmonic maps. We impose the condition that h is locally constant on $\partial\Delta_R$, i.e., takes two different constant values on the two boundary components of this annulus; by translation we assume that these two values are 0 and $h_0 > 0$.

We now pass to the induced mappings from the universal cover $M = \{w = w_1 + \mathrm{i}w_2 \in \mathbb{C} \,:\, 0 < w_2 < 1\}$ of Δ_R. The (holomorphic) covering map is

$$\varphi : M \longrightarrow \Delta_R, \qquad \varphi(w) = \exp\left(-\mathrm{i}\,(\log R)\, w\right).$$

Now $\hat{h} := h \circ \varphi : M \to \mathbb{R}$ is harmonic, and we normalize by assuming that $\hat{h}|_{w_2=0} \equiv 0$, $\hat{h}|_{w_2=1} \equiv h_0$. Since it is bounded in the lateral directions, $\hat{h}(w) = h_0 \,\mathrm{Im}\, w$; this is the imaginary part of the holomorphic function $z = h_0 w$ which, following notation in [6], we write as $z = -\theta + it$. This is defined on the strip $M_{h_0} = \{w = w_1 + \mathrm{i}w_2 \in \mathbb{C} \,:\, 0 < w_2 < h_0\}$.

We now bring in the fact that $\mathbb{A}$ is a surface of revolution, so we can write

$$\hat{F}(\theta, t) := (F \circ \varphi \circ z^{-1})(\theta, t) = r(t)\, e^{is(\theta)} \tag{1}$$

for some smooth functions s and r. The Hopf differential equals $\hat{\Phi} = \frac{1}{4}dz^2$. Since the pair $(\hat{F}, \hat{h}) = (\hat{F}, t)$ is conformal on M_{h_0}, we also have that

$$\|\hat{F}_t\|^2_{\mathbb{H}^2} + 1 = \|\hat{F}_\theta\|^2_{\mathbb{H}^2},$$

which becomes

$$\frac{4}{(1 - r(t)^2)^2}\left(s'(\theta)^2 r(t)^2 - r'(t)^2\right) = 1.$$

Rearranging this we obtain

$$s'(\theta)^2 = \frac{(1 - r(t)^2)^2}{4r(t)^2} + \left(\frac{r'(t)}{r(t)}\right)^2,$$

hence each of the two sides must equal the same constant κ^2. Write $s'(\theta) = \kappa > 0$ and

$$\kappa^2 = \frac{(1 - r(t)^2)^2}{4r(t)^2} + \left(\frac{r'(t)}{r(t)}\right)^2. \tag{2}$$

The harmonicity of $\hat{F}$ yields the equation

$$\hat{F}_{z\bar{z}} + 2(\log \rho \circ \hat{F})_u \hat{F}_z \hat{F}_{\bar{z}} = 0, \tag{3}$$

where u is a (holomorphic) coordinate of $\mathbb{H}^2$ and the metric $d\rho^2$ on $\mathbb{H}^2$ equals $\rho(u)^2|du|^2$. We compute that

$$2(\log \rho \circ \hat{F})_u = 2\overline{\hat{F}}/(1 - |\hat{F}|^2),$$

and hence the left hand side of (2) becomes

$$\frac{1}{4}e^{is(\theta)}\left(r''(t)-r(t)\left(s'(\theta)^2-is''(\theta)\right)\right)+$$
$$\frac{r(t)e^{is(\theta)}\left(r(t)^2s'(\theta)^2-r'(t)^2\right)}{2\left(r(t)^2-1\right)} \tag{4}$$

Substituting from (2) and using that $s''(\theta)=0$, we arrive at the expression

$$\hat{F}_{z\bar{z}}+2(\log\rho\circ\hat{F})_u\hat{F}_z\,\hat{F}_{\bar{z}}=\frac{e^{is(\theta)}\left(4r(t)r''(t)-4r'(t)^2+r(t)^4-1\right)}{16r(t)}.$$

We conclude finally that (3) is equivalent to

$$4r(t)r''(t)-4r'(t)^2+r(t)^4-1=0. \tag{5}$$

This is the equation obtained in a different way by Nelli and Rosenberg in [6].

From the fact that (2) is a first integral of (5), it is easy to read off that solutions $r_\kappa(t)$ are defined on the entire real line and are periodic in t, oscillating between two values:

$$\sqrt{\kappa^2+1}-\kappa\le r_\kappa(t)\le\sqrt{\kappa^2+1}+\kappa.$$

Noting that $\sqrt{\kappa^2+1}-\kappa<1<\sqrt{\kappa^2+1}+\kappa$, we see that $(\hat{F},\hat{h})$ extends to a map from $\mathbb{C}$ to $\mathbb{R}^3$, with image a complete surface of revolution which we write as U_κ; this has the conformal type of $\mathbb{C}^*$. This family of surfaces is similar in many ways to the classical family of Delaunay unduloids.

Proposition 1 *The surfaces U_κ converge to the cylinder $\{|z|=1\}\times\mathbb{R}$, as $\kappa\to 0$, and to a foliation of $\mathbb{R}^3$ by parallel planes as $\kappa\to\infty$. The connected components of $U_\kappa\cap\{|z|<1\}$, which are all identified with one another by appropriate vertical translations, are copies of the standard catenoid of revolution, which we write as C_κ, in $\mathbb{H}^2\times\mathbb{R}$, see Fig. 1.*

The surface C_κ is conformally equivalent to a proper annulus Δ_R where $R=R_\kappa\in(\mathrm{e}^{-\pi},1)$. The height $h=h(\kappa)$ of C_κ decreases monotonically from π to 0 as κ increases from 0 to ∞.

The connected components of $U_\kappa\cap\{|z|>1\}$ are related to the previous ones by the transformation:

$$I_\kappa:\mathbb{C}\times\mathbb{R}\longrightarrow\mathbb{C}\times\mathbb{R}$$

$$(z,t)\mapsto\left(\frac{1}{z},t+h(\kappa)\right).$$

Remark 1 We may also consider an alternative model of the hyperbolic plane consisting of the exterior of the unit disk in $\overline{\mathbb{C}}$:

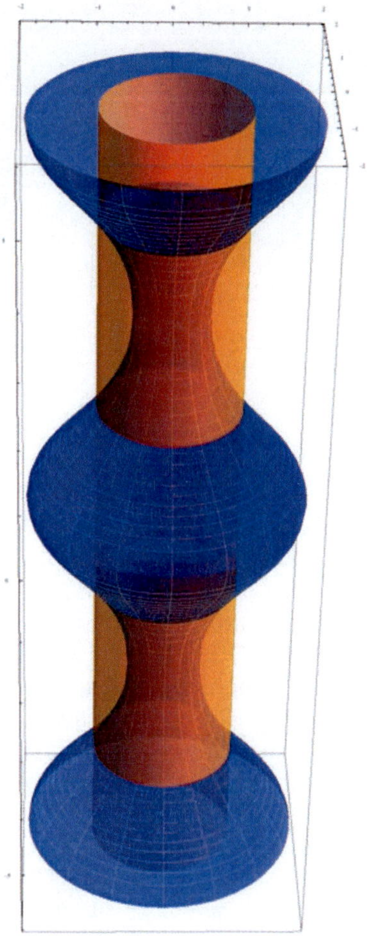

Fig. 1 One of the unduloid-type surfaces U_κ whose intersection with the solid cylinder $\{(x, y, t) \in \mathbb{R}^3 : x^2 + y^2 < 1\}$ gives the catenoids in $\mathbb{H}^2 \times \mathbb{R}$

$$\mathbb{E} = \{z \in \mathbb{C} : |z| > 1\} \cup \{\infty\},$$

with the metric

$$d\rho^2 = 4\frac{|dz|^2}{(1 - |z|^2)^2}.$$

Then

$$I_\kappa|_{U_\kappa \cap \{|z|<1\}} : U_\kappa \cap \{|z| < 1\} \longrightarrow U_\kappa \cap \{|z| > 1\}$$

is an isometry between minimal surfaces.

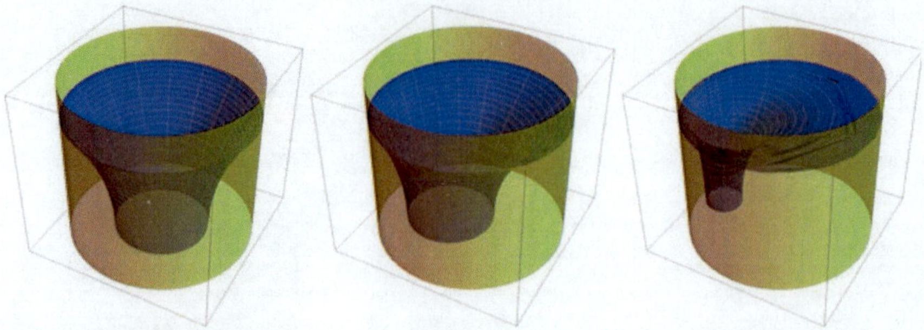

Fig. 2 The upper half of three surfaces in the sequence $\mathcal{C}_j$. The limit is the union of two horizontal disks

4 Parabolic Catenoids

We next consider a family of surfaces obtained via a particular limit of horizontal dilations of the catenoids $\mathcal{C}_\kappa$.

For any point $p \in \mathbb{H}^2$ and $h \in (0, \pi)$, let $\mathcal{C}_{h,p}$ denote the catenoid in $\mathbb{H}^2 \times \mathbb{R}$ which is rotationally symmetric around the axis $\{p\} \times \mathbb{R}$, symmetric with respect to reflections across $t = 0$ and has height $h \in (0, \pi)$. Observe that $\mathcal{C}_{h(\kappa),p}$ is obtained applying an horizontal dilation to $\mathcal{C}_\kappa$.

Now take a sequence of these catenoids, $\mathcal{C}_j := \mathcal{C}_{h_j,p_j}$ such that $p_j \to q \in \partial\mathbb{H}^2$ and h_j remains bounded away from both 0 and π. Then $\mathcal{C}_j$ converges locally in $\mathcal{C}^\infty$ on any compact set of $\mathbb{H}^2 \times \mathbb{R}$ to two horizontal disks $\mathbb{H}^2 \times \{t_j\}$, $j = 1, 2$, where $|t_2 - t_1| = \lim h_j$, see Fig. 2. (If $h_j \to 0$, then $\mathcal{C}_j$ converges to one horizontal disk with multiplicity two.)

Suppose however that we let $h_j \nearrow \pi$. Depending on the rates at which $p_j \to q$ and $h_j \to \pi$, various possibilities can occur. We suppose that these sequences are balanced in such a way that $\mathcal{C}_j$ intersects a fixed compact set $K \subset \mathbb{H}^2 \times \mathbb{R}$ for every j; this is easy to arrange by an elementary argument. In this case, standard results imply that some subsequence of the $\mathcal{C}_j$ converges locally in $\mathcal{C}^\infty$ to a complete properly embedded minimal surface which we write as $\mathfrak{D}$. By definition, $\mathfrak{D}$ is a parabolic catenoid. It has asymptotic boundary equal to the union of two horizontal circles separated by distance π, together with a vertical segment joining these two circles. This class of surfaces was discovered independently by Hauswirth [3] and Daniel [1].

We can finesse the construction of the sequence $\mathcal{C}_j$: start with a sequence of catenoids $\mathcal{C}_{\kappa_j}$, rotationally symmetric around the axis $\{o\} \times \mathbb{R}$, with $\kappa_j \searrow 0$, and lying in the slab $|t| < \frac{1}{2}h_j$. Let γ denote the hyperbolic geodesic through o and converging to $q \in \partial\mathbb{H}^2$, and let σ_j denote a hyperbolic dilation toward q along this geodesic which has the property that, extending σ_j to an isometry of $\mathbb{H}^2 \times \mathbb{R}$ which acts only on the first factor, $\sigma_j(\mathcal{C}_{\kappa_j})$ is tangent to the axis $\{o\} \times \mathbb{R}$ at the point $(o, 0)$

Fig. 3 The upper half of four surfaces in the sequence $\sigma_j\left(\mathcal{C}_{\kappa_j}\right)$. The limit as $\kappa_j \searrow 0$ is the parabolic catenoid

(i.e., on the central circle of the catenoid), with the neck of the catenoid lying between this point and q.

This sequence certainly intersects a fixed compact set for all j, hence converges to a parabolic catenoid which we denote by $\mathfrak{D}_q$, see Fig. 3. Its height is precisely equal to π. The symmetries of the catenoid and the naturality of the construction easily imply that these various Daniel surfaces are all related to one another by rotations of $\mathbb{H}^2 \times \mathbb{R}$ around the axis $\{o\} \times \mathbb{R}$. We can also apply horizontal hyperbolic dilations to these surfaces, which is the same as choosing a slightly different normalization of the sequence of surfaces C_j above; denoting the dilation parameter by λ, we obtain the family of surfaces $\mathfrak{D}_{q,\lambda}$; $\lambda = 1$ corresponds to the identity dilation. The surfaces in this entire family are all mutually isometric by rotations and hyperbolic dilations. We take as the standard model the surface in this family with $\lambda = 1$. It is an embedded disk with asymptotic boundary the two horizontal circles $S^1 \times \{0\} \sqcup S^1 \times \{\pi\}$ and the vertical segment $\{q\} \times [0, \pi]$ (Fig. 4).

Remark 2 Applying the limit process described above to the entire unduloid-type surfaces U_κ, one obtains a complete limiting surface $\mathcal{Q}_1$ in $\mathbb{R}^3$, see Fig. 5, described by the equation

$$\mathcal{Q}_1 = \left\{(x, y, t) \in \mathbb{R}^3 \,:\, 1 - x^2 - y^2 = \cos(t)((1 + x)^2 + y^2)\right\}.$$

The intersection

$$\mathcal{Q}_1 \cap \left\{(x, y, t) \in \mathbb{R}^3 \,:\, x^2 + y^2 < 1\right\}$$

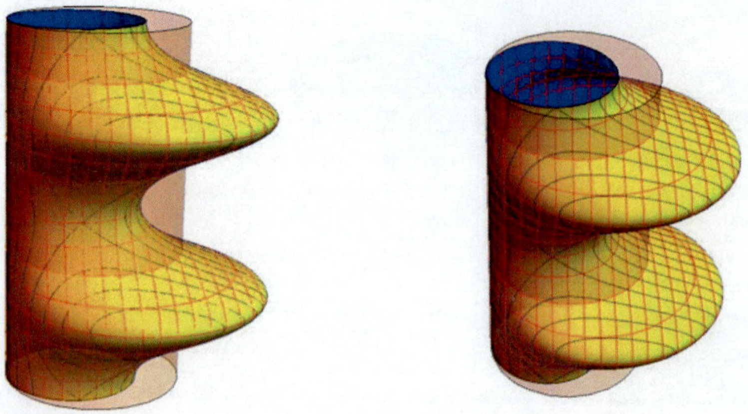

Fig. 4 Two different views of the surface $\mathcal{Q}_{1/2}$ in $\mathbb{R}^3$ whose intersection with the solid cylinder $\{(x, y, t) \in \mathbb{R}^3 : x^2 + y^2 < 1\}$ consists of translated copies of $\mathfrak{D}_{\infty,1/2}$

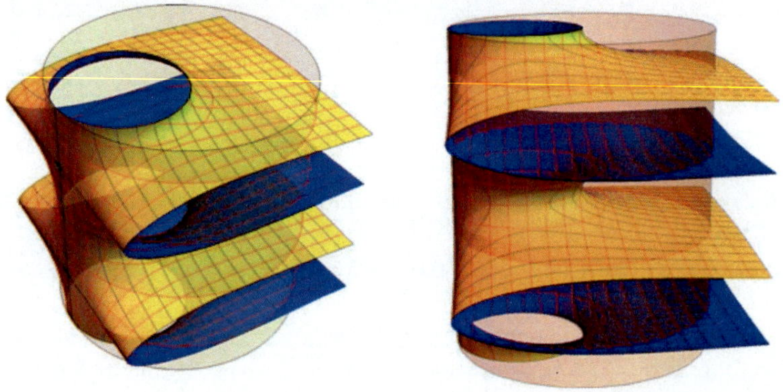

Fig. 5 Two different views of the surface $\mathcal{Q}_1$ in $\mathbb{R}^3$ whose intersection with the solid cylinder $\{(x, y, t) \in \mathbb{R}^3 : x^2 + y^2 < 1\}$ consists of translated copies of $\mathfrak{D}_{\infty,1}$

is an infinite union of (translated) copies of the parabolic catenoid. Similarly, the dilated surface $\mathfrak{D}_{q,\lambda}$ can be seen as the intersection of

$$\mathcal{Q}_\lambda = \left\{(x, y, t) \in \mathbb{R}^3 : \lambda(1 - x^2 - y^2) = \cos(t)\ ((1 + x)^2 + y^2)\right\}.$$

with the cylinder $\left\{(x, y, t) \in \mathbb{R}^3 : x^2 + y^2 < 1\right\}$, (see Figs. 4 and 6).

Note in particular that the surfaces $\mathcal{Q}_\lambda$ are invariant under the map:

$$I : \mathbb{C} \times \mathbb{R} \longrightarrow \mathbb{C} \times \mathbb{R},$$

$$(z, t) \mapsto \left(\frac{1}{z}, t + \pi\right).$$

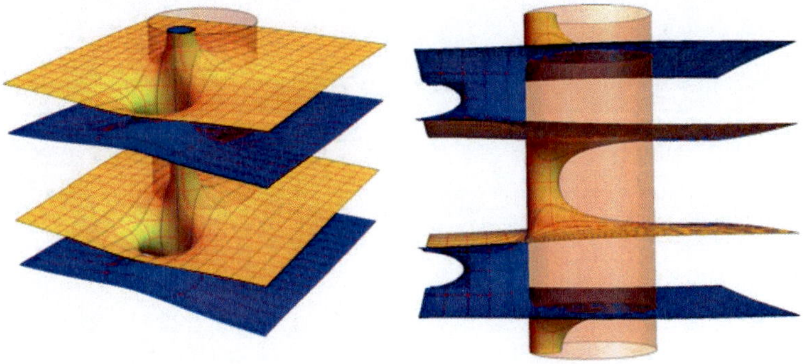

Fig. 6 Two different views of the surface $\mathcal{Q}_2$ in $\mathbb{R}^3$ whose intersection with the solid cylinder $\{(x, y, t) \in \mathbb{R}^3 \,:\, x^2 + y^2 < 1\}$ consists of translated copies of $\mathfrak{D}_{\infty,2}$

Similarly to what happens for catenoids, the map I induces an isometry

$$I : \mathcal{Q}_\lambda \cap \{(z, t) \,:\, |z| < 1\} \to \mathcal{Q}_\lambda \cap \{(z, t) \,:\, |z| > 1\}.$$

Observe that the surfaces $\mathcal{Q}_\lambda$, $\lambda \geq 1$, intersect the line $\{o\} \times \mathbb{R}$. As they are invariant under I, this means that $\mathcal{Q}_\lambda$ also contains horizontal straight lines. However, $\mathcal{Q}_\lambda \cap (\{o\} \times \mathbb{R}) = \varnothing$, for $\lambda \in (0, 1)$. Hence for these values of λ, the surface $\mathcal{Q}_\lambda$ is contained in a solid vertical cylinder.

The simplest analytic representation of $\mathfrak{D}_{q,\lambda}$ uses the upper half-space model for $\mathbb{H}^2$. Place q at infinity in $\mathfrak{H}$ and use the standard coordinates (x, y, t), $\mathfrak{H} \times \mathbb{R}$, where $y > 0$. The parabolic catenoid is invariant under parabolic translations, which in this representation take the form $(x, y, t) \mapsto (x + b, y, t)$ for any $b \in \mathbb{R}$, and it also intersects each horizontal slice $t =$ const. transversely. Therefore this surface is a sweep-out of some curve $(0, f(t), t)$, $0 < t < \pi$ by these parabolic translations. Searching for a minimal surface with these properties leads in a straightforward way to the family of solutions $f(t) = \lambda \sin t$ for any $\lambda \in \mathbb{R}^+$. The surface is the image of corresponding family of embeddings of the strip $M_\pi = \mathbb{R} \times (0, \pi)$ given by

$$\Psi_\lambda(x, t) = (\lambda x, \lambda \sin t, t). \tag{6}$$

For simplicity, we write $\Psi = \Psi_1$.

5 Tall Rectangles

The final family of surfaces in $\mathbb{H}^2 \times \mathbb{R}$ we consider here is the family of properly embedded minimal disks, described by Sa-Earp and Toubiana in [7]. These have ideal boundary consisting of two parallel arcs $\sigma \times \{\pm h/2\}$, where $h > \pi$ is arbitrary and σ is an arc in $\partial\mathbb{H}^2$ with endpoints q_1 and q_2, together with the vertical segments $\{q_1\} \times$

Fig. 7 The tall rectangle of height $h = \frac{9}{5}\pi$

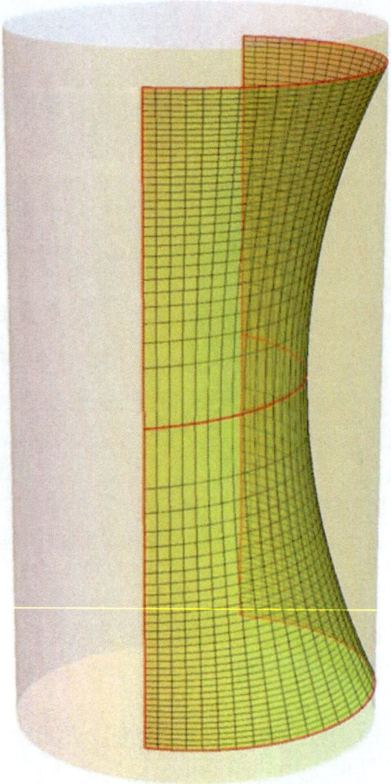

$[-h/2, h/2]$ and $\{q_2\} \times [-h/2, h/2]$, see Fig. 7. Each of these surfaces, denoted $\Sigma_{\sigma,h}$, is area minimizing. (A complete surface is area minimizing if any compact piece is area-minimizing among all the surfaces with the same boundary.) The intersections $\Sigma_{\sigma,h} \cap (\mathbb{H}^2 \times \{t\})$ foliate $\Sigma_{\sigma,h}$ by a family of curves which, if all projected down to $\mathbb{H}^2$, all have the same endpoints q_1 and q_2 and are equidistant to the geodesic $[q_1, q_2]$.

Following [7], we construct these surfaces as follows. Using the Poincaré disk model, consider the vertical plane $\mathcal{P} = \gamma \times \mathbb{R}$ where γ is the geodesic $\{\operatorname{Im} z = 0\} \subset \mathbb{D}$, parametrized either by $x \in (-1, 1)$ or by signed geodesic distance ρ from $\{o\}$. The relationship between the two parameters is $x = \tanh(\rho/2)$, $\rho = \log\left(\frac{1+x}{1-x}\right)$. Next, fix $0 < d < 1$ and consider the curve $\sigma_d \subset \mathcal{P}$ given as a bigraph of the two functions $t = \pm\lambda_d(\rho)$, i.e.,

$$\sigma_d = \left\{(\rho, \pm\lambda_d(\rho)) : \rho \geq \cosh^{-1}(1/d)\right\}.$$

We then determine the conditions under which the surface swept out by this curve with respect to horizontal hyperbolic dilations along the geodesic from $-i$ to i is minimal. A standard computations shows that minimality is equivalent to

$$\frac{d\lambda_d}{d\rho}(\rho) = \frac{1}{\sqrt{d^2 \cosh^2 \rho - 1}} \tag{7}$$

(N.B. the treatment in [7] uses $e = 1/d$ as a parameter instead.) By the chain rule,

$$\frac{d\lambda_d}{dx} = \frac{2}{\sqrt{d^2(1+x^2)^2 - (1-x^2)^2}}. \tag{8}$$

The lower bound $\rho > \cosh^{-1}(1/d)$ transforms to $x > d_1 := \frac{\sqrt{1-d}}{\sqrt{1+d}}$. Note that the derivative is infinite at $x = d_1$, so this graph together with its reflection cross the x-axis is at least $\mathcal{C}^1$. A closer analysis shows that this bigraph is in fact $\mathcal{C}^\infty$.

In summary, the complete properly embedded minimal disk Σ_d is the surface swept out by the curve σ_d, where

$$\lambda_d(x) := \int_{d_1}^{x} \frac{2\,dv}{\sqrt{d^2(1+v^2)^2 - (1-v^2)^2}}. \tag{9}$$

It is straightforward to check that:

$$\lambda_d(x) = -\frac{2}{1-d}\,\mathrm{Im}\left(F\left(\arcsin(d_1 x)\,\middle|\,\frac{1}{d_1^2}\right)\right),$$

where

$$F(\phi|z) := \int_0^{\phi}\left(1 - z\sin^2(\theta)\right)^{-1/2} d\theta, \quad -\frac{\pi}{2} < \phi < \frac{\pi}{2}$$

is the classical elliptic integral of the first kind. The surface itself is parametrized as a bigraph

$$\Upsilon_d : (d_1, 1) \times (-1, 1) \to \mathbb{H}^2 \times \mathbb{R},$$

$$\Upsilon_d(x, y) = \left(\frac{x - xy^2}{x^2y^2+1}, \frac{(x^2+1)\,y}{x^2y^2+1}, \pm\lambda_d(x)\right). \tag{10}$$

Note that the top and bottom halves $\Sigma_d^{\pm} = \Sigma_d \cap \{\pm t \geq 0\}$ are each graphs over the "lunette" region $L_d \subset \mathbb{D}$ lying between the circular arc γ_d passing through $\pm i$ and d_1, and the arc on the circumference joining $-i$ to i and passing through 1.

The curve σ_d and surface Σ_d are contained in the region where

$$|t| < \frac{1}{2}h_d = \int_{d_1}^{1} \frac{2\,dv}{\sqrt{d^2(v^2+1)^2 - (v^2-1)^2}} < \infty.$$

It is not hard from this expression to check that the height h_d of Σ_d increases monotonically in d and with the following asymptotic behavior:

- As $d \to 0$, $h_d \searrow \pi$ and Σ_d diverges to infinity. Denoting by $\widetilde{T}_d$ the horizontal dilation along γ which maps the point d_1 to 0, then the family of disks $Y_d := (\widetilde{T}_d \times \mathrm{Id}_{\mathbb{R}})(\Sigma_d)$ converges to the parabolic catenoid passing through the origin.

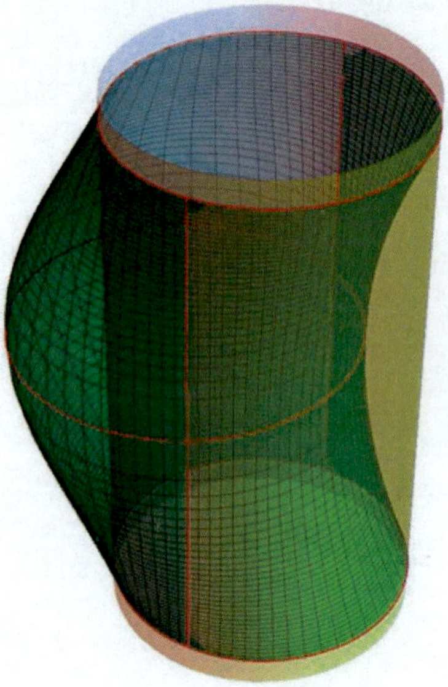

Fig. 8 The annulus $\Upsilon_d((d_1, 1/d_1) \times (-1, 1))$

- As $d \to 1$, $h_d \to \infty$ and Σ_d limits to the vertical plane $(-\mathrm{i}, \mathrm{i}) \times \mathbb{R}$.

Observe that, for a fixed $d \in (0, 1)$, the parametrisation Υ_d is defined on $(d_1, 1/d_1) \times (-1, 1)$ and the image of this extension is an annulus in $\mathbb{R}^3$, see Fig. 8. Applying an 180°-degree rotation around the line (geodesic) passing through the points $(-\mathrm{i}, h_d/2)$ and $(\mathrm{i}, h_d/2)$, then we get another annulus of height $2h_d$ which is the fundamental piece of a singly periodic surface in $\mathbb{R}^3$. The intersection of this periodic surface with the solid cylinder $\{(x, y, t) \in \mathbb{R}^3 \ : \ x^2 + y^2 < 1\}$ consists of an infinite number copies of the tall rectangle Σ_d, see Fig. 9.

Remark 3 As in the previous cases, the cylindrical surface in Fig. 9 is invariant under the map

$$I_d : \mathbb{C} \times \mathbb{R} \longrightarrow \mathbb{C} \times \mathbb{R},$$

$$(z, t) \mapsto \left(\frac{1}{z}, t + h_d\right).$$

Observe that when the tall rectangle intersects the vertical line $\{o\} \times \mathbb{R}$, the surface in $\mathbb{R}^3$ has "flat ends" similar to those in the surfaces $\mathcal{Q}_\lambda$, $\lambda \leq 1$ from Sect. 4.

6 Jacobi Fields on Parabolic Catenoids

In the remaining sections of this paper we initiate the study of the Jacobi operator on a parabolic catenoid. This is a necessary step before studying the broader space

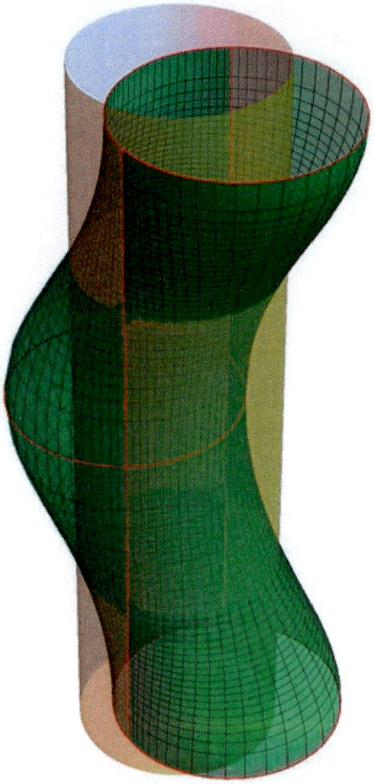

Fig. 9 Tall rectangles can also be seem as the intersection of a cylindrical surface of $\mathbb{R}^3$ with the cylinder $\{(x, y, t) \in \mathbb{R}^3 : x^2 + y^2 < 1\}$

of minimal disks with a similar asymptotic structure. In this section we consider the Jacobi fields on $\mathfrak{D}$ which are generated by deformations into catenoids or tall rectangles. The next section is a brief introduction to some more general aspects of the analysis of the Jacobi operator.

6.1 Deformations to Catenoids

We first analyse the deformations of $\mathfrak{D}$ into catenoids $\mathcal{C}_{q,h}$. A thorough analysis of the Jacobi operator and local deformation theory for catenoids $\mathcal{C}_\kappa$ appears in our earlier paper [2], but we consider here this limiting case and find an explicit expression for the Jacobi field on the parabolic catenoid arising from the 'regeneration' of this surface to the degenerating sequence of catenoids. This is a bit complicated because the rotational symmetry of the catenoids is lost in the limit. For this reason, we do all computations in the rectangular coordinates (x, y, t) on $\mathfrak{H} \times \mathbb{R}$.

Recall from Sect. 3 the conformal parametrization

Fig. 10 Half a catenoid is a graph over the annulus $A(r_0, 1)$

$$X_\kappa(\theta, t) = \left(r_\kappa(t)\mathrm{e}^{\mathrm{i}\sqrt{\kappa}\,\theta}, t\right), \quad \kappa > 0,$$

where

$$\kappa = \frac{(1 - r_\kappa(t)^2)^2}{4r_\kappa(t)^2} + \left(\frac{r'_\kappa(t)}{r_\kappa(t)}\right)^2. \tag{11}$$

We choose the solution for which $r < 1$ when $t \in \left(-\frac{h_\kappa}{2}, \frac{h_\kappa}{2}\right)$. We also know that

$$\min r_\kappa := r_0 = r_\kappa(0) = \sqrt{\kappa + 1} - \sqrt{\kappa} \leq r_\kappa(t) < 1.$$

Now write the catenoid as a bigraph over the planar annulus $A(r_0, 1)$, see Fig. 10, with t a function of r. Then

$$\kappa = \frac{(1 - r^2)^2}{4r^2} + \left(\frac{1}{r\,t'(r)}\right)^2,$$

or equivalently, $t'(r) = 2(4\kappa r^2 - (1 - r^2)^2)^{-1/2}$, whence

$$t_\kappa(r) = \int_{r_0}^{r} \frac{2\,du}{\sqrt{4\kappa u^2 - (1 - u^2)^2}}.$$

Before differentiating this with respect to κ, it is convenient to set $u = (r - r_0)s + r_0$, which yields

$$t_\kappa(r) = 2(r - r_0) \int_0^1 \frac{ds}{\sqrt{4\kappa((r - r_0)s + r_0)^2 - (1 - ((r - r_0)s + r_0)^2)^2}}.$$

We next transform to the upper half-space times $\mathbb{R}$, using the conformal diffeomorphism $g : \mathbb{D} \to \mathfrak{H}$ from the preliminaries. Write

$$\mu_0 := -\mathrm{i}\, g(r_0) = \frac{1 + \sqrt{\kappa} - \sqrt{1 + \kappa}}{1 - \sqrt{\kappa} + \sqrt{1 + \kappa}} \in (0, 1).$$

We obtain a representation of $\mathcal{C}_\kappa$ as a bigraph over the complement of a disk in the half-plane

$$\Omega_\kappa := \mathfrak{H} \setminus D(\mathrm{i}\,(\mu_0 + \mu_0^{-1})/2, (\mu_0^{-1} - \mu_0)/2) \subset \mathfrak{H},$$

where $D(z_0, r) := \{z \in \mathbb{C} : |z - z_0| < r\}$.

Next apply the horizontal dilation T_{1/μ_0} on $\mathfrak{H}$ which carries $z \mapsto \frac{1}{\mu_0} z$, and write $\widehat{\mathcal{C}}_\kappa := \left(T_{1/\mu_0} \times \mathrm{Id}_\mathbb{R}\right)(\mathcal{C}_\kappa)$. As described earlier, the parabolic catenoid is the limit of the family $\widehat{\mathcal{C}}_\kappa$, as $\kappa \to 0$. Note that

$$\lim_{\kappa \to 0} T_{1/\mu_0}(\Omega_\kappa) = \widetilde{M} = \{z = x + \mathrm{i}y \in \mathfrak{H} \;:\; 0 < y \leq 1\}.$$

Given a point $z \in \widetilde{M}$, set $r = \left\|g^{-1}(T_{\mu_0}(z))\right\|$; the corresponding point on the dilated catenoid is $(x, y, t_\kappa(r))$. This gives the family of minimal immersions

$$\Phi_\kappa(x, y) = (x, y, t_\kappa(x, y)).$$

To compute $d\Phi_\kappa/d\kappa|_{\kappa=0}$, it is first necessary to compute $dt_\kappa/d\kappa|_{\kappa=0}$. Write the integral formula for t_κ above as $t_\kappa(x, y) = \int_0^1 G(x, y, s, \kappa)ds$ and expand $G(x, y, s, \kappa)$ in powers of $\sqrt{\kappa}$:

$$G(x, y, s, \kappa) = a_0(x, y, s) + a_1(x, y, s)\sqrt{\kappa} + a_2(x, y, s)\kappa + o(\kappa^{3/2}).$$

These first few coefficients are given by

$$a_0(x, y, s) = \frac{1 - y}{\sqrt{s(1 - y)(2 - s + sy)}},$$

$$a_1(x, y, s) = \frac{-s^2y^3 + 3s^2y^2 - 3s^2y + s^2 - 3sy^2 + 6sy - 3s + y^2 - 2y + 1}{2(sy - s + 2)\sqrt{s(1-y)(sy-s+2)}}$$

and

$$a_2(x, y, s) = \frac{-2s^4(y-1)^5 + 2s^3(y-5)(y-1)^4 + s^2(10y-13)(y-1)^3}{8(sy-s+2)^2\sqrt{s(1-y)(sy-s+2)}} + \frac{2s(y-1)\left(y\left(x^2+8y-6\right)-2\right) + y\left(4x^2+(5-3y)y+7\right) - 9}{8(sy-s+2)^2\sqrt{s(1-y)(sy-s+2)}}$$

A straightforward but increasingly tedious computation gives

$$\int_0^1 a_0(x, y, s)ds = \arccos y, \tag{12}$$

$$\int_0^1 a_1(x, y, s)ds = 0, \tag{13}$$

$$\int_0^1 a_2(x, y, s)ds = \frac{1}{4}\left(\frac{x^2y}{\sqrt{1-y^2}} - \arccos y\right). \tag{14}$$

These yield

$$\lim_{\kappa \to 0} \Phi_\kappa(x, y) = \Phi_0(x, y) = (x, y, \arccos y),$$

and

$$\left.\frac{d\Phi_\kappa}{d\kappa}\right|_{\kappa=0}(x, y) = \left(0, 0, \frac{1}{4}\left(\frac{x^2y}{\sqrt{1-y^2}} - \arccos y\right)\right).$$

Finally, if we parametrize half of the parabolic catenoid as $\Phi_0(x, y) = (x, y, \arccos y)$, then the Gauss map is given by $\nu(x, y) = (0, -y^2, -\sqrt{1-y^2})$. Hence, the Jacobi field that we are looking for is

$$w(x, y) = \nu \cdot \left(\left.\frac{d\Phi_\kappa}{d\kappa}\right|_{\kappa=0}\right) = \frac{1}{4}(\sqrt{1-y^2}\arccos y - x^2y).$$

Using the parametrization $\widehat{F} : \mathbb{R} \times \left(-\frac{\pi}{2}, \frac{\pi}{2}\right) \to \mathbb{H}^2 \times \mathbb{R}$,

$$\widehat{F}(x, t) = (x, \cos t, t),$$

then this Jacobi field equals $w(x, t) = \frac{1}{4}(t \sin t - x^2 \cos t)$.

Remark 4 The parabolic catenoid in (6) and the one above differ by the vertical translation $t \mapsto t - \pi/2$. The Jacobi field on (6) is

$$w(x,t) = \frac{1}{8}\left((\pi - 2t)\cos(t) - 2x^2\sin(t)\right).$$

6.2 *Deformations to Tall Rectangles*

We next compute the Jacobi field on a parabolic catenoid associated to the variation of this surface into the family of tall rectangles. More specifically, we compute the variation associated to the family Y_d defined at the end of Sect. 5 which converge to a given parabolic catenoid $\mathfrak{D}$.

Recall the parametrization (10) for Σ_d. The top and bottom halves of Σ_d are graphs over the lunette $L_d \subset \mathbb{D}$ between the circular arc γ_d passing through $\pm i$ and d_1, and the boundary arc γ_1 passing through $\pm i$, 1. Set $\widehat{L}_d = g(L_d)$; this is the region in the half-plane $\mathfrak{H}$ between the segment $[-1, 1]$ and the circular arc passing through ± 1 and μ_1, where

$$\mu_1 = \mathrm{i}\,\frac{\sqrt{1+d} - \sqrt{1-d}}{\sqrt{1+d} + \sqrt{1-d}}.$$

Defining T_s as before, then the limit of the domains $T_{1/\mu_1}(\widehat{L}_d)$ as $d \to 0$ is the strip M where $0 < y < 1$.

Given $z = (x, y) \in M$, $z \in T_{1/\mu_1}(\widehat{L}_d)$ for all $d < d'$ if d' is small enough. The transformation of Σ_d is parametrized by the function

$$G_d(x, y) = (x, y, \lambda_d(X)),$$

where $X = X(x, y, d)$ is the positive solution to the quadratic system:

$$\begin{cases} \frac{X(1-Y^2)}{1+X^2Y^2} = \mu_1 x \\ \\ \frac{Y(1+X^2)}{1+X^2Y^2} = \mu_1 y \end{cases}$$

Now recall the formula (9) for $\lambda_d(X)$, and change variables in it by setting $r = (X - d_1)s + d_1$. This gives

$$\lambda_d(X) = \int_0^1 H(x, y, s, d)ds.$$

where

$$H(x, y, s, d) = \frac{2\,(X - d_1)ds}{\sqrt{d^2(((X-d_1)s+d_1)^2+1)^2 - (((X-d_1)s+d_1)^2-1)^2}}$$

Now proceed as in the previous subsection by expanding

$$H(x, y, s, \sqrt{d}) = h_0(x, y, s) + h_1(x, y, s)\sqrt{d} + h_2(x, y, s)d + o(d^{3/2}),$$

where

$$h_0(x, y, s) = \frac{1-y}{\sqrt{s(1-y)(2-s+sy)}},$$

$$h_1(x, y, s) = \frac{(1-y)^{3/2}\left(s^2y - s^2 + 3\,s - 1\right)}{2\sqrt{s}(sy-s+2)^{3/2}},$$

and

$$h_2(x, y, s) = \frac{-2\,s^4(y-1)^5 + 2\,s^3(y-5)(y-1)^4 + s^2(10y-17)(y-1)^3}{8(s(y-1)+2)^2\sqrt{s(1-y)(s(y-1)+2)}} + \frac{2\,s(y-1)\left(-\left(x^2+14\right)y + 8y^2 + 6\right) - y\left(4x^2 + y(3y-5) + 9\right) + 7}{8(s(y-1)+2)^2\sqrt{s(1-y)(s(y-1)+2)}}.$$

By a straightforward computation,

$$\int_0^1 h_1(x, y, s)ds = 0, \quad \text{and}$$

$$\int_0^1 h_2(x, y, s)ds = \frac{1}{4}\left(\arccos(y) - \frac{x^2y}{\sqrt{1-y^2}}\right).$$

Therefore, just as at the end of the previous subsection, the corresponding Jacobi field is

$$\widehat{w}(x, y) = -\frac{1}{4}(\sqrt{1-y^2}\arccos y - x^2y).$$

Note the unexpected fact that this is equal to $-w(x, y)$, where w is the Jacobi field associated to the deformation of $\mathfrak{D}$ into catenoids.

7 The Jacobi Operator on Parabolic Catenoids

We finally turn to the task of describing the beginnings of the analytic theory of the Jacobi operator on the parabolic catenoid $\mathfrak{D}_{\infty,1}$ given by parametrization (6) for $\lambda = 1$.

Coordinate vector fields and metric: Via the parametrization Ψ, the coordinates (x, t) on $\mathfrak{D}_{\infty,1}$ induce the coordinate vector fields

$$\Psi_*(\partial_x) = X_1 = (1, 0, 0), \qquad \Psi_*(\partial_t) = X_2 = (0, \cos t, 1).$$

The metric coefficients are

$$g_{11} = X_1 \cdot X_1 = 1/\sin^2 t, \quad g_{12} = X_1 \cdot X_2 = 0,$$

$$\text{and} \quad g_{22} = X_2 \cdot X_2 = (\cos^2 t/\sin^2 t) + 1 = 1/\sin^2 t,$$

(thus displaying the conformality of Ψ).

Jacobi operator: The unit normal to $\mathfrak{D}_{\infty,1}$ at $\Psi(x, t)$ is

$$\nu(x, t) = (0, \sin^2 t, -\cos t),$$

whence $\mathrm{Ric}_{\mathbb{H}^2\times\mathbb{R}}(\nu, \nu) = -g_{\mathbb{H}^2}((0, \sin^2 t), (0, \sin^2 t)) = -\sin^2 t$.

Remark 5 Given $q \in \partial\mathbb{H}^2$, $\tau \in \mathbb{R}$ and $\lambda > 0$, we are going to denote $\mathfrak{D}_{q,\tau,\lambda}$ the generalized catenoid whose vertical segment at infinity is $\{q\} \times [\tau, \tau + \pi]$ and is dilated by ratio λ. Noticed that $\mathfrak{D}_{q,\lambda} = \mathfrak{D}_{q,0,\lambda}$.

To compute the Jacobi operator we can avoid computing $|A|^2$ directly by the following observation.

We first compute the Jacobi field corresponding to the family $\lambda \mapsto \mathfrak{D}_{\infty,0,\lambda}$. Indeed,

$$\left.\frac{d}{d\lambda}\right|_{\lambda=1} \Psi_\lambda(x, t) = (x, \sin t, 0),$$

so the normal component of this, which is the Jacobi field we seek, equals

$$\psi = \nu \cdot (x, \sin t, 0) = \sin t.$$

This vanishes simply at both $t = 0$ and $t = \pi$ and is L^2 on any portion $|x| \leq C$ since the measure equals $dxdt/\sin^2 t$, but is not L^2 on Σ. Now, $\Delta_g = \sin^2 t(\partial_x^2 + \partial_t^2)$ and the Jacobi operator equals

$$L = -\left(\Delta_g + \mathrm{Ric}(\nu, \nu) + |A|^2\right).$$

Writing out the equality $L\psi = 0$ with $\psi = \sin t$, we obtain

$$\sin^2 t(-\sin t) + (-\sin^2 t + |A|^2)\sin t = 0 \Rightarrow |A|^2 = 2\ \sin^2 t$$

This shows that

$$L = -\sin^2 t(\partial_x^2 + \partial_t^2 + 1).$$

This is a nonnegative operator: indeed, $L = \sin^2 t L_0$, where $L_0 = -\partial_x^2 - \partial_t^2 - 1$, and its action on $L^2((\sin t)^{-2}dxdt)$ is equivalent to the action of L_0 on $L^2(dxdt)$ with Dirichlet boundary conditions at $t = 0, \pi$. This latter operator is nonnegative since $-\partial_t^2 - 1$ with Dirichlet conditions is nonnegative on $[0, \pi]$. The fact that the spectrum of L lies in $\mathbb{R}^+$ also follows from the existence of the nonnegative solution ψ to $L\psi = 0$.

The function $\tilde{u}(x, t) = x \sin t$ is another non-L^2 solution to $L\tilde{u} = 0$. It is not hard to show that u and $\tilde{u}$ span the full space of tempered solutions to the Jacobi equation which vanish at $t = 0, \pi$.

Mapping properties of the Jacobi operator: We next describe some aspects of the mapping properties of L on the infinite strip $S = \mathbb{R}_x \times [0, \pi]_t$. The remarks here are meant to be preparatory to a deeper study of the deformation theory of tall rectangles, to which we shall return in a work in progress. We consider two main questions:

(i) Find classes of functions $\phi_\pm(x)$ such that the problem $Lu = 0$ on S, $u(x, \pi) = \phi_+(x)$, $u(x, 0) = \phi_-(x)$ is solvable;
(ii) Find a class of functions f on S for which we can solve $Lu = f$ with $u = 0$ at $t = 0, \pi$, and $u \to 0$ as $x \to \pm\infty$.

We analyze these questions using the Fourier transform in x. Writing the dual variable as ξ, then question i) leads to the study of the two families of problems

$$\begin{aligned} \hat{L}_\xi \hat{u}(\xi, t) &:= \sin^2 t(-\partial_t^2 + \xi^2 - 1)\hat{u} = 0, \\ \hat{u}(\xi, \pi) &= \hat{\phi}_+(\xi),\ \hat{u}(\xi, 0) = \hat{\phi}_-(\xi) \end{aligned} \tag{15}$$

and

$$\hat{L}_\xi \hat{u}(\xi, t) = \hat{f}(\xi, t), \qquad \hat{u}(\xi, 0) = \hat{u}(\xi, \pi) = 0. \tag{16}$$

Consider (15) first. For $\xi \neq 0$, there exist two functions $v_\pm(\xi, t)$ which satisfy $\hat{L}_\xi v_\pm(\xi, t) = 0$ and

$$v_+(\xi, 0) = 0,\ v_+(\xi, \pi) = 1, \quad v_-(\xi, 0) = 1,\ v_-(\xi, \pi) = 0,$$

namely

$$\begin{cases} v_+(\xi, t) = \sin((1 - \xi^2)^{1/2} t)/\sin((1 - \xi^2)^{1/2}\pi), \\ v_-(\xi, t) = \sin((1 - \xi^2)^{1/2}(\pi - t))/\sin((1 - \xi^2)^{1/2}\pi) \end{cases}$$

when $|\xi| < 1$,

$$v_+(\pm 1, t) = t/\pi,\ v_-(\pm 1, t) = 1 - t/\pi,$$

and

$$\begin{cases} v_+(\xi, t) = \sinh((\xi^2 - 1)^{1/2}t)/\sinh((\xi^2 - 1)^{1/2}\pi), \\ v_-(\xi, t) = \sinh((\xi^2 - 1)^{1/2}(\pi - t))/\sinh((\xi^2 - 1)^{1/2}\pi) \end{cases}$$

for $|\xi| > 1$. These functions are clearly holomorphic when $\xi \in \mathbb{C} \setminus \{0, \pm 1\}$, but although they appear to be branched at $\xi = \pm 1$, they are single-valued at these points so are holomorphic away from $\xi = 0$, where they have a double pole.

Now suppose, for example, that $\hat{\phi}_\pm(\xi)$ are functions in L^1 such that $\phi_\pm(\xi)\xi^{-2} \in L^1_{\mathrm{loc}}$. Then the solution to problem i) is

$$u(x, t) = \mathcal{F}^{-1}\left(\hat{\phi}_+(\xi)v_+(\xi, t) + \hat{\phi}_-(\xi)v_-(\xi, t)\right),$$

where $\mathcal{F}$ is the Fourier transform. It is not hard to check that under these hypotheses, $u(x, t)$ is continuous on $\mathbb{R} \times [0, \pi]$, $u(x, t) \to 0$ uniformly in $t \in [0, \pi]$ as $x \to \pm\infty$, and furthermore, that

$$\int_{\mathbb{R}} u(x, \pi)\, dx = \int_{\mathbb{R}} u(x, 0)\, dx = 0.$$

It is clearly possible to choose functions $\phi_\pm$ satisfying these constraints but so that $u(x, \pi) \neq u(x, 0)$ for every $x \in \mathbb{R}$. This implies that there are infinitesimal deformations where the difference of the heights of the two boundary curves may vary, though the average of the difference of their heights equals π.

There are also some interesting constraints on Jacobi fields. Indeed, let

$$S_r = \{\Psi(x, t) \ : \ (x, t) \in [-r, r] \times [0, \pi]\}$$

denote the truncated surface. Consider the basic Jacobi field $u(x, t) = \sin t$, and suppose that w is any other Jacobi field, i.e., $Lw = 0$, which has sufficient decay as $|x| \to \infty$ for the following computations to make sense (such Jacobi fields certainly exist by virtue of the preceding calculations.) We then compute that

$$0 = \int_{S_r} ((Lu)w - u(Lw)) \frac{1}{\sin^2 t} dxdt =$$
$$\int_{-r}^{r} (w(x, \pi) + w(x, 0))\, dx + \int_0^\pi \sin t\ (w_x(r, t) - w_x(-r, t))\, dt.$$

We conclude that Jacobi fields (at least the well-behaved ones) must satisfy the 'moment condition'

$$\int_{-\infty}^{\infty} (w(x, \pi) + w(x, 0))dx + \int_0^\pi \sin t\ (w_x^+(t) - w_x^-(t))dt = 0,$$

where $w_x^{\pm}(t) := \lim_{x\to\pm\infty} w_x(x,t)$. The precise geometric meaning of this is not evident.

To convert these infinitesimal statements into statements about minimal surfaces near to $\mathfrak{D}_{q,\tau,\lambda}$, it is necessary to solve the inhomogeneous problem (ii). The details of this proceed in an unsurprising fashion: passing to the Fourier transform again, there is a Green function $\hat{G}(\xi,t,t')$ for $\hat{L}_\xi$ for $\xi \neq 0$, and this can be used to solve $Lu = f$ for a broad collection of functions f. By this linear theory and standard use of the implicit function theorem, the Jacobi fields discussed earlier can be integrated to nearby minimal surfaces.

The point of all of this is the following. There exist deformations of $\mathfrak{D}_{q,\tau,\lambda}$ which deform the top and bottom boundary curves to $\Gamma_\pm$, but which fix the vertical line connecting them. Although it might be natural to conjecture that any such deformation has top and bottom boundary separated at exactly distance π, i.e., $\Gamma_+(x) - \Gamma_-(x) = \pi$ for all x, we have shown that this is not true.

References

1. Daniel, B.: Isometric immersions into $\mathbb{S}^n \times \mathbb{R}$ and $\mathbb{H}^n \times \mathbb{R}$ and applications to minimal surfaces. Trans. A.M.S. **361**, 6255–6282 (2009)
2. Ferrer, L., Martín, F., Mazzeo, R., Rodríguez, M.: Properly embedded minimal annuli in $\mathbb{H}^2 \times \mathbb{R}$. Math. Ann. **375**, 541–594 (2019)
3. Hauswirth, L.: Minimal surfaces of Riemann type in three-dimensional product manifolds. Pacific J. Math. **224**, 91–117 (2006)
4. Kloeckner, B., Mazzeo, R.: On the asymptotic behavior of minimal surfaces in $\mathbb{H}^2 \times \mathbb{R}$. Indiana Math J. **66**(2), 631–658 (2017)
5. Mazet, L., Rodríguez, M., Rosenberg, H.: The Dirichlet problem for the minimal surface equation -with possible infinite boundary data- over domains in a Riemannian surface. Proc. Lond. Math. Soc. **102**(3), 985–023 (2011). https://doi.org/10.1112/plms/pdq032. arXiv:0806.0498
6. Nelli, B., Rosenberg, H.: Minimal surfaces in $\mathbb{H}^2 \times \mathbb{R}$. Bull. Braz. Math. Soc. (N.S.) **33**(2), 263–292 (2002)
7. Sa Earp, R., Toubiana, E.: An asymptotic theorem for minimal surfaces and existence results for minimal graphs in $\mathbb{H}^2 \times \mathbb{R}$. Math. Ann. **342** 309–331 (2008)

Jenkins–Serrin Graphs in $M \times \mathbb{R}$

Eddygledson S. Gama, Esko Heinonen, Jorge H. de Lira, and Francisco Martín

Abstract The so called Jenkins–Serrin problem is a kind of Dirichlet problem for graphs with prescribed mean curvature that combines, at the same time, continuous boundary data with regions of the boundary where the boundary values explodes either to $+\infty$ or to $-\infty$. We give a survey on the development of Jenkins–Serrin type problems over domains on Riemannian manifolds. The existence of this type of graphs imposes restrictions on the geometry of the boundary of these domains. We also improve some earlier results by proving Theorem 8, and prove the existence of translating Jenkins–Serrin graphs (Theorem 9).

Keywords Jenkins–Serrin solutions · Mean curvature flow solitons · Minimal surfaces

E. S. Gama—Supported by Capes/PDSE/88881.132464/2016-01.

E. Heinonen—Supported by the Finnish Academy of Science and Letters and by the Centre of Excellence in Analysis and Dynamics Research (Academy of Finland, decision 271983).

J. H. de Lira—Supported by PRONEX/FUNCAP/CNPq PR2-0101-00089.01.00-15. and CNPq/Edital Universal 409689/2016-5.

F. Martín—Partially supported by the MINECO/FEDER grant MTM2017-89677-P and by the Leverhulme Trust grant IN-2016-019.

E. S. Gama (✉) · J. H. de Lira
Universidade Federal do Ceará, Fortaleza, Ceará 60455-760, Brazil
e-mail: eddygledson@gmail.com

J. H. de Lira
e-mail: jorge.lira@mat.ufc.br

E. S. Gama · E. Heinonen · F. Martín
Departamento de Geometría y Topología, Universidad de Granada, 18071 Granada, Spain
e-mail: ea.heinonen@gmail.com

F. Martín
e-mail: fmartin@ugr.es

E. Heinonen
Department of Mathematics and Statistics, University of Helsinki, Helsinki, Finland

T. Hoffmann et al. (eds.), *Minimal Surfaces: Integrable Systems and Visualisation*, Springer Proceedings in Mathematics & Statistics 349,
https://doi.org/10.1007/978-3-030-68541-6_6

1 Introduction

Let (M, σ) be a Riemannian n-manifold (not necessarily complete) and let $\Omega \subset M$ be a domain whose boundary is piecewise smooth. The graph $\Sigma = \text{Graph}(u)$ of a smooth function $u : \Omega \to \mathbb{R}$ is a smooth hypersurface in the product manifold $M \times \mathbb{R}$ endowed with the product metric $g_0 := \sigma + dt^2$. Moreover, it is standard to check that if $H : M \to \mathbb{R}$ is a smooth function, then Σ has prescribed mean curvature H if and only if

$$\text{div}\left(\frac{\nabla u}{\sqrt{1 + |\nabla u|^2}}\right) = H, \tag{1}$$

for all $x \in \Omega$, where ∇ and div are the gradient and divergence with respect to σ.

In this paper we will work in the cases when either Σ is minimal, or Σ has constant mean curvature, or Σ is a translating soliton of the mean curvature flow. Moreover, we will adopt the convention that *the mean curvature is just the trace of the second fundamental form.*

First of all, let us explain what we mean by a Jenkins–Serrin solution of (1). Given any Riemannian manifold M of dimension $n \geq 2$, let $\Omega \subset M$ be a bounded Lipschitz domain whose boundary can be written as

$$\partial\Omega = \Gamma_0 \cup \Gamma_1 \cup \Gamma_2 \cup N$$

where $\Gamma_0, \Gamma_1, \Gamma_2$ are open disjoint subsets of class C^1 and N is a closed subset such that the $(n-1)$-dimensional Hausdorff measure $\mathcal{H}_{n-1}(N) = 0$. The Jenkins–Serrin solution of (1) with continuous data $f_0 : \Gamma_0 \to \mathbb{R}$ is a smooth function $u : \Omega \to \mathbb{R}$ that satisfies (1), $u|_{\Gamma_0} = f_0$ and

i. $u \to +\infty$ as $x \to \Gamma_1$;
ii. $u \to -\infty$ as $x \to \Gamma_2$.

As we will see later, the conditions i and ii impose important restrictions on the extrinsic geometry of Γ_1 and Γ_2, in general.

Probably, the best known example of Jenkins–Serrin solutions is given by the classical Scherk's minimal surface, that is the graph of the function $f(x, y) = \ln(\cos(x)/\cos(y))$ over the domain $\Omega = (-\pi/2, \pi/2) \times (-\pi/2, \pi/2)$. This idea of construction was generalized by Jenkins and Serrin in [14], but before stating their theorem, we need some notation. Let $\Omega \subset \mathbb{R}^2$ be a bounded domain whose boundary consists of a finite number of open straight arcs $A_1, \ldots, A_r$ (the set Γ_1), $B_1 \ldots, B_s$ (the set Γ_2), and a finite number of convex arcs $C_1, \ldots, C_t$ (the set Γ_0) with respect to Ω and the end points of these arcs (the set N). Let $\mathcal{P}$ be a polygon whose vertices are chosen among the vertices of Ω; this polygon will be called *an admissible polygon*. We are going to use also the following notation:

$$\alpha(\mathcal{P}) = \sum_{A_i \subset \mathcal{P}} ||A_i||, \ \beta(\mathcal{P}) = \sum_{B_i \subset \mathcal{P}} ||B_i|| \text{ and } \ell(\mathcal{P}) = \text{perimeter}(\mathcal{P}),$$

where $||A||$ denotes the arc length of A.

Theorem 1 (Jenkins–Serrin) *Let $\Omega \subset \mathbb{R}^2$ be as above. Assume also that no two arcs A_i and no two arcs B_i have a common endpoint. Given any continuous data $f_i\colon C_i \to \mathbb{R}$, then there exists a Jenkins–Serrin solution $u\colon \Omega \to \mathbb{R}$ for the minimal graph equation with continuous data $u|_{C_i} = f_i$ if, and only if, for any admissible polygon $\mathcal{P}$ we have*

$$2\alpha(\mathcal{P}) < \ell(\mathcal{P}) \text{ and } 2\beta(\mathcal{P}) < \ell(\mathcal{P}). \tag{2}$$

If $\{C_i\} = \varnothing$, we also require that $\alpha(\partial\Omega) = \beta(\partial\Omega)$ for $\mathcal{P} = \partial\Omega$. Moreover, if $\{C_i\} \neq \varnothing$, then u is unique, and if $\{C_i\} = \varnothing$, then u is unique up to adding a constant.

In [14], Jenkins and Serrin proved also that the existence of minimal Jenkins–Serrin solutions imposes several restrictions over the domain.

Theorem 2 (Jenkins–Serrin) *Let $\Omega \subset \mathbb{R}^2$ be a bounded domain. Suppose that $u\colon \Omega \to \mathbb{R}$ is a solution of the minimal graph equation satisfying $u \to \pm\infty$ as $x \to \gamma$, where γ is a smooth connected component of $\partial\Omega$. Then γ is a geodesic.*

Spruck [26] extended Jenkins' and Serrin's work for the CMC case. He proved the existence of CMC Jenkins–Serrin solutions for very specific domains in $\mathbb{R}^2$. The setting in [26] was the following:

Let $\Omega \subset \mathbb{R}^2$ be a bounded domain whose boundary consists of finite number of open arcs $A_1, \ldots, A_r$, $B_1 \ldots, B_s$, $C_1, \ldots, C_t$ and the endpoints of these arcs. Suppose that the curvatures of these arcs satisfy

$$\kappa(A_i) = H, \quad \kappa(B_i) = -H, \quad \kappa(C_i) \geq H. \tag{3}$$

Now, we say that $\mathcal{P}$ is *admissible* if $\mathcal{P}$ is a simple closed curvilinear polygon whose sides consists of circular arcs of curvature $\kappa = \pm H$ and whose vertices belong to the set of end points of the arcs $\{A_i\}$ and $\{B_i\}$. As before, α and β, respectively, denotes the total length of the arcs A_i and B_i, and ℓ denotes the perimeter of $\mathcal{P}$.

Theorem 3 (Spruck) *Let $\Omega \subset \mathbb{R}^2$ be a domain as above satisfying (3) with $\{C_i\} \neq \varnothing$ and $|B_i| < \pi/H$. Assume also that no two arcs A_i and no two arcs B_i have a common endpoint and that the domain Ω^* formed by reflection[1] of the arcs B_i is contained in a disk of radius $1/H$. Given any continuous data $f_i\colon C_i \to \mathbb{R}$, there exists a unique Jenkins–Serrin solution $u\colon \Omega \to \mathbb{R}$ with mean curvature H and satisfying $u|_{C_i} = f_i$, if and only if, for any admissible polygon $\mathcal{P}$ we have*

$$2\alpha(\mathcal{P}) < \ell(\mathcal{P}) + H\,\mathrm{Area}(\mathcal{P}) \text{ and } 2\beta(\mathcal{P}) < \ell(\mathcal{P}) - H\,\mathrm{Area}(\mathcal{P}).$$

In the case $\{C_i\} = \varnothing$, he obtained that the necessary and sufficient conditions for the existence of the solution are

$$2\alpha(\mathcal{P}) < \ell(\mathcal{P}) + H\,\mathrm{Area}(\mathcal{P}) \text{ and } 2\beta(\mathcal{P}) < \ell(\mathcal{P}) - H\,\mathrm{Area}(\mathcal{P})$$

[1] We reflect B_i with respect to the line passing through the end points. In this way we obtain an arc B_i^* with curvature $\kappa(B_i^*) = H$.

and

$$\alpha(\mathcal{P}) = \beta(\mathcal{P}) + H \operatorname{Area}(\Omega).$$

In higher dimensions the question is more difficult but Spruck proved the following local existence of CMC Jenkins–Serrin solution.

Theorem 4 (Spruck) *Let S be a C^2 hypersurface in $\mathbb{R}^n$ with mean curvature $H(>0)$. Let p be an interior point of S, and $B_\varepsilon(p)$ a ball with center p and radius ε. For small ε, let D^+ and D^- be the domains bounded by S and $\partial B_\varepsilon(p)$ so that the mean curvature $H_S = H$ with respect to D^+ and $H_S = -H$ with respect to D^-. Then for ε small enough, there exists a Jenkins–Serrin solution u in D^+ (respectively D^-) with boundary values $+\infty$ (respectively $-\infty$) on S and prescribed continuous data f on the remainder of the boundary.*

Moreover, he obtained an analogous result to Theorem 2 making a clever use of suitable barriers.

Theorem 5 (Spruck) *Let $\Omega \subset \mathbb{R}^n$ be a domain that is bounded in part by a C^2 hypersurface S. Suppose that $u\colon \Omega \to \mathbb{R}$ is a solution of the CMC equation with constant mean curvature $H(>0)$ satisfying $u \to \infty$ (respectively $u \to -\infty$) as $x \to S$. Then $H_S = H$ (respectively $-H$), where H_S denotes the mean curvature of S in $\mathbb{R}^n$.*

Further results about Jenkins–Serrin solutions in higher dimensions can be found in [10, 16].

Nelli and Rosenberg [21] extended the Jenkins–Serrin theory for domains in the hyperbolic plane $\mathbb{H}^2$, and after that, Pinheiro [22] to more general Riemannian surfaces.

Although we are only considering the case of bounded domains, we would like to mention that Mazet et al. [17] proved that, in the minimal case, it is possible to get Jenkins–Serrin solutions over non-compact domains in the hyperbolic plane $\mathbb{H}^2$. Their result generalizes previous results obtained by Collin and Rosenberg in [6]. To state the result, we use the following notation.

We say that a domain $\Omega \subset \mathbb{H}^2$ is an ideal Scherk domain if $\partial_\infty \Omega$ consists of a finite number of geodesic arcs A_i, B_i, a finite number of convex arcs C_i (with respect to Ω), a finite number of open arcs D_i at $\partial_\infty \mathbb{H}^2$, together with their endpoints (called the vertices of Ω) and no two A_i and no two B_i have a common end point. Now let Ω be an ideal Scherk domain; a polygon $\mathcal{P} \subset \Omega$ is said to be an admissible polygon if its vertices are among the vertices of Ω. For any ideal vertex p_i of Ω at $\partial_\infty \mathbb{H}^2$, let H_i be a horocycle at p_i such that for any $i \neq j$ we have $H_i \cap H_j = \varnothing$, and H_i does not intersect the bounded edges of $\partial \Omega$. Finally, given any admissible polygon $\mathcal{P} \subset \Omega$, denote by $\Gamma(\mathcal{P})$ the part of $\partial \mathcal{P}$ outside the horocycles H_i and call

$$\alpha(\mathcal{P}) = \sum_{A_i \subset \mathcal{P}} ||A_i \cap \Gamma(\mathcal{P})|| \text{ and } \beta(\mathcal{P}) = \sum_{B_i \subset \mathcal{P}} ||B_i \cap \Gamma(\mathcal{P})||.$$

With this notation, we can state their theorem.

Theorem 6 (Mazet–Rodríguez–Rosenberg) *Suppose that there is at least one edge C_i or D_i in $\partial_\infty \Omega$. Then there exists a Jenkins–Serrin solution with continuous data $f_i : C_i \to \mathbb{R}$ and $g_i : D_i \to \mathbb{R}$ if, and only if, the horocycles H_i can be chosen so that*

$$2\alpha(\mathcal{P}) < ||\Gamma(\mathcal{P})|| \text{ and } 2\beta(\mathcal{P}) < ||\Gamma(\mathcal{P})||.$$

for any admissible polygon $\mathcal{P}$ in Ω.

Around the same time, Gálvez and Rosenberg [9] proved similar result for simply connected and complete Riemannian surfaces whose Gaussian curvature is bounded from above by a negative constant.

Concerning the CMC case, Hauswirth et al. [11] proved the existence of Jenkins–Serrin solutions in $\mathbb{H}^2$ and $\mathbb{S}^2$. Later Folha and Rosenberg [8] generalized these results to the case of Hadamard surfaces.

We would like to point out that Folha and Rosenberg also proved the following theorem that provides a relationship between the mean curvature of the graph and the mean curvature of the boundary of the domain.

Theorem 7 (Folha–Rosenberg) *Let M be a Hadamard surface and $\Omega \subset M$ be a bounded domain. Suppose that $u : \Omega \to \mathbb{R}$ is a solution of CMC equation with constant mean curvature $H (> 0)$ satisfying $u \to \infty$ (respectively $u \to -\infty$) as $x \to \gamma$, where γ is a smooth connected component of $\partial\Omega$. Then $H_\gamma = H$ (respectively $-H$), where H_γ denotes the mean curvature of γ in M.*

Notice that a fact repeatedly appears in Theorems 2, 5 and 7. Namely, they showed that if we have a Jenkins–Serrin solution over a domain Ω, then the mean curvature of $\partial\Omega$ and the mean curvature of the surface given by the graph of the Jenkins–Serrin solution are equal, up to a sign. This suggests the following question:

Question 1 *How does the existence of complete graphs with prescribed mean curvature over a regular domain affect the geometrical structure of its boundary?*

In relation to the translating graphs, Shahriyari [24] obtained a complete result for translating graphs over regular domains in $\mathbb{R}^2$. The main idea in Shahriyari's proofs is to use Schoen's type estimates for stable minimal surfaces in 3-dimensional manifolds, an important tool in Colding & Minicozzi theory, to get information about the behaviour of the graph as we approach a component γ of the boundary, like Folha and Rosenberg do in [8]. Finally, she concluded that γ is a geodesic or has constant principal curvature in a CMC case. The main drawback of their method is that their proofs do not work in higher dimension.

Using a quite different strategy Eichmair and Metzger [7] were able to improve earlier result by Spruck (Theorem 3), in the case where $\{C_i\} = \emptyset$, into Riemannian manifolds of dimension $2 \leq n \leq 7$. Namely, they didn't need the existence of an auxiliary domain Ω^* nor a priori assumptions about existence of subsolutions, see [7, Sect. 6.1]. On the other hand, inspired by a classical compactness results of stable hypersurfaces by Schoen and Simon [23], Eichmair and Metzger gave a very simple proof that makes it possible to get an extension of Theorems 2, 5 and 7 for any

n-dimensional Riemannian manifold for $2 \leq n \leq 7$. We will show that the same method can be applied for any Riemannian manifold without any restriction on the dimension. More precisely, we prove the following theorem.

Theorem 8 *Let M be a complete Riemannian manifold and $\Omega \subset M$ be a domain with piecewise smooth boundary. Let $\Lambda \subset \partial\Omega$ be a smooth connected region and let Σ be the graph of a smooth function u on Ω in $M \times \mathbb{R}$. Suppose Σ is complete as we approach Λ. Then, we have*

a. If Σ is a translating soliton, then $H_\Lambda = 0$;
b. If Σ has constant mean curvature H_0 (with respect to the upward pointing normal vector field), then $H_\Lambda = H_0$, up to the sign.

The crucial ingredient in our proof is a local area estimate for the components of the graph in closed balls around the cylinder over the boundary. These area estimates, combined with a suitable compactness theorems (Theorem 10 in the minimal case and Theorem 11 for CMC hypersurfaces), yield precise information of the geometric structure of the domain through a surprisingly simple local analysis.

In the last section we consider the existence of Jenkins–Serrin type solutions to the translating soliton equation and prove the following result.

Theorem 9 *Let $\Omega \subset M$ be an admissible domain with $\{B_i\} = \varnothing$. Given any continuous data $f_i \colon C_i \to \mathbb{R}$, there exists a Jenkins–Serrin solution $u \colon \Omega \to \mathbb{R}$ for translating equation with continuous data $u|_{C_i} = f_i$ if for any admissible polygon $\mathcal{P}$ we have*

$$2\alpha(\mathcal{P}) < \ell(\mathcal{P}).$$

2 Compactness Theorem for Stable Integral Varifolds

In this section, we give a brief review of the results that we need about varifolds. We refer the reader to [25] for an elegant introduction on the subject. Before we state the compactness theorem for varifolds, recall the following concepts of weak and strong convergence.

Definition 1 Let $\{S_i\}$ be a sequence of varifolds in a smooth manifold $\bar{M}$. We say that $\{S_i\}$ converges weakly to a varifold S, if for every smooth function $\varphi \colon \bar{M} \to \mathbb{R}$ with compact support we have

$$\lim_{i\to\infty} \int_{\bar{M}} \varphi \, \mathrm{d}\mu_{S_i} = \int_{\bar{M}} \varphi \, \mathrm{d}\mu_S,$$

where μ_{S_i} is the Radon measure in $\bar{M}$ associated to varifold S_i. If the sequence $\{S_i\}$ converges weakly to S, then we will write $S_i \rightharpoonup S$.

Let S be a hypersurface in a Riemannian manifold $\bar{M}$. Given $p \in S$ and $r > 0$ we denote by $B_p(r)$ the tangent ball in $T_pS \subset T_p\bar{M}$ centered at $0 \in T_pS$ with radius r.

$$B_r(p) = \{v \in T_pS \colon |v| < r\}$$

Let N be a unit normal vector to S at p. Fixed a sufficiently small $\varepsilon > 0$, let $W_{r,\varepsilon}(p)$ be the solid cylinder around p, that is,

$$W_{r,\varepsilon}(p) := \{\exp_p(q + tN) \colon q \in B_r(p) \text{ and } |t| < \varepsilon\},$$

where exp is the exponential map of the manifold $\bar{M}$. Given a smooth function $f : B_r(p) \to (-\varepsilon, \varepsilon)$, the set

$$\text{Graph}[f] := \{\exp_p(q + u(q)N) \colon q \in B_r(p)\},$$

is called the graph of f over $B_r(p)$. Finally, the convergence in C^∞-topology can be defined as follows.

Definition 2 Let $\{S_i\}$ be a sequence of smooth codimension one submanifolds of a smooth manifold $\bar{M}$. We say that $\{S_i\}$ converges in C^∞-topology with finite multiplicity to a smooth embedded submanifold S if

a. S consists of accumulation points of $\{S_i\}$, that is, for each $p \in S$ there exists a sequence $\{p_i\}$ such that $p_i \in S_i$, for each $i \in \mathbb{N}$, and $p = \lim_i p_i$;
b. For every $p \in S$ there exist $r, \varepsilon > 0$ such that $S \cap W_{r,\varepsilon}(p)$ can be represented as a graph of a function f over $B_r(p)$;
c. For i large enough, the set $S_i \cap W_{r,\varepsilon}(p)$ consists of a finite number, say k, independent of i, of graphs of functions $f_i^1, \dots, f_i^k$ over $B_r(p)$ that converge smoothly to f.

The multiplicity of a given point $p \in S$ is defined by k. If $\{S_i\}$ converges smoothly to S, then we will write $S_i \to S$.

In this setting, the strong compactness theorem for integral varifolds can be stated as follows.

Theorem 10 (Strong compactness theorem) *Let $\{S_i\}$ be a sequence of stable minimal hypersurfaces in Riemannian manifold $\bar{M}^{n+1}$, without boundary, such that*

$$\limsup \mu_{S_i}(\bar{M}) < \infty,$$

where μ_{S_i} is the (Riemannian) measure associated to S_i. Suppose that there exists a sequence $\{p_i\}$ so that $p_i \in S_i$ and $p_i \to p_\infty \in \bar{M}$. Then there exist a closed set, sing S_∞, *and a stationary integral varifold, S_∞, such that a subsequence of $\{S_i\}$ converges smoothly to S_∞ away from* sing S_∞ *and $S_i \rightharpoonup S_\infty$ with $p_\infty \in \text{spt}\, S_\infty$. Moreover, the set* sing S_∞ *has Hausdorff dimension at most $n - 7$. Hence, it is empty for $n < 7$.*

Remark 1 This version is a consequence of regularity and compactness results due to Schoen and Simon in [23] and Corollary 17.8 in [25]. See also [27].

In very recent works about regularity and compactness of stable CMC integral varifolds Bellettini and Wickramasekera, in [2–4], have extended the previous theorem to the class of all stable CMC integral varifolds in the following way:

Theorem 11 (General compactness theorem) *Let $\{S_i\}$ be a sequence of stable CMC hypersurfaces in Riemannian manifold $\bar{M}^{n+1}$, without boundary, satisfying*

$$\limsup\{\mu_{S_i}(\bar{M}) + |H_i|\} < \infty,$$

where μ_{S_i} is the (Riemannian) measure associated to S_i and H_i is the mean curvature of S_i. Suppose that there exists a sequence $\{p_i\}$ so that $p_i \in S_i$ and $p_i \to p_\infty \in \bar{M}$. Then there exist a real number H_∞, a closed set $\operatorname{sing} S_\infty$, an integral varifold S_∞ and subsequence $\{S_{i_j}\}$ and subsequence of $\{H_{i_j}\}$ such that $S_i \rightharpoonup S_\infty$, $H_i \to H_\infty$ and $p_\infty \in \operatorname{spt} S_\infty$. Moreover $\operatorname{sing} S_\infty \setminus \operatorname{sing}_T S_\infty$ has Hausdorff dimension at most $n-7$, and $\operatorname{sing}_T S_\infty \subset \text{gen-reg}\, S_\infty$, $\operatorname{sing} S_\infty$ is locally contained in a smooth submanifold of dimension $n-1$, and $\text{gen-reg}\, S_\infty$ is a classical CMC immersion with mean curvature H_∞.

Remark 2 Both Theorems 10 and 11, as stated above, are weak versions of the compactness theorems by Bellettini and Wickramasekera.

In the statement of Theorem 11 the sets $\operatorname{sing}_C S_\infty$, $\operatorname{reg}_1 S$, $\operatorname{sing}_T S_\infty$ and $\text{gen-reg}\, S_\infty$ are defined as follows. Here $B_\varepsilon(x)$ denotes an open ball of radius ε and center x in $\bar{M}$.

Definition 3 (*Set of classical singularities*) $\operatorname{sing}_C S_\infty$ denotes the set of all $x \in \operatorname{spt} S$ such that, for some $\alpha \in (0, 1]$, there exists $\varepsilon > 0$ so that $\operatorname{spt} S \cap B_\varepsilon(x)$ is the union of three or more embedded $C^{1,\alpha}$ hypersurface with boundary meeting pairwise only along their common $C^{1,\alpha}$ boundary γ containing x and such that at least one pair of the hypersurfaces with boundary meet transversely everywhere along γ.

Definition 4 (*C^1 regular set*) $\operatorname{reg}_1 S$ denotes the set of all $x \in \operatorname{spt} S$ such that there exists $\varepsilon > 0$ so that $\operatorname{spt} S \cap B_\varepsilon(x)$ is an embedded hypersurface of $B_\varepsilon(x)$ of class C^1.

Definition 5 (*Set of touching singularities*) $\operatorname{sing}_T S_\infty$ denotes the set of all $x \in \operatorname{sing} S_\infty \setminus \operatorname{reg}_1 S_\infty$ so that $x \notin \operatorname{sing}_C S_\infty$ and there exists $\varepsilon > 0$ such that $\operatorname{spt} S_\infty \cap B_\varepsilon(x)$ is the union of two embedded $C^{1,\alpha}$ hypersurfaces of $B_\varepsilon(x)$ which touch at x.

Definition 6 (*Generalized regular set*) $\text{gen-reg}\, S_\infty$ denotes the set of all points $x \in \operatorname{spt} S_\infty$ so that either $x \in \operatorname{reg} S_\infty$ or $x \in \operatorname{sing}_T S_\infty$ and there exists $\varepsilon > 0$ such that $\operatorname{spt} S_\infty \cap B_\varepsilon(x)$ is the union of two embedded C^∞ hypersurfaces of $B_\varepsilon(x)$.

3 Translating Graphs

In this section we present some preliminaries on translating and minimal graphs. We will first study the behaviour of the translating graphs in the product space $M \times \mathbb{R}$

and in the end we prove our main theorem for translating and minimal graphs. Here M is a complete Riemannian manifold whose metric is denoted by σ. Following [15] we say that a hypersurface Σ in $M \times \mathbb{R}$ is a translating soliton with respect to the parallel vector field $X = \partial_t$ (with translation speed $c \in \mathbb{R}$) if

$$\mathbf{H} = c\, X^{\perp},$$

where $\mathbf{H}$ is the mean curvature vector field of Σ and $\perp$ indicates the projection onto the normal bundle of Σ. In particular, if N is a normal vector field along Σ, then we have

$$H = c\langle X, N\rangle, \tag{4}$$

where $\langle\cdot,\cdot\rangle$ denotes the Riemannian product metric in $M \times \mathbb{R}$.

Although translating solitons seem to define a completely new class of hypersurfaces, the next result due to Ilmanen [13] tells that they are actually minimal hypersurfaces with respect to the so-called Ilmanen's metric

$$g_c = e^{\frac{2c}{m}t}(\sigma + \mathrm{d}t^2) \tag{5}$$

which is a metric conformal to the Riemannian product metric $g_0 = \sigma + \mathrm{d}t^2$ in $M \times \mathbb{R}$, see also [15].

Hence (4) can be interpreted as the fact that the mean curvature of Σ with respect to g_c vanishes, i.e.

$$e^{\frac{c}{m}t}\widetilde{H} = H - c\langle X, N\rangle = 0. \tag{6}$$

This is the Euler–Lagrange equation (for normal variations) of the volume functional

$$\mathcal{A}_c[\Sigma] = \int_\Sigma e^{c\eta}\mathrm{d}\mu_\Sigma$$

where $\eta = t|_\Sigma$ and $e^{c\eta}\,\mathrm{d}\mu_\Sigma$ is the volume element in Σ induced by g_c.

Remark 3 In geometric measure theory, $\mathcal{A}_c[\Sigma]$ is usually called the mass of the varifold associate to Σ. Indeed, the standard notation to $\mathcal{A}_c[\Sigma]$ is $M_c(\Sigma)$, where we put the index c to indicate the dependence of the metric g_c. However, we will use the notation $\mathcal{A}_c[\Sigma]$ to denote the integral above.

Lemma 1 (T. Ilmanen) *Translating solitons with translation speed $c \in \mathbb{R}$ are minimal hypersurfaces in the product $M \times \mathbb{R}$ with respect to the Ilmanen's metric $g_c = e^{\frac{2c}{m}t}(\sigma + \mathrm{d}t^2)$.*

With this interpretation, it is natural to consider the second variation of the volume and the corresponding Jacobi operator

$$L_c[v] = \Delta v + c\langle X, \nabla v\rangle + (|A|^2 + \overline{\mathrm{Ric}}(N, N))v, \quad v \in C^2(\Sigma),$$

where $|A|$ is the norm of the second fundamental form and $\overline{\text{Ric}}$ is the Ricci curvature of $M \times \mathbb{R}$, both calculated with respect to the Riemannian product metric. We refer to [1, 15, 29] for further details. Taking derivatives on both sides of (4) we obtain

$$\nabla H = -c\, AT, \tag{7}$$

where $T = X^{\top}$ is the tangential component of X. Since X is parallel in $M \times \mathbb{R}$ with respect to the product metric g_0, we also have

$$\nabla_V T = \frac{H}{c} AV \tag{8}$$

for any $V \in \Gamma(T\Sigma)$. Hence, using Codazzi equation for any $U, V \in \Gamma(T\Sigma)$ we have

$$\begin{aligned}\langle \nabla_U \nabla H, V\rangle &= -c\langle (\nabla_U A)T, V\rangle - H\langle AU, AV\rangle \\ &= -c\langle (\nabla_T A)U, V\rangle + c\langle \bar{R}(U, T)N, V\rangle - H\langle AU, AV\rangle.\end{aligned}$$

Therefore

$$\Delta H = -c\langle \nabla H, T\rangle + c\,\text{Ric}_{\bar{M}}(T, N) - |A|^2 H.$$

Since

$$0 = \overline{\text{Ric}}(X, N) = \overline{\text{Ric}}(T, N) + \langle X, N\rangle \overline{\text{Ric}}(N, N) = \overline{\text{Ric}}(T, N) + \frac{1}{c} H\overline{\text{Ric}}(N, N)$$

we conclude that

$$L_c[H] = \Delta H + c\langle X, \nabla H\rangle + (|A|^2 + \text{Ric}_{\bar{M}}(N, N))H = 0. \tag{9}$$

Therefore, the function $h := \langle X, N\rangle$ satisfies $L_c[h] = 0$.

Outside the points where a translating soliton is vertical, it can be described locally in non-parametric terms as a graph

$$\Sigma = \{(x, u(x)) : x \in \Omega\}$$

of a smooth function u defined in a domain $\Omega \subset M$ with regular boundary (possibly empty.) In this case, we denote $\Sigma = \text{Graph}[u]$ and we refer to those solitons as *translating graphs*. From (4) we can check that u satisfies the following partial differential equation

$$\text{div}\left(\frac{\nabla u}{W}\right) = \frac{1}{W}, \tag{10}$$

where $W := \sqrt{1 + |\nabla u|^2}$, and the gradient and divergence operators are taken with respect to the Riemannian metric σ of M. In this case, Σ can be oriented by the normal vector field

$$N = \frac{1}{W}(X - \nabla u) \tag{11}$$

with ∇u translated from $x \in \Omega$ to the point $(x, u(x)) \in \Sigma$. Proceeding as in [29] we prove the following result.

Lemma 2 (Shahriyari–Zhou) *All translating graphs are stable in $M \times \mathbb{R}$ endowed with Ilmanen's metric g_c.*

Remark 4 If Σ is a *graphical* translator and N is the (upward-pointing) unit-normal vector field to Σ, then $\langle \partial_t, N \rangle$ is a positive Jacobi field. Consequently, Σ is a stable g_c-minimal surface. Therefore a sequence of translating graphs will converge, subsequentially, to a translator. Moreover the vertical translates of Σ are also g_c-minimal and foliate a cylinder $\Omega \times \mathbb{R}$, where Ω is the region over which Σ is a graph. As a consequence, Σ is a g_c-area minimizing surface in $\Omega \times \mathbb{R}$.

The proof of the last assertion in the previous remark is quite simple, as we will see in the next lemma. Recall that a complete hypersurface is called area-minimizing if any compact piece is area-minimizing among all the hypersurfaces with the same boundary.

Lemma 3 *Let $\Omega \subset M$ be a bounded domain and Σ a translating graph over $\bar{\Omega}$. For any hypersurface Σ' in $\bar{\Omega} \times \mathbb{R}$ with $\partial\Sigma = \partial\Sigma'$, we have*

$$\mathcal{A}_c[\Sigma] = \int_\Sigma e^{ct}\, \mathrm{d}\mu_\Sigma \leq \int_{\Sigma'} e^{ct}\, \mathrm{d}\mu_{\Sigma'} = \mathcal{A}_c[\Sigma'],$$

and the equality holds if, and only if, $\Sigma = \Sigma'$.

Proof Let $\Psi_\tau(x, t) = \Psi((x, t), \tau) = (x, t + \tau)$ the flow generated by $X = \partial_t$ in $M \times \mathbb{R}$. Consider the one-parameter family of translated copies of Σ given by

$$\Sigma_\tau = \Psi_\tau(\Sigma)$$

and let N_τ be the vector field in $\bar{\Omega} \times \mathbb{R}$ given by

$$N_\tau(\Psi_\tau(x, t)) = \Psi_{\tau *}(x, t) \cdot N(x, t),$$

where N is a unit normal vector field along Σ. It is obvious that N_τ is a normal vector field along Σ_τ. Consider the vector field in $\bar{\Omega} \times \mathbb{R}$ defined by

$$Y = e^{ct} N_\tau. \tag{12}$$

Let $\{\mathsf{e}_i\}_{i=1}^m$ be a local orthonormal frame tangent to Σ_τ. Therefore, denoting the Riemannian connection and the divergence in $(M \times \mathbb{R}, g_0)$ by $\bar{\nabla}$ and div, we obtain

$$\begin{aligned}
\operatorname{div} Y|_{\Sigma_\tau} &= ce^{ct}\langle \bar{\nabla} t, N_\tau\rangle + e^{ct}\operatorname{div} N_\tau \\
&= ce^{ct}\langle X, N_\tau\rangle + e^{ct}\sum_{i=1}^{m}\langle \bar{\nabla}_{\mathsf{e}_i} N_\tau, \mathsf{e}_i\rangle + e^{ct}\langle \bar{\nabla}_{N_\tau} N_\tau, N_\tau\rangle \\
&= ce^{ct}\langle X, N\rangle - e^{ct} H_\tau,
\end{aligned}$$

where we used the fact that $|N_\tau|^2 = 1$. Here $H_\tau(\Psi_\tau(x,t)) = H(x,t)$ is the mean curvature of Σ. Using (4), we conclude that

$$\operatorname{div} Y = e^{ct}(H - c\langle X, N\rangle) = 0 \tag{13}$$

in $\bar{\Omega} \times \mathbb{R}$. First, suppose that $\Sigma' \subset \bar{\Omega} \times \mathbb{R}$ lies in one side of Σ with $\partial\Sigma = \partial\Sigma'$ and denote by U the domain bounded by $\Sigma \cup \Sigma'$. We have

$$\begin{aligned}
0 = \int_U \operatorname{div} Y &= \int_\Sigma \langle Y, N\rangle \, \mathrm{d}\mu_\Sigma - \int_{\Sigma'} \langle Y, N'\rangle \, \mathrm{d}\mu_{\Sigma'} \\
&= \int_\Sigma e^{ct} \, \mathrm{d}\mu_\Sigma - \int_{\Sigma'} e^{ct}\langle N, N'\rangle \, \mathrm{d}\mu_{\Sigma'} \\
&\geq \int_\Sigma e^{ct} \, \mathrm{d}\mu_\Sigma - \int_{\Sigma'} e^{ct} \, \mathrm{d}\mu_{\Sigma'},
\end{aligned}$$

where N' and $\mathrm{d}\mu_{\Sigma'}$ define the orientation and the volume element in Σ', respectively. This completes the proof in this particular case. The general case follows by breaking up the hypersurface Σ' into regions that lie on one side of Σ and applying the previous argument to get the inequality in each one of these regions. □

Remark 5 Similar result was proved by Xin [28] for translating graphs in $\mathbb{R}^{n+1}$.

Remark 6 Notice that by Lemma 3 translating graphs are, in fact, area-minimizing with respect to g_c.

Next we provide a proof for the first equality in (6), that is, the relation between the mean curvatures with respect to g_c and g_0. The Riemannian connections $\widetilde{\nabla}$ and ∇ for the metrics g_c and g_0, respectively, are related by

$$\widetilde{\nabla}_V W = \nabla_V W + \frac{c}{m}\big(\langle V, \partial_t\rangle W + \langle W, \partial_t\rangle V - \langle V, W\rangle \partial_t\big).$$

If N is a unit normal vector field along Σ with respect to g_0, here it is not necessary to suppose that Σ is a graph. Then the normal vector field with respect to g_c is given by $\widetilde{N} = e^{-\frac{c}{m}t} N$. Therefore, the second fundamental forms $\widetilde{II}$ and II of Σ with respect to g_c and g_0 are related by

$$\widetilde{II} = e^{\frac{c}{m}t}\left(II - \frac{c}{m}\langle \partial_t, N\rangle g_0|_\Sigma\right)$$

Taking traces with respect to $g_c|_\Sigma$ one gets

$$\widetilde{H} = e^{-\frac{c}{m}t}(H - c\langle \partial_t, N\rangle) \tag{14}$$

as we have stated above.

As an application of (14), we have the next lemma.

Lemma 4 *Suppose that Λ is a hypersurface in M. Then the mean curvature $\widetilde{H}_{\Lambda\times\mathbb{R}}$ of $\Lambda \times \mathbb{R}$ in $(M \times \mathbb{R}, g_c)$ is given by*

$$\widetilde{H}_{\Lambda\times\mathbb{R}}(x,t) = e^{-\frac{c}{m}t} H_\Lambda(x). \tag{15}$$

for all $(x,t) \in \Lambda \times \mathbb{R}$. Here H_Λ is the mean curvature of Λ in (M, σ).

Proof First, observe that

$$H_{\Lambda\times\mathbb{R}}(x,t) = H_\Lambda(x)$$

for all $(x,t) \in \Lambda \times \mathbb{R}$, where both mean curvatures are calculated with respect to g_0. Moreover, $\langle \partial_t, N\rangle = 0$ along $\Lambda \times \mathbb{R}$ since N is merely the horizontal lift of the unit normal vector field along Λ in M. Hence, (14) yields (15). □

Before proving the main theorem of this section, we recall what we mean by a complete graph.

Definition 7 Let $\Omega \subset M$ be a domain, not necessarily regular, and let $\Lambda \subset \partial\Omega$ be a smooth open set. We say that a smooth function $u\colon \Omega \to \mathbb{R}$ is complete as we approach Λ, if

$$\lim_{x\to x_0} u(x) = \pm\infty, \text{ for any } x_0 \in \Lambda.$$

Theorem 12 *Let M be a complete Riemannian manifold and $\Omega \subset M$ be a domain (not necessarily regular). Let $\Lambda \subset \partial\Omega$ be a smooth open set and Σ a translating or minimal graph of a smooth function $u\colon \Omega \to \mathbb{R}$ that is complete as we approach Λ. Then $H_\Lambda = 0$.*

Proof Fix $x_0 \in \Lambda$ and take a sequence $\{x_i\}$ in Ω with $x_i \to x_0$. Assume that $u(x_i) \to \infty$ and define the sequence of hypersurfaces $\{\Sigma_i := \text{Graph}[u - u(x_i)]\}$ in $M \times \mathbb{R}$ endowed with the Ilmanen's metric g_c.

Fix a closed ball B around $(x_0, 0)$ in $M \times \mathbb{R}$ so that B does not intersect $\partial\Lambda \times \mathbb{R}$. Notice that each Σ_i intersects B for i sufficiently large. So, up to subsequence, we can suppose that each Σ_i intersects B. For each i let S_i be the connected component of $\Sigma_i \cap B$ so that $(x_i, 0) \in S_i$. By Lemma 3

$$\mathcal{A}_c[S_i] \leq \mathcal{A}_c[\partial B],$$

for all i. Hence, by Theorem 10, up to a subsequence, we may assume $S_i \to S_\infty$ in int $B \setminus \text{Sing}(S_\infty)$ and $S_i \rightharpoonup S_\infty$ in int B. Note also that $(x_0, 0) \in (\Lambda \times \mathbb{R}) \cap \text{spt}\, S_\infty$. Then we have two possibilities for $(x_0, 0)$: either it is a regular point or not.

Suppose first that $(x_0, 0)$ is a regular point of S_∞. We claim that a small neighbourhood of $(x_0, 0)$ in S_∞ lies on the cylinder $\Lambda \times \mathbb{R}$. In fact, if this does not hold for any small neighbourhood of $(x_0, 0)$, then we could choose a small domain S in S_∞ near $(x_0, 0)$ and away from the singular set, such that S lies in $\Omega \times \mathbb{R}$. Therefore, we could take a small compact cylinder C through S in $M \times \mathbb{R}$ such that it does not touch $\partial\Omega \times \mathbb{R}$. Since $S_i \to S_\infty$ in int $B \setminus \mathrm{Sing}(S_\infty)$, it follows that, for i sufficiently large, S_i must intersect C. However the assumption $u(x_i) \to \infty$ implies that, for sufficiently large i, $S_i \cap C = \varnothing$, which is a contradiction. Therefore a neighbourhood of $(x_0, 0)$ in the minimal surface S_∞ lies on $\Lambda \times \mathbb{R}$ and, in particular, we conclude that $\widetilde{H}_{\Lambda\times\mathbb{R}}(x_0, 0) = 0$, that is, the mean curvature of $\Lambda \times \mathbb{R}$ with respect to g_c vanishes at $(x_0, 0) = 0$.

If $(x_0, 0)$ is not a regular point of S_∞, take a neighborhood W of $(x_0, 0)$ in S_∞. As the Hausdorff dimension of $\mathrm{Sing}(\Sigma_\infty)$ is less than $n - 7$, we know that $W \setminus \mathrm{Sing}(S_\infty)$ is an open dense subset of W. Furthermore, we can apply the previous argument to prove that any connected component of $W \setminus \mathrm{Sing}(S_\infty)$, which is regular, must lie on $\Lambda \times \mathbb{R}$. Hence, we can take a sequence $\{y_i\} \subset \Lambda \times \mathbb{R}$ such that $y_i \to (x_0, 0)$ and $\widetilde{H}_{\Lambda\times\mathbb{R}}(y_i) = 0$. By continuity $\widetilde{H}_{\Lambda\times\mathbb{R}}(x_0, 0) = 0$. Therefore in both cases, Lemma 4 implies that we must have $H_\Lambda = 0$ on Λ. □

Finally, using Theorem 11 and the idea of the proof of Theorem 12, we can obtain a proof of a result that was obtained by M. Eichmair, J. Metzger for $2 \le n \le 7$ [7, Appendix B].

Remark 7 Let M be a complete Riemannian manifold and $\Omega \subset M$ be a domain whose boundary is not necessarily regular. Let Σ be a graph of a smooth function $u\colon \Omega \to \mathbb{R}$ with constant mean curvature $H_0 > 0$. Let $\Lambda \subset \partial\Omega$ be a smooth open set and suppose that u is complete as we approach Λ.

a. If $u \to \infty$ on Λ, then $H_\Lambda = H_0$ with respect to the inward normal to $\partial\Omega$.
b. If $u \to -\infty$ on Λ, then $H_\Lambda = -H_0$ with respect to the inward normal to $\partial\Omega$.

The idea of the proof (which was already present in [7]) is the following. We fix $x_0 \in \Lambda$ and take a sequence $\{x_i\}$ in Ω with $x_i \to x_0$. Assume first that $u(x_i) \to \infty$ and define the sequence of hypersurfaces $\{\Sigma_i := \mathrm{Graph}[u - u(x_i)]\}$ in $M \times \mathbb{R}$. Fix a closed ball B around $(x_0, 0)$ in $M \times \mathbb{R}$ so that B does not intersect $\partial\Lambda \times \mathbb{R}$, and suppose that each Σ_i intersects B. For each i let S_i be the connected component of $\Sigma_i \cap B$ so that $(x_i, 0) \in S_i$. Reasoning as in Lemma 3, we deduce that

$$\mathcal{A}_0[S_i] \le \mathcal{A}_0[\partial B] + H_0 \,\mathrm{Vol}(B),$$

for all i. Moreover, it is easy to check that S_i is stable for all i. Hence Theorem 11 implies that, up to a subsequence, we may assume $S_i \rightharpoonup S_\infty$ in int B and $(x_0, 0) \in (\Lambda \times \mathbb{R}) \cap \mathrm{spt}\, S_\infty$.

We claim that $\mathrm{spt}\, S_\infty \subset \Lambda \times \mathbb{R}$. Indeed, suppose that $y_0 \notin \Lambda \times \mathbb{R}$ and take any small ball B' around y_0 so that it does not intersect $\Lambda \times \mathbb{R}$. If φ is any smooth function with support in B', our definition of S_i give us that

$$\int_B \varphi \mathrm{d}\mu_{S_i} = 0$$

for all sufficiently large i. Hence,

$$\int_B \varphi \mathrm{d}\mu_{S_\infty} = \lim_i \int_B \varphi \mathrm{d}\mu_{S_i} = 0.$$

This proves that $y_0 \notin \operatorname{spt} S_\infty$. Consequently we must have $\operatorname{spt} S_\infty \subset \Lambda \times \mathbb{R}$, and by regularity of S_∞ according to Theorem 11, we can argue as in Theorem 12 and conclude that $H_\Lambda = H_0$. On the other hand, if $u(x_i) \to -\infty$, we can argue as above and get $-H_\Lambda = H_0$, where $-H_\Lambda$ is the mean curvature with respect to the outward normal to Ω. So $H_\Lambda = -H_0$ with respect to the inward normal to Ω.

3.1 The Euclidean Case

In the Euclidean case $M = \mathbb{R}^n$ endowed with the Euclidean metric $\langle \cdot, \cdot \rangle$, we can obtain a better result. In this particular case, it is very natural to impose on a translators $\Sigma \subset \mathbb{R}^{n+1}$ the condition that

$$\sup_{x \in \mathbb{R}^{n+1},\, r>0} \left(\frac{\operatorname{Area}(\Sigma \cap B(x,r))}{r^n} \right) < \infty. \tag{16}$$

In particular, if Σ arise as a blow up of some mean curvature flow, then it has to satisfy (16) by [5, Corollary 2.13].

The theorem is then:

Theorem 13 *Suppose $u \colon \Omega \subset \mathbb{R}^n \to \mathbb{R}$ is a smooth function whose graph is a complete translator satisfying* (16)*. Then the following holds.*

(i) If $n < 8$, then $\partial\Omega$ is a smooth minimal hypersurface.
(ii) For general n, $\partial\Omega$ is a smooth minimal hypersurface except for a closed singular set of Hausdorff dimension at most $n - 8$.

Proof Fix $x_0 \in \partial\Omega$ and take a sequence $\{x_i\}$ in Ω with $x_i \to x_0$. As before, we define the sequence of hypersurfaces $\{\Sigma_i := \operatorname{Graph}[u - u(x_i)]\}$ in $\mathbb{R}^{n+1}$ endowed with the Ilmanen's metric g_c. The idea of the proof is that the result of Schoen and Simon [23] shows that a weak limit V of smooth n-dimensional stable minimal hypersurfaces, is smooth except for an $(n-7)$-dimensional singular set, assuming we have local area bounds. These local area bounds are provided by the assumption (16).

Furthermore, any point of V where the tangent cone is a plane (possibly with multiplicity) is a regular point. Now we can apply the Schoen–Simon theorem to $V = (\partial\Omega) \times \mathbb{R}$, which is the limit of our sequence Σ_i. At first glance, it seems that we get a singular set of dimension at most $n - 7$, not $n - 8$. But since V is translation-invariant in the vertical direction, in fact one can get one dimension better. One way to see that is the following.

Since $(\partial\Omega) \times \mathbb{R}$ is stable (and has a small singular set) for the Ilmanen metric, it follows that for any sufficiently small ball B in $\mathbb{R}^n$, $(\partial\Omega) \cap B$ is stable for the

Euclidean metric. Hence the Schoen–Simon theorem implies that the singular set of $(\partial \Omega) \cap B$ has dimension at most $n - 8$. □

Remark 8 Under additional hypotheses on the geometry of the translator, it is possible to prove that $\partial \Omega$ is not only minimal, but totally geodesic (see [12]). In the case of $\mathbb{R}^n$, this means that $\partial \Omega$ consists of (disjoint) affine hyperplanes.

4 Jenkins–Serrin Translating Solitons

In this section we study the existence of Jenkins–Serrin solutions for the translating soliton equation on Riemannian surfaces M. To be more precise, we will prove an existence theorem for type 1 Jenkins–Serrin solutions for the translating soliton equation. Let us recall that u is a solution of the translating soliton equation if

$$\operatorname{div}\left(\frac{\nabla u}{W}\right) = \frac{1}{W},$$

where $W = \sqrt{1 + |\nabla u|^2}$, and the gradient and divergence are taken with respect to the Riemannian metric σ of M. The principal concepts we are going to need are the following:

Definition 8 (*Nitsche curve*) Let Ω be a domain in M and $\Gamma \subset M \times \mathbb{R}$ be a Jordan curve. We say that Γ is a Nitsche curve, if it admits a parametrization $\Gamma(t) = \{(\alpha(t), \beta(t)) : t \in \mathbb{S}^1\}$ such that $\alpha(t)$ is a monotone parametrization of $\partial \Omega$. This means that $\alpha : \mathbb{S}^1 \to \partial \Omega$ is continuous and monotone, and there exist closed disjoint intervals $J_1, \ldots, J_v$ such that $\alpha|_{J_i}$ is constant for all i and $\alpha|_{\mathbb{S}^1 \setminus \cup J_i}$ is one-to-one and smooth.

Definition 9 (*Admissible domain*) Let Ω be a connected domain in M. We say that Ω is an admissible domain if it is geodesically convex and $\partial \Omega$ is a union of geodesic arcs $A_1, \ldots, A_s, B_1 \ldots, B_r$, convex arcs $C_1, \ldots, C_t$, the end points of these arcs and that no two arcs A_i and no two arcs B_i have a common endpoint.

Definition 10 (*Admissible polygon*) Let Ω be an admissible domain. We say that $\mathcal{P}$ is an admissible polygon if $\mathcal{P} \subset \Omega$ and the vertices of $\mathcal{P}$ are chosen among the vertices of Ω.

Let Γ be a Nitsche curve over the boundary $\partial \Omega$ of an admissible domain Ω. By a translating soliton with boundary Γ we mean a translating soliton in $\Omega \times \mathbb{R}$ that is a graph over Ω. Using classical results about the solvability of the Plateau problem, we can prove that any Nitsche curve over an admissible domain admits a unique translating soliton with it as the boundary.

Theorem 14 (Local existence) *Let Ω be an admissible domain in M and Γ a Nitsche curve over $\partial \Omega$. Then there exists a unique translating soliton with boundary Γ.*

Proof The proof is similar to that of [22] and therefore we skip some details. First we note that $\Omega \times \mathbb{R}$ is homogeneous in the sense of Meeks and Yau [18, 19] and Morrey [20]. Since the boundary $\partial\Omega$ consists of geodesic and convex arcs, the boundary $\partial(\Omega \times \mathbb{R})$ is mean convex. By Lemma 4 it remains mean convex also when we change the metric to Ilmanen's metric g_c. Therefore there exists an embedded minimal (w.r.t. g_c) disk $\Sigma \subset \Omega \times \mathbb{R}$ with boundary Γ. It remains to prove that $\text{int}(\Sigma)$ is a graph over Ω.

First we show that, for all $p \in \text{int}(\Sigma)$, $T_p\Sigma$ is not a vertical plane. On the contrary, suppose that there exists a point $p \in \text{int}(\Sigma)$ such that $p \in M \times \{c\}$ for some $c \in \mathbb{R}$ and that the tangent plane Π to Σ is vertical in $\Omega \times \mathbb{R}$. Take a basis $\{\partial_t, \ v\}$ tangent to Π at p, where ∂_t is tangent to Σ and v is tangent to $M \times \{c\}$ with $||v|| = 1$. Let γ be the unique geodesic in $M \times \{c\}$ such that $\gamma(0) = p$ and $\gamma'(0) = v$. Note that γ intersects $\partial(\Omega \times \mathbb{R})$ exactly in two points.

Now $\gamma \times \mathbb{R}$ is a totally geodesic surface, in particular minimal, in $\Omega \times \mathbb{R}$ and by Lemma 4 it is minimal also with respect to the metric g_c. Moreover, we have $T_p(\gamma \times \mathbb{R}) = \Pi$ and therefore, near p, $I := \Sigma \cap (\gamma \times \mathbb{R})$ contains at least two curves that intersect transversally at p. If there exists a closed curve α in $I \setminus \partial\Sigma$, then α is the boundary of a minimal disk D in Σ. Thus we could choose a geodesic curve β in D so that the totally geodesic surface $\beta \times \mathbb{R}$ touches D at an interior point. But this is impossible by the maximum principle.

Since I does not contain a closed curve, each of the branches leaving p must go to $\partial\Sigma$. Moreover, γ intersects $\partial\Omega$ at two points so at least two of these branches must go to the same point or vertical segment of $\partial\Sigma$. However, this yields again closed curve that bounds a minimal surfaces and we get a contradiction with the maximum principle. Therefore $T_p\Sigma$ is not a vertical plane,

Finally, the same argument as in [22] shows that each vertical line in $\Omega \times \mathbb{R}$ intersects Σ exactly at one point and therefore Σ is a graph over the interior of Ω. □

To prove our main theorem, we will need the following maximum principle.

Proposition 1 (Maximum principle) *Let $\Omega \subset M$ be an admissible domain. Suppose that u_1 and u_2 satisfy*

$$\text{div}\left(\frac{\nabla u_1}{\sqrt{1+|\nabla u_1|^2}}\right) \geq \text{div}\left(\frac{\nabla u_2}{\sqrt{1+|\nabla u_2|^2}}\right),$$

and $\liminf(u_2 - u_1) \geq 0$ *for any approach of $\partial\Omega$, with possible exception of finite numbers of points $\{q_1, \ldots, q_r\} =: E \subset \partial\Omega$. Then $u_2 \geq u_1$ on $\partial\Omega \setminus E$ with strict inequality unless $u_2 = u_1$.*

Proof The proof follows similar arguments as in [26]. Let K and ε be positive constants, with K large enough and ε small enough. Define a function

$$\varphi := \begin{cases} K - \varepsilon, & \text{if } u_1 - u_2 \geq K; \\ u_1 - u_2 - \varepsilon, & \text{if } \varepsilon < u_1 - u_2 \leq K; \\ 0, & \text{if } u_1 - u_2 \leq \varepsilon. \end{cases}$$

Notice that φ is Lipschitz with $0 \leq \varphi \leq K$, and $\nabla\varphi = \nabla u_1 - \nabla u_2$ in the set $\{\varepsilon < u_1 - u_2 < K\}$ and $\nabla\varphi = 0$ almost everywhere in the complement of $\{\varepsilon < u_1 - u_2 < K\}$. Around any point $q_i \in E$, consider an open geodesic disk $B_\varepsilon(q_i)$ of radius ε and center q_i. Let $\Omega_\varepsilon := \Omega \setminus \cup B_\varepsilon(q_i)$, and suppose that $\partial\Omega_\varepsilon = \tau_\varepsilon \cup \rho_\varepsilon$, where $\rho_\varepsilon = \cup(\partial B_\varepsilon(q_i) \cap \Omega)$ and $\tau_\varepsilon = \partial\Omega_\varepsilon \cap \partial\Omega$. Since $\liminf(u_2 - u_1) \geq 0$ in $\partial\Omega \setminus E$, we have $\varphi \equiv 0$ in a neighbourhood of τ_ε.

Define

$$J := \int_{\rho_\varepsilon} \varphi \left\{ \left\langle \frac{\nabla u_1}{W_1}, \nu \right\rangle - \left\langle \frac{\nabla u_2}{W_2}, \nu \right\rangle \right\}, \tag{17}$$

where ν is the unit outer conormal to Ω_ε and $W_i = \sqrt{1 + |\nabla u_i|^2}$. From (17) and $0 \leq \varphi \leq K$ we obtain

$$J \leq 2K \sum_{i=1}^{r} ||\partial B_\varepsilon(q_i)||, \tag{18}$$

where $||\partial B_\varepsilon(q_i)||$ denotes the length of $\partial B_\varepsilon(q_i)$. On ther other hand, since φ is Lipschitz, we have

$$\operatorname{div}\left(\varphi\left\{\frac{\nabla u_1}{W_1} - \frac{\nabla u_2}{W_2}\right\}\right) = \nabla\varphi\left\{\frac{\nabla u_1}{W_1} - \frac{\nabla u_2}{W_2}\right\} + \varphi\left\{\operatorname{div}\left(\frac{\nabla u_1}{W_1}\right) - \operatorname{div}\left(\frac{\nabla u_2}{W_2}\right)\right\},$$

almost everywhere in Ω. By Stokes theorem we get

$$\begin{aligned} J &= \int_{\Omega_\varepsilon} \left\{ \left\langle \nabla\varphi, \left(\frac{\nabla u_1}{W_1} - \frac{\nabla u_2}{W_2}\right) \right\rangle + \varphi\left(\operatorname{div}\left(\frac{\nabla u_1}{W_1}\right) - \operatorname{div}\left(\frac{\nabla u_2}{W_2}\right)\right) \right\} \\ &\geq \int_{\Omega_\varepsilon} \left\langle \nabla\varphi, \left(\frac{\nabla u_1}{W_1} - \frac{\nabla u_2}{W_2}\right) \right\rangle. \end{aligned} \tag{19}$$

Now if we consider the unit normals (choosing $X = \partial_t$ in (11))

$$N_i := \frac{\partial_t}{W_i} - \frac{\nabla u_i}{W_i},$$

then

$$\left\langle \nabla u_1 - \nabla u_2, \left(\frac{\nabla u_1}{W_1} - \frac{\nabla u_2}{W_2}\right)\right\rangle = \langle N_1 - N_2, W_1 N_1 - W_2 N_2\rangle$$
$$= W_1 - (W_1 + W_2)\langle N_1, N_2\rangle + W_2$$
$$= \frac{1}{2}(W_1 + W_2)||N_1 - N_2||^2. \tag{20}$$

From (18), (19) and (20) we get

$$2K \sum_{i=1}^{r} ||\partial B_\varepsilon(q_i)|| \geq \frac{1}{2}\int_{\Omega_\varepsilon \cap \{0<u_1-u_2<K\}} (W_1 + W_2)||N_1 - N_2||^2 \geq 0.$$

In particular, when $\varepsilon \to 0$ we obtain

$$\int_{\{0<u_1-u_2<K\}} (W_1 + W_2)||N_1 - N_2||^2 = 0.$$

Therefore $N_1 = N_2$ in $\{0 < u_1 - u_2 < K\}$, so $\nabla u_1 = \nabla u_2$ in $\{0 < u_1 - u_2 < K\}$. As K was arbitrary we may conclude $\nabla u_1 = \nabla u_2$ in the set $\{0 < u_1 - u_2\}$. Suppose now that $\{0 < u_1 - u_2\}$ contains a connected component with non-empty interior. By the previous argument $u_1 = u_2 + c$, where c is a positive constant, so by the maximum principle $u_1 = u_2 + c$ in Ω. On the other hand, as $\liminf(u_2 - u_1) \geq 0$ for any approach of $\partial\Omega \setminus E$, then c is a non-positive constant, which is impossible. This finishes the proof. □

Finally, we can prove the main theorem of this section.

Theorem 15 (Existence of Jenkins–Serrin Solution) *Let $\Omega \subset M$ be an admissible domain with $\{B_i\} = \varnothing$. Given any continuous data $f_i : C_i \to \mathbb{R}$, there exists a Jenkins–Serrin solution $u : \Omega \to \mathbb{R}$ for the translating soliton equation with continuous data $u|_{C_i} = f_i$, if for any admissible polygon $\mathcal{P}$ we have*

$$2\alpha(\mathcal{P}) < \ell(\mathcal{P}). \tag{21}$$

Proof Define a Nitsche curve $\Gamma_n = (\alpha_n, \beta_n)$ by setting $\beta_n = n$ on $\{A_i\}$ and $\beta_n = \min\{f_i, n\}$ on C_i for all i. By Theorem 14, for all $n \in \mathbb{N}$, there exists $u_n : \Omega \to \mathbb{R}$ so that Graph$[u_n]$ is a translating soliton in $\Omega \times \mathbb{R}$ with boundary Γ_n. Notice that if $n > m$ we have $u_n \geq u_m$ on $\partial\Omega$, so $u_n > u_m$ in Ω by comparison principle. Hence $\{u_n\}$ is a monotone sequence. Taking into account Pinheiro's results [22], (21) guarantees that there exists a Jenkins–Serrin solution $v : \Omega \to \mathbb{R}$ for the minimal graph equation with continuous data f_i. Since

$$\operatorname{div}\left(\frac{v}{\sqrt{1+|v|^2}}\right) = 0 < \frac{1}{\sqrt{1+|u_n|^2}} = \operatorname{div}\left(\frac{u_n}{\sqrt{1+|u_n|^2}}\right)$$

and $\liminf(v - u_n) \geq 0$ on $\partial\Omega \setminus E$, where E is the set of vertices of Ω, Proposition 1 implies $v > u_n$ for all n. Therefore $\lim u_n = u$ exists and u satisfies

$$\operatorname{div}\left(\frac{u}{\sqrt{1+|\nabla u|^2}}\right) = \frac{1}{\sqrt{1+|u|^2}}$$

in Ω. Clearly $u|_{C_i} = f_i$, by construction, and $u \to \infty$ as we approach A_i for all i. □

Acknowledgements We would like to thank Brian White for his valuable comments, in particular concerning Theorem 13. E. S. Gama is very grateful to the Institute of Mathematics at the University of Granada for its hospitality during the time the research and preparation of this article were conducted.

References

1. Alías, L., Lira, J.H., Rigoli, M.: Mean curvature flow solitons in the presence of conformal vector fields (2017). arXiv:1707.07132v2
2. Bellettini, C., Wickramasekera, N.: Stable CMC integral varifolds of codimension 1: regularity and compactness (2018). arXiv:1802.00377
3. Bellettini, C., Wickramasekera, N.: Stable prescribed-mean-curvature integral varifolds of codimension 1: regularity and compactness. In preparation
4. Bellettini, C., Wickramasekera, N.: The Allen–Cahn equation and the existence of prescribed mean-curvature hypersurfaces. In preparation
5. Colding, T.H., Minicozzi, W.: Generic mean curvature flow I; generic singularities. Ann. Math. **175**(2), 755–833 (2012)
6. Collin, P., Rosenberg, H.: Construction of harmonic diffeomorphisms and minimal graphs. Ann. Math. **1879–1906** (2010)
7. Eichmair, M., Metzger, J.: Jenkins–Serrin-type results for the Jang equation. J. Differ. Geom. **102**(2), 207–242 (2016)
8. Folha, A., Rosenberg, H.: The Dirichlet problem for constant mean curvature graphs in $M \times \mathbb{R}$. Geom. Topol. **16**(2), 1171–1203 (2012)
9. Gálvez, J.A., Rosenberg, H.: Minimal surfaces and harmonic diffeomorphisms from the complex plane onto certain Hadamard surfaces. Am. J. Math. **132**(5), 1249–1273 (2010)
10. Giusti, E.: Minimal surfaces and functions of bounded variation, vol. 80. Birkhäuser, Boston (1984)
11. Hauswirth, L., Rosenberg, H., Spruck, J.: Infinite boundary value problems for constant mean curvature graphs in $\mathbb{H} \times \mathbb{R}$ and $\mathbb{S} \times \mathbb{R}$. Am. J. Math. **131**(1), 195–226 (2009)
12. Hoffman, D., Ilmanen, T., Martín, F., White, B.: Graphical translators for mean curvature flow (2018). arXiv: 1805.10860
13. Ilmanen, T.: Elliptic regularization and partial regularity for motion by mean curvature. Mem. Am. Math. Soc. **520** (1994)
14. Jenkins, H., Serrin, J.: Variational problems of minimal surface type II. Boundary value problems for the minimal surface equation. Arch. Rat. Mech. Anal. **21.4**, 321–342 (1966)
15. Lira, J.H., Martín, F.: Translating solitons in Riemannian products. J. Differ. Equ. **266**(12), 7780–7812 (2019)
16. Massari, U.: Problema di Dirichlet per l'equazione delle superfici di curvatura media assegnata con dato infinito. Annali dell'Università di Ferrara **23**(1), 111–141 (1977)

17. Mazet, L., Rodríguez, M.M., Rosenberg, H.: The Dirichlet problem for the minimal surface equation, with possible infinite boundary data, over domains in a Riemannian surface. Proc. Lond. Math. Soc. **102**(6), 985–1023 (2011)
18. Meeks III, W., Yau, S.T.: The classical Plateau problem and the topology of three-dimensional manifolds: the embedding of the solution given by Douglas-Morrey and an analytic proof of Dehn's Lemma. Topology **21**(4), 409–442 (1982)
19. Meeks III, W., Yau, S.T.: The existence of embedded minimal surfaces and the problem of uniqueness. Mathematische Zeitschrift **179**(2), 151–168 (1982)
20. Morrey, C.B.: The problem of Plateau on a Riemannian manifold. Ann. Math. **49**, 807–851 (1948)
21. Nelli, B., Rosenberg, H.: Minimal surfaces in $\mathbb{H}^2 \times \mathbb{R}$. Bull. Braz. Math. Soc. **33**, 263–292 (2002)
22. Pinheiro, A.L.: A Jenkins–Serrin theorem in $M^2 \times \mathbb{R}$. Bull. Braz. Math. Soc. **40**(1), 117–148 (2009)
23. Schoen, R., Simon, L.: Regularity of stable minimal hypersurfaces. Commun. Pure Appl. Math. **34**(6), 741–797 (1981)
24. Shahriyari, L.: Translating graphs by mean curvature flow. Geom. Dedicata **175**, 57–64 (2015)
25. Simon, L.: Lectures on geometric measure theory. The Australian National University, Mathematical Sciences Institute, Centre for Mathematics and its Applications (1983)
26. Spruck, J.: Infinite boundary value problems for surface of constant mean curvature. Arch. Rat. Mech. Anal. **49**(1), 1–31 (1972)
27. Wickramasekera, N.: A general regularity theory for stable codimension 1 integral varifolds. Ann. Math. **179**(3), 843–1007 (2014)
28. Xin, Y.L.: Translating solitons of the mean curvature flow. Calc. Var. Part. Differ. Equ. **54**(2), 1995–2016 (2015)
29. Zhou, H.: The boundary behavior of domains with complete translating, minimal and CMC graphs in $N^2 \times \mathbb{R}$. Sci. China Math. 1–12

Survey on the Asymptotic Dirichlet Problem for the Minimal Surface Equation

Esko Heinonen

Abstract We give a survey on the development of the study of the asymptotic Dirichlet problem for the minimal surface equation on Cartan–Hadamard manifolds. Part of this survey is based on the introductory part of the doctoral dissertation [29] of the author. The paper is organised as follows. First we introduce Cartan–Hadamard manifolds and the concept of asymptotic Dirichlet problem, then discuss about the development of the results and describe the methods used in the proofs. In the end we mention some results about the nonsolvability of the asymptotic Dirichlet problem.

Keywords Minimal surfaces · Asymptotic dirichlet problem · Hadamard manifolds

1 Preliminaries

1.1 Cartan–Hadamard Manifolds

A *Cartan–Hadamard* (also *Hadamard*) manifold M is a complete simply connected Riemannian manifold whose all sectional curvatures satisfy

$$K_M \leq 0.$$

The most simple examples of such manifolds are the Euclidean space $\mathbb{R}^n$, with zero curvature, and the hyperbolic space $\mathbb{H}^n$, with constant negative curvature -1. The name of these manifolds has its origin in the Cartan–Hadamard theorem which states that the exponential map is a diffeomorphism in the whole tangent space at every point of M.

Supported by a grant of the Finnish Academy of Science and Letters.

E. Heinonen (✉)
Departamento de Geometría y Topología, Universidad de Granada, 18071 Granada, Spain
e-mail: ea.heinonen@gmail.com

T. Hoffmann et al. (eds.), *Minimal Surfaces: Integrable Systems and Visualisation*, Springer Proceedings in Mathematics & Statistics 349,
https://doi.org/10.1007/978-3-030-68541-6_7

Another important examples of Cartan–Hadamard manifolds are given by the rotationally symmetric model manifolds with radial curvature function f satisfying $f'' \geq 0$. We recall that a model manifold is $\mathbb{R}^n$ equipped with the rotationally symmetric metric that can be written as

$$g = dr^2 + f(r)^2 d\vartheta^2,$$

where r (as in the rest of the paper) is the distance to a pole o and $d\vartheta$ is the standard metric on the unit sphere $\mathbb{S}^{n-1}$. To justify the requirement $f'' \geq 0$, we note that the sectional curvatures of a model manifold can be obtained from the radial curvature function, namely we have

$$K_{M_f}(P_x) = -\frac{f''\big(r(x)\big)}{f\big(r(x)\big)} \cos^2 \alpha + \frac{1 - f'\big(r(x)\big)^2}{f\big(r(x)\big)^2} \sin^2 \alpha, \tag{1}$$

where α is the angle between $\nabla r(x)$ and the 2-plane $P_x \subset T_x M$. In the case of the radial sectional curvature the formula simplifies to

$$K_{M_f} = -\frac{f''}{f}.$$

For the verification of these formulae one could see e.g. [4, 49] where more general formula was given for the sectional curvatures of warped product manifolds.

The radial curvature function f is an example of solutions to the Jacobi equation that is defined as follows. Given a smooth function $k\colon [0, \infty) \to [0, \infty)$ we denote by $f_k\colon [0, \infty) \to \mathbb{R}$ the smooth non-negative solution to the initial value problem (Jacobi equation)

$$\begin{cases} f_k(0) = 0, \\ f_k'(0) = 1, \\ f_k'' = k^2 f_k. \end{cases}$$

These functions play an important role in estimates involving curvature bounds since the resulting model manifolds can be used in various comparison theorems, e.g. Hessian and Laplace comparison (see [26]).

1.2 Asymptotic Dirichlet Problem on Cartan–Hadamard Manifolds

Cartan–Hadamard manifolds can be compactified by adding the *asymptotic boundary* (also *sphere at infinity*) $\partial_\infty M$ and equipping the resulting space $\bar{M} := M \cup \partial_\infty M$ with the *cone topology*. This makes $\bar{M}$ homeomorphic to the closed unit ball. The asymptotic boundary $\partial_\infty M$ consists of equivalence classes of geodesic rays under the equivalence relation

$$\gamma_1 \sim \gamma_2 \quad \text{if} \quad \sup_{t\geq 0} \operatorname{dist}\big(\gamma_1(t), \gamma_2(t)\big) < \infty.$$

Equivalently it can be considered as the set of geodesic rays emitting from a fixed point $o \in M$, when each ray corresponds to a point on the unit sphere of $T_o M$, and this justifies the name sphere at infinity.

The basis for the cone topology in $\bar{M}$ is formed by cones

$$C(v, \alpha) := \{y \in M \setminus \{x\} \colon \sphericalangle(v, \dot{\gamma}_0^{x,y}) < \alpha\}, \quad v \in T_x M, \ \alpha > 0,$$

truncated cones

$$T(v, \alpha, R) := C(v, \alpha) \setminus \bar{B}(x, R), \quad R > 0,$$

and all open balls in M. Here $\gamma^{x,y}$ denotes the unique geodesic joining x to y, $\dot{\gamma}_0$ is the initial unit vector of geodesic γ, and $\sphericalangle(\cdot, \cdot)$ the angle between two vectors. Cone topology was first introduced by Eberlein and O'Neill in [23].

Now we can define the *asymptotic Dirichlet problem* (also *Dirichlet problem at infinity*). Even though we will consider the minimal surface equation

$$\operatorname{div} \frac{\nabla u}{\sqrt{1 + |\nabla u|^2}} = 0, \tag{2}$$

we formulate the problem for a general quasilinear elliptic operator Q. Concerning terminology, the minimal surfaces given as graphs of solutions to (2) will be called also *minimal graphs*.

Problem 1 *Let $\theta \colon \partial_\infty M \to \mathbb{R}$ be a continuous function. Does there exist a continuous function $u \colon \bar{M} \to \mathbb{R}$ with*

$$\begin{cases} Q[u] = 0 & \text{in } M; \\ u|\partial_\infty M = \theta, \end{cases}$$

and if yes, is the function u unique?

In the case such function u exists for every $\theta \in C(\partial_\infty M)$, we say that the asymptotic Dirichlet problem in $\bar{M}$ is *solvable*. We will see that the solvability of this problem depends heavily on the geometry of the manifold M, but the uniqueness of the solutions depends also on the operator Q. Namely, for the usual Laplace, $\mathcal{A}$-harmonic and minimal surface operators we have the uniqueness but in the case of more complicated operators, that do not satisfy maximum principles, the uniqueness of solutions will be lost (see e.g. [10]).

2 Overview of the Results

The study of the asymptotic Dirichlet problem has its origin in the question of the existence of entire bounded nonconstant harmonic functions on Cartan–Hadamard manifolds. Namely, one way to prove the existence is to solve the asymptotic Dirichlet problem with arbitrary continuous boundary data on $\partial_\infty M$. This question gained lot of interest after the seminal work by Greene and Wu [26], where they conjectured that if the sectional curvatures of a Cartan–Hadamard manifold M satisfy

$$K_M \leq \frac{C}{r^2}, \quad C > 0,$$

outside a compact set, then there exists an entire bounded nonconstant harmonic function on M.

From now on we will focus on the mean curvature equation; a brief discussion about the results concerning harmonic and $\mathcal{A}$-harmonic functions can be found e.g. in [29]. We will mention also some results that are not directly about the asymptotic Dirichlet problem but that have motivated the study or have had an important role in the proofs of other results.

2.1 Minimal Surfaces

To begin with bounded domains $\Omega \subset \mathbb{R}^n$, we mention that the usual Dirichlet problem for the minimal graphs was solved in 1968 by Jenkins and Serrin [34] under the assumption that $\partial\Omega$ has nonnegative mean curvature. Serrin [47] generalised this for the graphs with prescribed mean curvature, and more recently Guio and Sa Earp [28] considered similar Dirichlet problem in the hyperbolic space.

First result about the existence of entire minimal graphs is due to Nelli and Rosenberg. In [39], in addition to constructing catenoids, helicoids and Scherk-type surfaces in $\mathbb{H}^2 \times \mathbb{R}$, they proved the following theorem using the disk model of $\mathbb{H}^2$.

Theorem 1 *Let Γ be a continuous rectifiable Jordan curve in $\partial_\infty \mathbb{H}^2 \times \mathbb{R}$, that is a vertical graph. Then, there exists a minimal vertical graph on $\mathbb{H}^2$ having Γ as asymptotic boundary. The graph is unique.*

In 2005 Meeks and Rosenberg [36] developed the theory of properly embedded minimal surfaces in $N \times \mathbb{R}$, where N is a closed orientable Riemannian surface but the existence of entire minimal surfaces in product spaces $M \times \mathbb{R}$ really draw attention after the papers by Collin and Rosenberg [17] and Gálvez and Rosenberg [24]. In [17] Collin and Rosenberg constructed a harmonic diffeomorphism from $\mathbb{C}$ onto $\mathbb{H}^2$ and hence disproved the conjecture of Schoen and Yau [45]. Gálvez and Rosenberg generalised this result to Hadamard surfaces whose curvature is bounded from above by a negative constant. A key tool in their constructions was to solve the Dirichlet problem on unbounded ideal polygons with alternating boundary values

$\pm\infty$ on the sides of the ideal polygons (so-called Jenkins-Serrin problem). In the end of [24] Gálvez and Rosenberg obtain also a counterpart of the Theorem 1 under the assumption $K_M \le C < 0$.

Sa Earp and Toubiana [44] constructed minimal vertical graphs over unbounded domains in $\mathbb{H}^2 \times \mathbb{R}$ taking certain prescribed finite boundary data and certain prescribed asymptotic boundary data. Espírito-Santo and Ripoll [22] considered the existence of solutions to the exterior Dirichlet problem on simply connected manifolds with negative sectional curvature. Here the idea is to find minimal hypersurfaces on unbounded domains with compact boundary assuming zero boundary values.

Espírito-Santo et al. [21] proved the solvability of the asymptotic Dirichlet problem on Riemannian manifold M whose sectional curvatures satisfy $K_M \le -k^2$, $k > 0$, and under the assumption that there exists a point $p \in M$ such that the isotropy group at p of the isometry group of M acts transitively on the geodesic spheres centred at p.

Rosenberg et al. [43] studied minimal hypersurfaces in $N \times \mathbb{R}_+$ with N complete Riemannian manifold having non-negative Ricci curvature and sectional curvatures bounded from below. They proved so-called half-space properties both for properly immersed minimal surfaces and for graphical minimal surfaces. In the latter, a key tool was a global gradient estimate for solutions of the minimal surface equation.

Ripoll and Telichevesky [42] showed the existence of entire bounded nonconstant solutions for slightly larger class of operators, including minimal surface operator, by studying the strict convexity (SC) condition of the manifold. Similar class of operators was studied also by Casteras et al. [12] but instead of considering the SC condition, they solved the asymptotic Dirichlet problem by using similar barrier functions as in [33]. Both of these gave the existence of minimal graphic functions under the sectional curvature assumption

$$- r(x)^{-2-\varepsilon} e^{2kr(x)} \le K_M(P_x) \le -k^2 \tag{3}$$

outside a compact set, and the latter also included the assumption

$$- r(x)^{2(\phi-2)-\varepsilon} \le K(P_x) \le -\frac{\phi(\phi-1)}{r(x)^2}, \tag{4}$$

$r(x) \ge R_0$, for some constants $\phi > 1$ and ε, $R_0 > 0$.

In Casteras et al. [14] adapted a method that was earlier used by Cheng [15] for the study of harmonic functions and by Vähäkangas [51] for $\mathcal{A}$-harmonic functions, and proved the following.

Theorem 2 *Let M be a Cartan–Hadamard manifold of dimension $n \ge 3$ and suppose that*

$$-\frac{\big(\log r(x)\big)^{2\bar{\varepsilon}}}{r(x)^2} \le K_M(P_x) \le -\frac{1+\varepsilon}{r(x)^2 \log r(x)} \tag{5}$$

holds for all 2-planes $P_x \subset T_xM$ and for some constants $\varepsilon > \tilde{\varepsilon} > 0$ and r large enough. Then the asymptotic Dirichlet problem for the minimal surface equation (2) *is uniquely solvable.*

The proof was based on the usage of Sobolev and Caccioppoli-type inequalities together with complementary Young functions.

Telichevesky [48] considered the Dirichlet problem on unbounded domains Ω proving the existence of solutions provided that $K_M \leq -1$, the ordinary boundary of Ω is mean convex and that Ω satisfies the SC condition at infinity. The SC condition was studied by Casteras et al. also in [13] and they proved that the manifold M satisfies the SC condition under very general curvature assumptions. As special cases they obtain the bounds (5) and

$$-ce^{(2-\varepsilon)r(x)}e^{e^{r(x)/e^3}} \leq K_M \leq -\phi e^{2r(x)} \tag{6}$$

for some constants $\phi > 1/4$, $\varepsilon > 0$ and $c > 0$. In addition to the asymptotic Dirichlet problem, Casteras, Holopainen and Ripoll applied the SC condition to prove also the solvability of the asymptotic Plateau problem.

Adapting Vähäkangas' method Casteras, Heinonen and Holopainen showed that, as in the case of $\mathcal{A}$-harmonic functions, the curvature lower bound can be replaced by a pointwise pinching condition obtaining the following result.

Theorem 3 *Let M be a Cartan–Hadamard manifold of dimension $n \geq 2$ and let $\phi > 1$. Assume that*

$$K(P_x) \leq -\frac{\phi(\phi-1)}{r(x)^2}, \tag{7}$$

holds for all 2-planes $P_x \subset T_xM$ containing the radial vector $\nabla r(x)$, with $x \in M \setminus B(o, R_0)$. Suppose also that there exists a constant $C_K < \infty$ such that

$$|K(P_x)| \leq C_K|K(P'_x)| \tag{8}$$

whenever $x \in M \setminus B(o, R_0)$ and $P_x, P'_x \subset T_xM$ are 2-planes containing the radial vector $\nabla r(x)$. Moreover, suppose that the dimension n and the constant ϕ satisfy the relation

$$n > \frac{4}{\phi} + 1.$$

Then the asymptotic Dirichlet problem for the minimal surface equation (2) *is uniquely solvable.*

It is worth to point out that choosing $\phi > 4$ in Theorem 3, the result holds for every dimension $n \geq 2$ and if $n \geq 5$, ϕ can be as close to 1 as we wish. Moreover, if M is 2-dimensional, the condition (8) is trivially satisfied and the asymptotic Dirichlet problem can be solved by assuming only the curvature upper bound (7) and $\phi > 4$.

In the case of rotationally symmetric manifolds it is possible to obtain solvability results without any assumptions on the lower bound of the sectional curvatures. Ripoll and Telichevesky [41] solved the asymptotic Dirichlet problem on 2-dimensional surfaces, and later Casteras et al. [9] gave a proof that holds in any dimension $n \geq 2$, obtaining the following.

Theorem 4 *Let M be a rotationally symmetric n-dimensional Cartan–Hadamard manifold whose radial sectional curvatures outside a compact set satisfy the upper bounds*

$$K(P_x) \leq -\frac{1+\varepsilon}{r(x)^2 \log r(x)}, \quad \text{if } n = 2$$

and

$$K(P_x) \leq -\frac{1/2+\varepsilon}{r(x)^2 \log r(x)}, \quad \text{if } n \geq 3.$$

Then the asymptotic Dirichlet problem for the minimal surface equation (2) *is uniquely solvable.*

We point out that the curvature assumptions in Theorem 4 are most likely optimal (even the constants in the numerators). Namely, March [35] gave an if and only if result for the existence of bounded entire nonconstant harmonic functions under the same curvature assumptions (see also the discussion in Sect. 4).

2.2 Other Prescribed Mean Curvature Surfaces

As we have already mentioned, the asymptotic Dirichlet problem has been previously considered also for other type of operators (starting from the Laplacian) but the class of surfaces, that is very closely related to minimal surfaces, are the surfaces of prescribed mean curvature. It is well known that the graph of a function $u\colon M \to \mathbb{R}$ has prescribed mean curvature H if u satisfies

$$\operatorname{div} \frac{\nabla u}{\sqrt{1+|\nabla u|^2}} = nH,$$

where $H\colon M \to \mathbb{R}$ is a given function. As it is reasonable to believe, in this case the solvability of the asymptotic Dirichlet problem will depend also on the function H.

A very special type of prescribed mean curvature surfaces are the so-called f-minimal surfaces that are obtained by replacing the function nH by

$$\langle \bar{\nabla} f, \nu_u \rangle,$$

where $f\colon M \times \mathbb{R} \to \mathbb{R}$ and ν_u is the downward pointing unit normal to the graph of u. In this case the mean curvature will depend, not only on the point of the manifold, but also on the solution itself. The name f-minimal comes from the fact that these surfaces are minimal in the weighted manifolds $(M, g, e^{-f} d\,\mathrm{vol}_M)$, where (M, g) is a complete Riemannian manifold with volume element $d\,\mathrm{vol}_M$.

In [10] Casteras et al. studied these f-minimal graphs and solved first the usual Dirichlet problem under suitable assumptions and then applied it to solve the asymptotic Dirichlet problem under curvature assumptions similar to those that appeared in [12], i.e. (3) and (4). We point out that in [10] it was necessary to assume that the function $f \in C^2(M \times \mathbb{R})$ is of the form

$$f(x,t) = m(x) + r(t).$$

Other key assumptions are related to the decay of $|\bar{\nabla} f|$ compared to the curvature upper bound of the manifold. For example, it is required that

$$\sup_{\partial B(o,r)\times\mathbb{R}} |\bar{\nabla} f| = o\left(\frac{f_a'(r)}{f_a(r)} r^{-\varepsilon-1}\right),$$

where f_a is the Jacobi solution related to the curvature upper bound. Note that $(n-1)f_a'(r)/f_a(r)$ is the mean curvature of a sphere of radius r on a model manifold with curvature $-a^2$. The assumptions related to the decay of $|\bar{\nabla} f|$ are really necessary in view of a result due to Pigola et al. [40].

In [11] Casteras et al. studied the asymptotic Dirichlet problem for Killing graphs with prescribed mean curvature on warped product manifolds $M \times_\varrho \mathbb{R}$. Here M is a complete n-dimensional Riemannian manifold and $\varrho \in C^\infty(M)$ is a smooth warping function. This means that the metric of $M \times_\varrho \mathbb{R}$ can be written in the form

$$(\varrho \circ \pi_1)^2 \pi_2^* dt^2 + \pi_1^* g,$$

where g is the metric on M and $\pi_1 \colon M \times \mathbb{R} \to M$ and $\pi_2 \colon M \times \mathbb{R} \to \mathbb{R}$ are the standard projections. In this setting $X = \partial_t$ is a Killing field with $|X| = \varrho$ on M. The norm of X is preserved along its flow lines and therefore ϱ can be extended to a smooth function $\varrho = |X| \in C^\infty(M \times_\varrho \mathbb{R})$. Then the Killing graph of a function $u \colon M \to \mathbb{R}$ is the hypersurface given by

$$\Sigma_u = \{\Psi(x, u(x)) \colon x \in M\},$$

where $\Psi \colon M \times \mathbb{R} \to M \times_\varrho \mathbb{R}$ is the flow generated by X. Such Killing graphs were first introduced in [19, 20] (see also [18]). In [19] it was shown that the Killing graph Σ_u has mean curvature H if u satisfies the equation

$$\operatorname{div}_{-\log \varrho} \frac{\nabla u}{\sqrt{\varrho^{-2} + |\nabla u|^2}} = nH, \tag{9}$$

where $\operatorname{div}_{-\log \varrho}(\cdot) = \operatorname{div}(\cdot) + \langle \nabla \log \varrho, (\cdot) \rangle$ is weighted divergence operator.

In the case of Killing graphs the solvability of the asymptotic Dirichlet problem depends on the geometry of M, on the warping function ϱ and on the prescribed mean curvature H. In [11] the authors consider the same curvature assumptions (3) and (4) on M as in [10, 12]. We point out that here it is necessary to assume that the warping function ϱ is convex since otherwise the whole warped product space would not be a Cartan–Hadamard manifold, see [4].

Depending on the curvature bounds on M and the warping function ϱ, it is possible to find entire Killing graphs with prescribed mean curvature H such that $H(x) \not\to 0$ as $r(x) \to \infty$. For example, the hyperbolic space $\mathbb{H}^{n+1}$ with constant curvature -1, can be written as a warped product $\mathbb{H}^{n+1} = \mathbb{H}^n \times_{\cosh} \mathbb{R}$ and in this case the natural bound for the mean curvature function is

$$|H| < 1.$$

This comes from the fact that, in order to solve the Dirichlet problem, the prescribed mean curvature must be bounded from above by the mean curvature of the Killing cylinder over the domain.

To finish this section, we mention also the article [5] that appeared after writing the first version of this survey. In [5] Bonorino et al. study the asymptotic Dirichlet problem in $\mathbb{H}^n \times \mathbb{R}$ with prescribed mean curvature $H\colon \mathbb{H}^n \times \mathbb{R} \to \mathbb{R}$ depending also on the height variable.

3 Different Strategies to Solve the Asymptotic Dirichlet Problem

The proof of the solvability of the asymptotic Dirichlet problem is done in two steps. Namely, first one needs to obtain an entire solution, and secondly, show that this solution has the correct behaviour at infinity. The first step is easy (at least for the minimal graphs) in a sense that it follows from the known theory of PDE's: The Dirichlet problem for the minimal surface equation is solvable on bounded domains if the mean curvature of the boundary is nonnegative. To obtain an entire solution this can be used as follows.

Let $\theta \in C(\partial_\infty M)$ be a continuous function defined on the asymptotic boundary. Fix a point $o \in M$ and extend θ to a function $\theta \in C(\bar{M})$. Then for each $k \in \mathbb{N}$ there exists a solution $u_k \in C^{2,\alpha}(B(0,k)) \cap C(\bar{B}(o,k))$ to

$$\begin{cases} \operatorname{div} \frac{\nabla u_k}{\sqrt{1+|\nabla u_k|}} = 0 & \text{in } B(o,k) \\ u_k | \partial B(o,k) = \theta. \end{cases}$$

Applying interior gradient estimate in compact subsets of M and using the diagonal argument, one finds a subsequence that converges locally uniformly with respect to C^2-norm to a solution u. Then the hard part is to show that u extends continuously to $\partial_\infty M$ and satisfies $u|\partial_\infty M = \theta$, that is

$$\lim_{x \to x_0} u(x) = \theta(x_0) \tag{10}$$

for any $x_0 \in \partial_\infty M$. In the case of operators mentioned in Sect. 2.2, it is also necessary to obtain global height estimates to get the converging subsequence, since constants are no more solutions.

In order to complete the proof of the correct behaviour at infinity, a couple of different strategies have been applied. One possibility is to construct local barriers at infinity and use them to show that the solution must have the desired behaviour. Another possibility is to use more PDE related techniques, such as Caccioppoli and Poincaré-type inequalities, in order to obtain integral estimates and to show that the difference of the solution and boundary data is pointwise bounded by these integrals that tend to 0 at infinity. It is this latter part where the geometry of the manifold, i.e. the curvature assumptions, play a key role.

Finally, the uniqueness of the solutions (for operators satisfying maximum principle) can be proved as follows. Assume that u and $\tilde{u}$ are solutions, continuous up to the boundary, and $u = \tilde{u}$ on $\partial_\infty M$. Assume that there exists $y \in M$ with $u(y) > \tilde{u}(y)$. Now denote $\delta = (u(y) - \tilde{u}(y))/2$ and let $U \subset \{x \in M : u(x) > \tilde{u}(x) + \delta\}$ be the component of M containing the point y. Since u and $\tilde{u}$ are continuous functions that coincides on the asymptotic boundary $\partial_\infty M$, it follows that U is relatively compact open subset of M. Moreover, $u = \tilde{u} + \delta$ on ∂U, which implies $u = \tilde{u} + \delta$ in U. This is a contradiction since $y \in U$.

3.1 Methods Related to Barriers

First way to show (10) is to use local barriers at the given point x_0, i.e. to bound the values of the solution u from above and from below by sub- and supersolutions that are approaching the desired value $\theta(x_0)$ when $x \to x_0$. The idea is as follows. Let $x_0 \in \partial_\infty M$ and $\varepsilon > 0$ be arbitrary. Then by the continuity of θ (the extended function in $\bar{M}$), there exists a neighbourhood W of x_0 such that

$$|\theta(x) - \theta(x_0)| < \varepsilon/2$$

for all $x \in W$.

If we have a supersolution ψ (in W) so that $\psi(x) \to 0$ as $x \to x_0$, the function $-\psi$ will be a subsolution and we aim to bound the solution u in W as

$$-\psi(x) + \theta(x_0) - \varepsilon \le u(x) \le \psi(x) + \theta(x_0) + \varepsilon.$$

This follows since, by comparison principle, we can bound every solution u_k (for k large enough) to be between these barriers. Finally it follows that

$$\limsup_{x \to x_0} |u(x) - \theta(x_0)| \le \varepsilon$$

since $\lim_{x \to x_0} \psi(x) = 0$. Because ε and $x_0 \in \partial_\infty M$ were arbitrary, this shows the claim.

Strict convexity condition. Motivated by Choi's [16] convex conic neighbourhood condition, that was used for the Laplacian, Ripoll and Telichevesky [42] introduced the following strict convexity condition to suit more general divergence form quasilinear elliptic PDE's.

Definition 1 A Cartan–Hadamard manifold M satisfies the strict convexity condition (SC condition) if, given $x \in \partial_\infty M$ and a relatively open subset $W \subset \partial_\infty M$ containing x, there exists a C^2 open subset $\Omega \subset M$ such that $x \in \operatorname{Int} \partial_\infty \Omega \subset W$, where $\operatorname{Int} \partial_\infty \Omega$ denotes the interior of $\partial_\infty \Omega$ in $\partial_\infty M$, and $M \setminus \Omega$ is convex.

They showed that if M satisfies the SC condition and the sectional curvatures of M are bounded from above by a negative constant, $K_M \leq -k^2$, it is possible to use the distance function, $s \colon \Omega \to \mathbb{R}$, to $\partial\Omega$ to construct barriers at x. They also proved that rotationally symmetric manifolds with $K_M \leq -k^2$ and manifolds satisfying (3) satisfy the SC condition.

In order to prove the SC condition under (3), Ripoll and Telichevesky generalise former constructions of Anderson [2] and Borbély [6]. The idea is that since $K_M \leq -k^2$, the principal curvatures of the geodesic spheres are at least k, and therefore it is possible to take out small pieces of the spheres such that the remaining set is still convex.

Later, Casteras et al. [13] proved the SC condition under the curvature bounds (5) and (6). In order to do this, they used slightly modified version of the local barrier function that will be introduced next.

Barrier from an angular function. In [33] Holopainen and Vähäkangas solved the asymptotic Dirichlet problem for the p-Laplacian under very general curvature conditions on the manifold M. A key tool was a local barrier function at infinity that was constructed by generalising the ideas that go back to Anderson and Schoen [3] and to Holopainen [31]. The idea was to take continuous function on the boundary $\partial_\infty M$, extend it to the whole manifold and after a smoothening procedure obtain sub- and supersolutions that can be used as barriers. The clever idea in [33] was that the smoothening procedure depends also on the curvature lower bound, and this allowed the more general curvature conditions (3) and (4).

The barrier function obtained in [33] has appeared to be very flexible and suit also other quasilinear elliptic PDE's. The same barrier has been used to solve the asymptotic Dirichlet problem for a large class of operators in [12], for f-minimal graphs in [10], and for Killing graphs in [11]. As mentioned in the previous section, modified version of this barrier was used also in [13].

We point out that since the smoothening procedure in [33] depends on the curvature bounds, the computations become very long and technical. However, to give a very brief idea, we describe some steps. The barrier function is constructed under curvature bounds

$$-(b \circ r)^2(x) \leq K(P_x) \leq -(a \circ r)^2(x),$$

where r is the distance to a fixed point $o \in M$ and $a, b\colon [0, \infty) \to [0, \infty)$ are smooth functions satisfying certain conditions (see [33]). These curvature bounds are used to control the first two derivatives of the "barrier" that is constructed for each boundary point $x_0 \in \partial_\infty M$.

The construction starts from an angular function h that is defined on the boundary $\partial_\infty M$, then extended inside the manifold, smoothened by integrating against certain kernel, and appropriately normalised. We denote this smooth function still by h. Then adding this function h to the distance function r with certain negative power and multiplying by a constant we will obtain a function ψ that is a supersolution in a sufficiently small neighbourhood of x_0 in the cone topology. Finally this supersolution can be used to construct local barriers in that neighbourhood.

Rotationally symmetric manifolds. On rotationally symmetric manifolds, with the metric

$$g^2 = dr^2 + f(r)^2 d\vartheta^2,$$

the solvability of the asymptotic Dirichlet problem under the optimal curvature bounds of Theorem 4 is obtained by proving first the following integral condition that can be used to construct global barrier functions.

Theorem 5 *Assume that*

$$\int_1^\infty \Big(f(s)^{n-3} \int_s^\infty f(t)^{1-n} dt \Big) ds < \infty. \tag{11}$$

Then there exist non-constant bounded solutions of the minimal surface equation and, moreover, the asymptotic Dirichlet problem for the minimal surface equation is uniquely solvable for any continuous boundary data on $\partial_\infty M$.

The condition (11) was earlier considered by March in [35] where, by studying the behaviour of the Brownian motion, he proved that entire bounded nonconstant harmonic functions exist if and only if (11) is satisfied. Similar conditions appeared also in [16, 37]. We will sketch the proof for the minimal surface equation from [9]. How the condition (11) gives the curvature bounds of Theorem 4 can be found from [35].

Proof (*Proof of Theorem 5 (Sketch)*)

Changing the order of integration, the condition (11) reads

$$\int_1^\infty \frac{\int_1^t f(s)^{n-3} ds}{f(t)^{n-1}} dt < \infty. \tag{12}$$

Now interpret $\partial_\infty M$ as $\mathbb{S}^{n-1}$ and let $b\colon \mathbb{S}^{n-1} \to \mathbb{R}$ be a smooth non-constant function and define $B\colon M \setminus \{o\} \to \mathbb{R}$,

$$B(\exp(r\vartheta)) = B(r, \vartheta) = b(\vartheta), \ \vartheta \in \mathbb{S}^{n-1} \subset T_o M.$$

Define also

$$\eta(r) = k \int_r^\infty f(t)^{-n+1} \int_1^t f(s)^{n-3} ds dt,$$

with $k > 0$ to be determined later, and note that by the assumption (12) $\eta(r) \to 0$ as $r \to \infty$. The idea in the proof is to use the functions η and B, and condition (12) to construct global barrier functions.

The minimal surface equation for $\eta + B$ can be written as

$$\begin{aligned} \operatorname{div} \frac{\nabla(\eta + B)}{\sqrt{1 + |\nabla(\eta + B)|^2}} &= \frac{\Delta(\eta + B)}{\sqrt{1 + |\nabla(\eta + B)|^2}} \\ &\quad + \left\langle \nabla(\eta + B), \nabla\Big(\frac{1}{\sqrt{1 + |\nabla(\eta + B)|^2}}\Big) \right\rangle, \end{aligned} \tag{13}$$

and we want to estimate the terms on the right hand side from above.

An important fact is that on the rotationally symmetric manifolds the Laplace operator can be written as

$$\Delta = \frac{\partial^2}{\partial r^2} + (n-1)\frac{f' \circ r}{f \circ r}\frac{\partial}{\partial r} + \frac{1}{(f \circ r)^2}\Delta^{\mathbb{S}}, \tag{14}$$

where $\Delta^{\mathbb{S}}$ is the Laplacian on the unit sphere $\mathbb{S}^{n-1} \subset T_o M$, and for the gradient of a function φ we have

$$\nabla\varphi = \frac{\partial\varphi}{\partial r}\frac{\partial}{\partial r} + \frac{1}{f(r)^2}\nabla^{\mathbb{S}}\varphi \tag{15}$$

and

$$|\nabla\varphi|^2 = \varphi_r^2 + f^{-2}|\nabla^{\mathbb{S}}\varphi|^2.$$

Here $\nabla^{\mathbb{S}}$ is the gradient on $\mathbb{S}^{n-1}$, $|\nabla^{\mathbb{S}}\varphi|$ denotes the norm of $\nabla^{\mathbb{S}}\varphi$ with respect to the Euclidean metric on $\mathbb{S}^{n-1}$, and $\varphi_r = \partial\varphi/\partial r$.

Therefore using (13), (14) and (15) we obtain, by somewhat long computation,

$$\operatorname{div} \frac{\nabla(\eta + B)}{\sqrt{1 + |\nabla(\eta + B)|^2}} \le 0 \tag{16}$$

when we choose r large enough and then $k \ge ||b||_{C^2}$ large enough. In particular, $\eta + B$ is a supersolution to the minimal surface equation in $M \setminus B(o, r_0)$ for some r_0.

To obtain a global upper barrier choose k so that (16) holds and $\eta > 2\max|B|$ on the geodesic sphere $\partial B(o, r_0)$. Then $a := \min_{\partial B(o,r_0)}(\eta + B) > \max B$. Since $\eta(r) \to 0$ as $r \to \infty$, the function

$$w(x) := \begin{cases} \min\{(\eta + B)(x), a\} & \text{if } x \in M \setminus B(o, r_0); \\ a & \text{if } x \in B(o, r_0) \end{cases}$$

is continuous in $\bar{M}$ and coincide with b on $\partial_\infty M$. Global lower barrier v can be obtained similarly by replacing η with $-\eta$. Then $v \le B \le w$ by construction and the same will hold also for the solution u. In the end, continuous boundary values can be handled by approximation. □

3.2 Sobolev and Poincaré Inequalities with Moser Iteration

The method of proving the correct boundary values at infinity without barriers goes back to [15] where Cheng considered the asymptotic Dirichlet problem for the harmonic functions. Cheng's approach was modified by Vähäkangas in [50, 51] for the case of $\mathcal{A}$-harmonic functions. It turned out that Cheng's and Vähäkangas' proofs could be further developed to suit also the minimal surface equation and weaker assumptions on the curvature [8, 14]. We will describe some steps of the proof of Theorem 3; the proof of Theorem 2 follows similar steps although details differ due to the different curvature assumptions.

Within this approach we will deal with weak solutions of the minimal surface Eq. (2) that are defined as follows. Let $\Omega \subset M$ be an open set. Then a function $u \in W^{1,1}_{\text{loc}}(\Omega)$ is a *(weak) solution of the minimal surface equation* if

$$\int_\Omega \frac{\langle \nabla u, \nabla \varphi \rangle}{\sqrt{1 + |\nabla u|^2}} = 0 \tag{17}$$

for every $\varphi \in C_0^\infty(\Omega)$. Note that the integral is well-defined since

$$\sqrt{1 + |\nabla u|^2} \ge |\nabla u| \quad \text{a.e.},$$

and thus

$$\int_\Omega \frac{|\langle \nabla u, \nabla \varphi \rangle|}{\sqrt{1 + |\nabla u|^2}} \le \int_\Omega \frac{|\nabla u||\nabla \varphi|}{\sqrt{1 + |\nabla u|^2}} \le \int_\Omega |\nabla \varphi| < \infty.$$

Now let θ be the boundary data function that is extended inside the manifold M and let u be the solution obtained via the method mentioned in the beginning of this section. In order to show (10), we construct a certain smooth homeomorphism $\varphi\colon [0, \infty) \to [0, \infty)$ (see (22)) and prove first a pointwise estimate

$$\sup_{B(x,s/2)} \varphi\big(|u - \theta|/\nu\big)^{n+1} \le c \int_{B(x,s)} \varphi\big(|u - \theta|/\nu\big), \tag{18}$$

where c and ν are constants. Second step is to show that the integral on the right hand side of (18) tends to 0 as $r(x) \to \infty$, which then implies that the pointwise difference of the boundary data and the solution tends to 0. This second step follows from an integral estimate (Poincaré-type inequality)

$$\int_M \varphi(|u-\theta|/\nu) \le c + c\int_M F(r|\nabla\theta|) + c\int_M F_1(r^2|\nabla\theta|^2) < \infty. \tag{19}$$

Namely, let now (x_i) be a sequence of points such that $x_i \to x_0 \in \partial_\infty M$. Since the integral of φ over the whole manifold is finite by (19), we must have

$$\int_{B(x_i,s)} \varphi(|u-\theta|/\nu) \to 0$$

as $x_i \to x_0$ and (10) follows.

Key tools to obtain the pointwise estimate (18) are the Sobolev inequality (see e.g. [30])

$$\left(\int_{B(x,r_s)} |\eta|^{n/(n-1)}\right)^{(n-1)/n} \le C_s \int_{B(x,r_s)} |\nabla\eta|, \tag{20}$$

$\eta \in C_0^\infty(B(x,r_s))$, and a Caccioppoli-type inequality [8, Lemma 3.1]

$$\int_B \eta^2\varphi'(|u-\theta|/\nu)\frac{|\nabla u|^2}{\sqrt{1+|\nabla u|^2}} \le C_\varepsilon \int_B \eta^2\varphi'(|u-\theta|/\nu)|\nabla\theta|^2 + (4+\varepsilon)\nu^2\int_B \frac{\varphi^2}{\varphi'}(|u-\theta|/\nu)|\nabla\eta|^2, \tag{21}$$

where $\eta \ge 0$ is a C^1 function. We note that the inequality (21) can be proved by using

$$f = \eta^2\varphi\left(\frac{(u-\theta)^+}{\nu}\right) - \eta^2\varphi\left(\frac{(u-\theta)^-}{\nu}\right)$$

as a test function in the weak form of the minimal surface Eq. (17). These inequalities can be then used to run a Moser iteration process that yields the estimate (18). The idea of the Moser iteration, that goes back to [38, 46] (see also [25]), is that if one has an estimate for L^p-norms, i.e. $L^{p'}$-norm can be suitably bounded by L^p-norm when $p' \ge p$, then it is possible to iterate this estimate and obtain a bound for the L^∞-norm in terms of the L^p-norm with finite p. In our case we first obtain an estimate that can be written as a recursion formula $I_{j+1} \le C^{1/\kappa^j}\kappa^{j/\kappa^j} I_j$, where

$$I_j = \left(\int_{B_j} \varphi(h)^{m_j}\right)^{1/\kappa^j},$$

$\kappa = n/(n-1)$, $j \in \mathbb{N}$, $m_j = (n+1)\kappa^j - n$ and the radii of the balls B_j converge to $s/2$ as $j \to \infty$, and this finally yields (18).

The Caccioppoli inequality (21), together with the curvature bound, play a central role also in the proof of the integral estimate (19). Another essential part is the Young's inequality

$$ab \le F(a) + G(b)$$

for special type of Young functions F and G that are constructed as follows. Let $H: [0, \infty) \to [0, \infty)$ be a certain homeomorphism (for details, see [8, Sect. 2.3]) and define $G(t) = \int_0^t H(s)ds$ and $F(t) = \int_0^t H^{-1}(s)ds$. Then

$$\psi(t) = \int_0^t \frac{ds}{G^{-1}(s)} \quad \text{and} \quad \varphi = \psi^{-1} \tag{22}$$

are homeomorphisms so that $G \circ \varphi' = \varphi$. Another pair of Young functions F_1 and G_1 are constructed similarly, and so that $G_1 \circ \varphi'' \approx \varphi$. The integrability of the functions F and F_1 in (19) follows from the construction and the curvature assumptions on M.

4 Non-solvability of the Asymptotic Dirichlet Problem

From the classical results on Bernstein's problem we know that if the graph of $u: \mathbb{R}^n \to \mathbb{R}$ is a minimal surface in $\mathbb{R}^{n+1}$, then u is affine for $n \le 7$, and so a bounded solution must be constant. Therefore it is clear, that if we wish to solve the asymptotic Dirichlet problem with any continuous boundary data, the curvature cannot be zero everywhere. On the other hand, the discussion about the rotationally symmetric case and the theorems replacing the sectional curvature lower bound with the pinching condition (8) raise a question about the necessity of the curvature lower bound. It turns out that for the solvability of the asymptotic Dirichlet problem for (2) on general non-rotationally symmetric Cartan–Hadamard manifold, some control on the negativity of the curvature is necessary (e.g. the condition (8) or a curvature lower bound).

The first result in this direction was proved in 1994 by Ancona in [1] where he constructed a manifold with $K_M \le -1$ so that the Brownian motion almost surely exits M at a single point on the asymptotic boundary, and therefore the asymptotic Dirichlet problem for the Laplacian is not solvable. Borbély [7] constructed similar manifold using analytic arguments and later Holopainen and Ripoll [32] generalised Borbély's example to cover also the minimal surface equation by proving the following.

Theorem 6 *There exists a 3-dimensional Cartan–Hadamard manifold M with sectional curvatures ≤ -1 such that the asymptotic Dirichlet problem for the minimal surface Eq. (2) is not solvable with any continuous nonconstant boundary data, but there are nonconstant bounded continuous solutions of (2) on M.*

It is also worth mentioning two closely related results by Greene and Wu [27] that partly answer the question about the optimal curvature upper bound. Firstly, in [27, Theorems 2 and 4] they showed that an n-dimensional, $n \neq 2$, Cartan–Hadamard manifold with asymptotically nonnegative sectional curvature is isometric to $\mathbb{R}^n$. Secondly, in [27, Theorem 2] they showed that an odd dimensional Riemannian manifold with a pole $o \in M$ and everywhere non-positive or everywhere nonnegative sectional curvature is isometric to $\mathbb{R}^n$ if $\liminf_{s\to\infty} s^2 k(s) = 0$, where $k(s) = \sup\{|K(P_x)|: x \in M,\ d(o,x) = s,\ P_x \in T_x M \text{ two-plane}\}$.

Above the asymptotically nonnegative sectional curvature means the following.

Definition 2 Manifold M has asymptotically nonnegative sectional curvature if there exists a continuous decreasing function $\lambda\colon [0,\infty) \to [0,\infty)$ such that

$$\int_0^\infty s\lambda(s)\,ds < \infty,$$

and that $K_M(P_x) \geq -\lambda(d(o,x))$ at any point $x \in M$.

In [9] Casteras et al. proved the following under the conditions of Definition 2.

Theorem 7 *Let M be a complete Riemannian manifold with asymptotically nonnegative sectional curvature and only one end. If $u\colon M \to \mathbb{R}$ is a solution to the minimal surface equation that is bounded from below and has at most linear growth, then it must be a constant.*

Compared to the results of Greene and Wu, here M does not need to be simply connected and the curvature is allowed to change sign. Similar result was proved also in [43] but assuming nonnegative Ricci curvature. To end the discussion, we point out that for example the curvature lower bound

$$K(P_x) \geq -\frac{C}{r(x)^2\big(\log r(x)\big)^{1+\varepsilon}}, \qquad C > 0,$$

satisfies Definition 2, and this should be compared to the bounds (5) and those of Theorem 4.

References

1. Ancona, A.: Convexity at infinity and Brownian motion on manifolds with unbounded negative curvature. Rev. Mat. Iberoamericana **10**(1), 189–220 (1994)
2. Anderson, M.T.: The Dirichlet problem at infinity for manifolds of negative curvature. J. Differ. Geom. **18**(4), 701–721, 1983 (1984)
3. Anderson, M.T., Schoen, R.: Positive harmonic functions on complete manifolds of negative curvature. Ann. Math. (2) **121**(3), 429–461 (1985)
4. Bishop, R.L., O'Neill, B.: Manifolds of negative curvature. Trans. Am. Math. Soc. **145**, 1–49 (1969)

5. Bonorino, L., Casteras, J.-B., Klaser, P., Ripoll, J., Telichevesky M.: On the asymptotic Dirichlet problem for a class of mean curvature type partial differential equations. Calc. Var. Partial Differ. Equ. **59**(4), 34, Paper No. 135 (2020)
6. Borbély, A.: A note on the Dirichlet problem at infinity for manifolds of negative curvature. Proc. Am. Math. Soc. **114**(3), 865–872 (1992)
7. Borbély, A.: The nonsolvability of the Dirichlet problem on negatively curved manifolds. Differ. Geom. Appl. **8**(3), 217–237 (1998)
8. Casteras, J.-B., Heinonen, E., Holopainen, I.: Solvability of minimal graph equation under pointwise pinching condition for sectional curvatures. J. Geom. Anal. **27**(2), 1106–1130 (2017)
9. Casteras, J.-B., Heinonen E., Holopainen, I.: Existence and non-existence of minimal graphic and p-harmonic functions. Proc. Roy. Soc. Edinburgh Sect. A **150**(1), 341–366 (2020)
10. Casteras, J.-B., Heinonen, E., Holopainen I.: Dirichlet problem for f-minimal graphs. J. Anal. Math. **138**(2), 917–950 (2019)
11. Casteras, J.-B., Heinonen, E., Holopainen, I., Lira, J.: Asymptotic Dirichlet problems in warped products. Math. Z. **295**(1–2), 211–248 (2020)
12. Casteras, J.-B., Holopainen, I., Ripoll, J.B.: On the asymptotic Dirichlet problem for the minimal hypersurface equation in a Hadamard manifold. Potent. Anal. **47**(4), 485–501 (2017)
13. Casteras, J.-B., Holopainen, I., Ripoll, J.B.: Convexity at infinity in Cartan-Hadamard manifolds and applications to the asymptotic Dirichlet and Plateau problems. Math. Z. **290**(1–2), 221–250 (2018)
14. Casteras, J.-B., Holopainen, I., Ripoll, J.B.: Asymptotic Dirichlet problem for $\mathcal{A}$-harmonic and minimal graph equations in a Cartan–Hadamard manifolds. Comm. Anal. Geom. **27**(4), 809–855 (2019)
15. Shiu Yuen Cheng: The Dirichlet problem at infinity for non-positively curved manifolds. Commun. Anal. Geom. **1**(1), 101–112 (1993)
16. Hyeong In Choi: Asymptotic Dirichlet problems for harmonic functions on Riemannian manifolds. Trans. Am. Math. Soc. **281**(2), 691–716 (1984)
17. Collin, P., Rosenberg, H.: Construction of harmonic diffeomorphisms and minimal graphs. Ann. Math. (2), **172**(3), 1879–1906 (2010)
18. Dajczer, M., de Lira, J.H., Ripoll, J.: An interior gradient estimate for the mean curvature equation of Killing graphs and applications. J. Anal. Math. **129**, 91–103 (2016)
19. Dajczer, M., Hinojosa, P.A., de Lira, J.H.: Killing graphs with prescribed mean curvature. Calc. Var. Part. Differ. Equ. **33**(2), 231–248 (2008)
20. Dajczer, M., Ripoll, J.: An extension of a theorem of Serrin to graphs in warped products. J. Geom. Anal. **15**(2), 193–205 (2005)
21. do Espírito Santo, N., Fornari, S., Ripoll, J.B.: The Dirichlet problem for the minimal hypersurface equation in $M \times \mathbb{R}$ with prescribed asymptotic boundary. J. Math. Pures Appl. (9) **93**(2), 204–221 (2010)
22. do Espírito-Santo, N., Ripoll, J.: Some existence results on the exterior Dirichlet problem for the minimal hypersurface equation. Ann. Inst. H. Poincaré Anal. Non Linéaire **28**(3), 385–393 (2011)
23. Eberlein, P., O'Neill, B.: Visibility manifolds. Pacific J. Math. **46**, 45–109 (1973)
24. Gálvez, J.A., Rosenberg, H.: Minimal surfaces and harmonic diffeomorphisms from the complex plane onto certain Hadamard surfaces. Am. J. Math. **132**(5), 1249–1273 (2010)
25. Gilbarg, D., Trudinger, N.S.: Elliptic partial differential equations of second order. Classics in Mathematics. Springer, Berlin (2001). Reprint of the 1998 edition
26. Greene, R.E., Hung, W.: Function theory on manifolds which possess a pole. Lecture Notes in Mathematics, vol. 699. Springer, Berlin (1979)
27. Greene, R.E., Hung, W.: Gap theorems for noncompact Riemannian manifolds. Duke Math. J. **49**(3), 731–756 (1982)
28. Guio, E.M., Sa Earp, R.: Existence and non-existence for a mean curvature equation in hyperbolic space. Commun. Pure Appl. Anal. **4**(3), 549–568 (2005)
29. Heinonen, E.: Dirichlet problems for mean curvature and p-harmonic equations on Cartan-Hadamard manifolds. Doctoral dissertation, University of Helsinki (2017)

30. Hoffman, D., Spruck, J.: Sobolev and isoperimetric inequalities for Riemannian submanifolds. Commun. Pure Appl. Math. **27**, 715–727 (1974)
31. Holopainen, I.: Asymptotic Dirichlet problem for the p-Laplacian on Cartan-Hadamard manifolds. Proc. Am. Math. Soc. **130**(11), 3393–3400 (electronic) (2002)
32. Holopainen, I., Ripoll, J.B.: Nonsolvability of the asymptotic Dirichlet problem for some quasilinear elliptic PDEs on Hadamard manifolds. Rev. Mat. Iberoam. **31**(3), 1107–1129 (2015)
33. Holopainen, I., Vähäkangas, A.: Asymptotic Dirichlet problem on negatively curved spaces. J. Anal. **15**, 63–110 (2007)
34. Jenkins, H., Serrin, J.: The Dirichlet problem for the minimal surface equation in higher dimensions. J. Reine Angew. Math. **229**, 170–187 (1968)
35. March, P.: Brownian motion and harmonic functions on rotationally symmetric manifolds. Ann. Probab. **14**(3), 793–801 (1986)
36. Meeks, W.H., Rosenberg, H.: The theory of minimal surfaces in $M \times \mathbb{R}$. Comment. Math. Helv. **80**(4), 811–858 (2005)
37. Milnor, J.: On deciding whether a surface is parabolic or hyperbolic. Amer. Math. Monthly **84**(1), 43–46 (1977)
38. Moser, J.: On Harnack's theorem for elliptic differential equations. Commun. Pure Appl. Math. **14**, 577–591 (1961)
39. Nelli, B., Rosenberg, H.: Minimal surfaces in $\mathbb{H}^2 \times \mathbb{R}$. Bull. Braz. Math. Soc. (N.S.) **33**(2), 263–292 (2002)
40. Pigola, S., Rigoli, M., Setti, A.G.: Some remarks on the prescribed mean curvature equation on complete manifolds. Pacific J. Math. **206**(1), 195–217 (2002)
41. Ripoll, J., Telichevesky, M.: Complete minimal graphs with prescribed asymptotic boundary on rotationally symmetric Hadamard surfaces. Geom. Dedicata **161**, 277–283 (2012)
42. Ripoll, J., Telichevesky, M.: Regularity at infinity of Hadamard manifolds with respect to some elliptic operators and applications to asymptotic Dirichlet problems. Trans. Am. Math. Soc. **367**(3), 1523–1541 (2015)
43. Rosenberg, H., Schulze, F., Spruck, J.: The half-space property and entire positive minimal graphs in $M \times \mathbb{R}$. J. Differ. Geom. **95**(2), 321–336 (2013)
44. Ricardo Sa Earp and Éric Toubiana: An asymptotic theorem for minimal surfaces and existence results for minimal graphs in $\mathbb{H}^2 \times \mathbb{R}$. Math. Ann. **342**(2), 309–331 (2008)
45. Schoen, R., Yau, S.-T.: Lectures on harmonic maps. In: Conference Proceedings and Lecture Notes in Geometry and Topology, II. International Press, Cambridge, MA (1997)
46. Serrin, J.: Local behavior of solutions of quasi-linear equations. Acta Math. **111**, 247–302 (1964)
47. Serrin, J.: The problem of Dirichlet for quasilinear elliptic differential equations with many independent variables. Philos. Trans. R. Soc. Lond. Ser. A **264**, 413–496 (1969)
48. Telichevesky, M.: A note on minimal graphs over certain unbounded domains of Hadamard manifolds. Pacific J. Math. **281**(1), 243–255 (2016)
49. Vähäkangas, A.: Bounded p-harmonic functions on models and Cartan–Hadamard manifolds. Unpublished licentiate thesis, Department of Mathematics and Statistics, University of Helsinki (2006)
50. Vähäkangas, A.: Dirichlet problem at infinity for $\mathcal{A}$-harmonic functions. Potent. Anal. **27**(1), 27–44 (2007)
51. Vähäkangas, A.: Dirichlet problem on unbounded domains and at infinity. Reports in Mathematics, Preprint 499, Department of Mathematics and Statistics, University of Helsinki (2009)

Generalized Whitham Flow and Its Applications

Lynn Heller

Abstract The generalized Whitham flow [12] is a technique to interpolate between (symmetric) solutions of differential equations on surfaces with differing topology by introducing boundary conditions. This is a survey article on applications of the flow to the harmonic maps and self-duality equations over Riemann surfaces. We also discuss conjectures arising from the long time existence of such a flow.

Keywords Computational geometry · Graph theory · Hamilton cycles

1 Introduction

The genesis of this survey article are two very influential papers by Hitchin [18, 19]. The first [19] studies harmonic maps from compact Riemann surfaces into the 3-sphere, including minimal surfaces and (non-conformal) harmonic maps into a totally geodesic 2-sphere inside S^3. The latter are the Gauss maps of constant mean curvature (CMC) surfaces into (euclidean or spherical) 3-space. In the gauge theoretic set up the harmonicity of the maps translates to

$$F^{\nabla} = [\Phi, \Phi^*]; \quad \bar{\partial}^{\nabla}\Phi = 0, \tag{1}$$

where F^{∇} is the curvature of a special unitary connection ∇ on a rank 2 complex bundle V over the compact Riemann surface M, and Φ is a $(1,0)$-form with values in the trace-free endomorphism bundle $\mathrm{End}_0(V)$. Adding a complex parameter—called spectral parameter—(1) is shown to be equivalent to the flatness of the whole $\mathbb{C}_*$-family of flat connections on V:

$$\lambda \in \mathbb{C}_* \longmapsto \tilde{\nabla}^{\lambda} := \nabla + \lambda^{-1}\Phi - \lambda\Phi^*. \tag{2}$$

L. Heller (✉)
Leibniz University Hannover, Institut for Differential Geometrie, Welfengarten 1, 30167 Hannover, Germany
e-mail: lynn.heller@math.uni-hannover.de

T. Hoffmann et al. (eds.), *Minimal Surfaces: Integrable Systems and Visualisation*, Springer Proceedings in Mathematics & Statistics 349,
https://doi.org/10.1007/978-3-030-68541-6_8

Hitchin [19], and independently Pinkall–Sterling [25] and Bobenko [6], made a deep and systematic investigation of these integrable surfaces using variants of the associated family of flat connections (2), with particular success for compact surfaces of genus one. The existence of the immersion on M (rather than on its universal covering) is equivalent to the family $\tilde{\nabla}^\lambda$ satisfying two properties (called intrinsic and extrinsic closing conditions):

- $\tilde{\nabla}^\lambda$ is unitary for $\lambda \in S^1$.
- there exist two points $\lambda_i \in S^1$ such that $\tilde{\nabla}^{\lambda_i}$ has trivial monodromy (for target space S^3).

Because the fundamental group of a torus (g = 1) is abelian, the monodromy of $\tilde{\nabla}^\lambda$ can be simultaneously diagonalized with respect to both generators of the fundamental group. Due to its unitarity along S^1, the monodromy of $\tilde{\nabla}^\lambda$ is then diagonalizable for generic $\lambda \in \mathbb{C}_*$. In other words, $\tilde{\nabla}^\lambda$ generically reduces to a direct sum of $\mathbb{C}_*$-connections. This is very different for (immersed) higher genus surfaces, where the connection $\tilde{\nabla}^\lambda$ is irreducible for generic $\lambda \in \mathbb{C}_*$ by [16]. On the other hand, the closing condition for harmonic maps demands the existence of two trivial connections for $\lambda = \lambda_i \in S^1$. Reducible flat connections are singular points of the moduli space of flat connections. Thus the complex curve into $\mathcal{M}_{dR}$ given by

$$\lambda \mapsto [\tilde{\nabla}^\lambda]$$

maps into the smooth part of $\mathcal{M}_{dR}$ for generic λ, but intersects the singular set at finitely many λ. This is one of the main reasons that compact CMC surfaces of more complicated topology are far less understood from the integrable systems perspective.

The generalized Whitham flow [12] is a technique to overcome these difficulties. Starting at a torus solution, for which the integrable systems theory is well-understood, the flow deforms $\tilde{\nabla}^\lambda$ over the torus towards symmetric connections over higher genus surfaces by introducing boundary conditions, see Fig. 1 for the simplest examples of Lawson surfaces [23]. The existence of such a flow has ramifications for the higher genus Willmore and Lawson conjectures and helps to create a systematic approach to construct higher genus CMC and constrained Willmore surfaces.

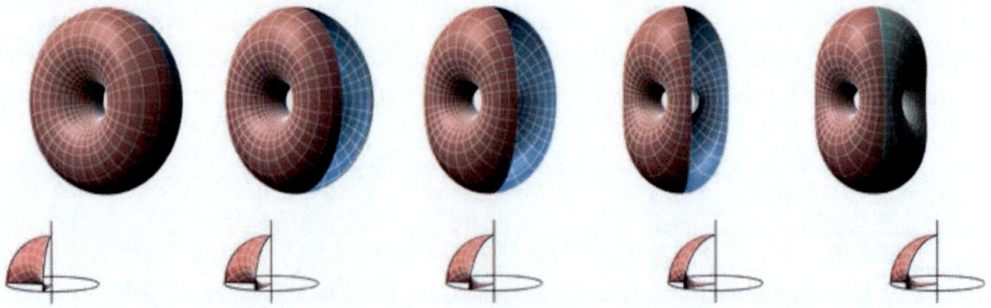

Fig. 1 The flow of minimal surfaces from the Clifford torus to Lawson's genus 2 surface $\xi_{1,2}$. The second row shows the Plateau solutions for the corresponding geodesic polygon

The other Hitchin paper mentioned above [18] deals with the reduction of the 4-dimensional self-dual Yang–Mills equations to Riemann surfaces. The resulting equations differ from those in (1) only by a sign:

$$F^{\nabla} = -[\Phi, \Phi^*]; \quad \bar{\partial}^{\nabla}\Phi = 0, \tag{3}$$

since solutions of the self-duality equations give rise to (equivariant) harmonic maps into the hyperbolic 3-space. In contrast to (1), solutions to (3) over tori are always reducible, yielding the classical abelian Hodge theory. For compact Riemann surfaces with $g \geq 2$ the moduli space $\mathcal{M}$ of solutions of the self-duality equations (modulo the gauge group) possesses a very rich geometric structure. It is a smooth manifold of dimension $12g - 12$ when restricted to irreducible solutions.

Irreducible solutions are uniquely determined by their holomorphic (Dolbeaut) data $(\bar{\partial}^{\nabla}, \Phi)$—called the Higgs pair. From this perspective Φ is a holomorphic $\mathrm{End}_0(V)$-valued 1-form of the holomorphic vector bundle $(V, \bar{\partial}^{\nabla})$. The irreducibility of the solution translates to the stability of the Higgs pair: Φ-invariant holomorphic line subbundles of V have strictly negative degree. Conversely, Hitchin has shown that every stable Higgs pair gives rise to an irreducible solution of the self-duality equations, giving a 1:1 correspondence between stable Higgs pairs and irreducible self-duality solutions—the Hitchin-Kobayashi correspondence. By construction, the moduli space $\mathcal{M}_{Dol}$ of stable Higgs bundles $(\bar{\partial}^{\nabla}, \Phi)$ is a holomorphic symplectic manifold. The cotangent space of the moduli space of stable holomorphic bundles is contained in $\mathcal{M}_{Dol}$ as an open dense subset. Through the Hitchin-Kobayashi correspondence $\mathcal{M}$ is equipped with a complex structure I.

As in the S^3-case considered in [19] a family of flat connections can be associated to a solution of (3)

$$\lambda \in \mathbb{C}_* \longmapsto \nabla^{\lambda} := \nabla + \lambda^{-1}\Phi + \lambda\Phi^*. \tag{4}$$

In stark contrast to the S^3-case Donaldson [8] showed that every irreducible flat $\mathbf{SL}(2, \mathbb{C})$-connection ∇ uniquely determines a solution of the self-duality equations such that $\nabla = \nabla^{\lambda=1}$ (up to gauge-equivalence). The moduli space of irreducible flat $\mathbf{SL}(2, \mathbb{C})$-connections $\mathcal{M}_{dR}$ is again a holomorphic symplectic manifold, and $\mathcal{M}$ inherits a second complex structure, denoted by J, turning $\mathcal{M}$ into a hyper-Kähler manifold.

Though both Eqs. (1) and (3) describe (equivariant) harmonic maps, the analytic properties of their solutions are determined by the curvature of the target space. In $\mathbb{H}^3$ the negative curvature yields existence and uniqueness of the harmonic map in each homotopy class relating them to representations of the fundamental group of the underlying Riemann surface. The failure of such results for S^3 on the other hand gives rise a cascade of examples of increasing complexity (even when fixing the genus).

A unified way to treat both theories is to consider the associated families as complex curves in the Deligne–Hitchin moduli space with different reality conditions. We give a comprehensive introduction to the space and its real sections in the second section. In the third section we revisit the construction of (branched) CMC surfaces of higher genus using the generalized Whitham flow [12]. Computer experiments indicate the longtime existence of the flow leading to experimental pictures of embedded CMC surfaces of genus 2, see Fig. 4. We then discuss the ramifications of the longtime existence of the flow for the higher genus Willmore and Lawson conjectures.

This technique can also be applied to the self-duality case [11] which allow one to flow between solutions of the self-duality solutions on Riemann surfaces of different genera. Moreover, new examples of real holomorphic sections of the Deligne–Hitchin moduli space are constructed in [11] which are counterexamples to a question raised by Simpson in [27]. The resulting solutions are harmonic maps in hyperbolic space away from certain codimension one singularity sets on the Riemann surface, and intersect the infinity boundary of $\mathbb{H}^3$ there. The behavior of these new solutions resembles the higher spectral genus solutions to the harmonic map equations (1) and they give smooth (equivariant) Willmore surfaces in S^3. We explain the main ideas behind these counterexamples in the last section. Moreover, we discuss possible applications of the involved ideas to study the singular strata of the moduli space of flat connections.

2 The Deligne–Hitchin Moduli Space

To every solution of (1) or (3) a $\mathbb{C}_*$-family of flat connections (2) and (4) can be associated. Those families give rise to complex curves in the moduli space of flat $\mathbf{SL}(2, \mathbb{C})$-connections $\mathcal{M}_{dR}$:

$$\mathcal{D}\colon \mathbb{C}_* \longrightarrow \mathcal{M}_{dR}, \quad \lambda \longmapsto [\tilde{\nabla}^\lambda] \text{ or } [\nabla^\lambda].$$

The space $\mathcal{M}_{dR}$ has a hyper-Kähler structure, i.e., it possesses three anti-commuting complex structures I, J and K which are Kähler with respect to the same Kähler metric. The smooth subset corresponds to irreducible connections. Although these different complex structures on $\mathcal{M}$ are naturally given, the transition between the Higgs pair and the flat connections picture is little understood. The construction of the Deligne–Hitchin moduli space [27, 28] is an approach to the problem which interpolates between these two pictures using λ-connections.

Definition 1 Let $\lambda \in \mathbb{C}$. An integrable λ-connection on the Riemann surface M is a pair $(\bar\partial, D)$ consisting of a holomorphic structure $\bar\partial$ and a first order linear differential operator D satisfying the λ-∂-Leibniz rule

$$D(fs) = fDs + \lambda\partial f \otimes s \quad \text{(for all functions } f \text{ and sections } s)$$

and the integrability condition

$$D\bar\partial + \bar\partial D = 0.$$

For $\lambda \neq 0$ a λ-connection gives rise to a flat connection via

$$(\lambda, \bar{\partial}, D) \mapsto \frac{1}{\lambda}D + \bar{\partial},$$

where flatness is resulting from the integrability condition. For $\lambda = 0$ we have that D is tensorial and the integrability gives in this case $D = \Phi \in \mathrm{H}^0(M; K\mathrm{End}(V))$. Thus the notion of λ-connections allows to smoothly interpolate between the space of Higgs pairs at $\lambda = 0$ and the space of flat connections at $\lambda = 1$. We restrict to $\mathbf{SL}(2, \mathbb{C})$-$\lambda$-connections in the following, i.e., the corresponding flat connection is $\mathbf{SL}(2, \mathbb{C})$ or the corresponding Φ is trace free and the underlying holomorphic structure has trivial determinant bundle.

2.1 *Stability*

The stability of Higgs bundles and irreducibility of connections are natural conditions to single out the smooth locus of the moduli space. In this paper we restrict to the case of the trivial $\mathbb{C}^2$-bundle V and $\mathbf{SL}(2, \mathbb{C})$-connections.

Definition 2 Let M be a compact Riemann surface. A $\mathbf{SL}(2, \mathbb{C})$-$\lambda$-connection $(\bar{\partial}, D)$ is called stable, if every $\bar{\partial}$-holomorphic subbundle $L \subset V = \underline{\mathbb{C}}^2$ with

$$D(\Gamma(M, L)) \subset \Omega^{(1,0)}(M, L)$$

satisfies

$$\deg(L) < 0$$

and semi-stable if

$$\deg(L) \leq 0.$$

All other λ-connections are called unstable.

For $\lambda \neq 0$, every λ-connection $(\bar{\partial}, D)$ is automatically semi-stable. Moreover, $(\bar{\partial}, D)$ is stable if and only if the connection $\nabla = \frac{1}{\lambda}D + \bar{\partial}$ is irreducible. For $\lambda = 0$ there exist unstable λ-connections and their gauge orbits are infinitesimally close to the gauge orbits of (certain) stable λ-connections.

Definition 3 Let M be a compact Riemann surface of genus $g \geq 2$. The Hodge moduli space $\mathcal{M}_{Hod} = \mathcal{M}_{Hod}(M)$ is the space of all semi-stable, $\mathbf{SL}(2, \mathbb{C})$-$\lambda$-connections on M modulo gauge transformations. In other words, $\mathcal{M}_{Hod}$ is the space of triples $[\lambda, \bar{\partial}, D]$.

The Hodge moduli space admits a holomorphic map

$$\pi = \pi_M : \mathcal{M}_{Hod} \longrightarrow \mathbb{C}; \quad [\lambda, \bar{\partial}, D] \longmapsto \lambda$$

whose fiber over $\lambda = 0$ is the Higgs moduli space $\mathcal{M}_{Dol}$, and over $\lambda = 1$ it is the deRham moduli space of flat SL(2, $\mathbb{C}$)-connections $\mathcal{M}_{dR}$, which we consider as complex analytic spaces endowed with their respective natural complex structures.

In order to obtain the Deligne–Hitchin moduli space $\mathcal{M}_{DH}$, the Hodge moduli space of M is glued with the Hodge moduli space of the complex conjugate Riemann surface $\bar{M}$ over $\mathbb{C}_*$ by identifying

$$[\lambda, \bar{\partial}, D]_M \sim \left[\frac{1}{\lambda}, \frac{1}{\lambda}D, \frac{1}{\lambda}\bar{\partial}\right]_{\bar{M}}.$$

Remark 1 The Deligne-gluing Ψ maps stable λ-connections on M to stable $\frac{1}{\lambda}$-connections on $\bar{M}$. Therefore, it maps the smooth locus of $\mathcal{M}_{Hod}(M)$ (consisting of stable λ-connections) to the smooth locus of $\mathcal{M}_{Hod}(\bar{M})$, and $\mathcal{M}_{DH}$ is equipped with a structure of a complex manifold at all of its stable points.

By construction we have a natural projection π from $\mathcal{M}_{DH}$ to $\mathbb{C}P^1$ and a section s of $\mathcal{M}_{DH}$ is a map

$$s\colon \mathbb{C}P^1 \to \mathcal{M}_{DH}, \quad \pi \circ s = \mathrm{id}_{\mathbb{C}P^1}.$$

The associated families ∇^λ and $\tilde{\nabla}^\lambda$ give rise to particularly simple sections of the Deligne–Hitchin moduli space, as

$$\lambda \in \mathbb{C}_* \longmapsto (\lambda, \bar{\partial}^\nabla \pm \lambda\Phi^*, \lambda\partial^\nabla + \Phi) \tag{5}$$

extends holomorphically to ∞. By the Hitchin-Kobayashi correspondence, the curve given by ∇^λ is determined by its Higgs pair $(0, \bar{\partial}, \Phi)$.

Definition 4 A holomorphic section s of $\mathcal{M}_{DH}$ is called stable, if the λ-connection $s(\lambda)$ is stable for all $\lambda \in \mathbb{C}^*$ and if the Higgs pairs $s(0)$ on M and $s(\infty)$ on $\bar{M}$ are stable.

It follows from the work of Hitchin [18] and Donaldson [8] that every stable point in $\mathcal{M}_{DH}$ uniquely determines a section s of the form (5) (with "+" -sign)—a twistor line. Thus a twistor line s is already stable if it is stable at some $\lambda_0 \in \mathbb{C}$.

2.2 Automorphisms of the Deligne–Hitchin Moduli Space

The Deligne–Hitchin moduli space admits some natural automorphisms. Let $N : \mathcal{M}_{DH} \to \mathcal{M}_{DH}$ be the map given by

$$[\nabla, \bar{\partial}, D] \longmapsto [-\lambda, \bar{\partial}, -D].$$

and $C: \mathcal{M}_{DH} \longrightarrow \mathcal{M}_{DH}$ be the continuation of the map

$$\tilde{C}: \mathcal{M}_{Hod}(M) \longrightarrow \mathcal{M}_{Hod}(\bar{M}), \quad \tilde{C}([\lambda, \bar{\partial}, D]_M) \longmapsto [\bar{\lambda}, \bar{\bar{\partial}}, \bar{D}]_{\bar{M}}.$$

To be more concrete, for

$$\bar{\partial} = \bar{\partial}^0 + \eta \quad \text{and} \quad D = \lambda(\partial^0) + \omega$$

where $d = \bar{\partial}^0 + \partial^0$ is the trivial connection, $\eta \in \Omega^{0,1}(M, \mathfrak{sl}(2, \mathbb{C}))$, and $\omega \in \Omega^{1,0}(M, \mathfrak{sl}(2, \mathbb{C}))$, we define the complex conjugate on the trivial $\mathbb{C}^2$-bundle over $\bar{M}$ to be

$$\bar{\bar{\partial}} = \partial^0 + \bar{\eta} \quad \text{and} \quad \bar{D} = \bar{\lambda}(\bar{\partial}^0) + \bar{\omega}.$$

The two involutions N and C commute. The real involution $\varrho = C$ covers the map

$$\lambda \in \mathbb{C}P^1 \longmapsto \bar{\lambda}^{-1} \in \mathbb{C}P^1$$

with fixed points given by $\lambda \in S^1$, while the real involution

$$\tau = CN$$

covers the fixed-point free involution $\lambda \mapsto -\bar{\lambda}^{-1}$ of $\mathbb{C}P^1$. A section s of $\mathcal{M}_{DH}$ is called τ-real and ϱ-real respectively if

$$\tau(s(-\bar{\lambda}^{-1})) = s(\lambda), \quad \varrho(s(\bar{\lambda}^{-1})) = s(\lambda).$$

Example 1 The associated family (2) of a harmonic maps into S^3 gives rise to a ϱ–real section of $\mathcal{M}_{DH}$ leading to $\tilde{\nabla}^\lambda$ being unitary along to unit circle. On the other hand, τ–real sections are given by the associated families of self-duality solutions (4) inducing twistorlines.

The space of real sections of the Deligne–Hitchin moduli space—or more generally of a twistor space—is of particular interest, as it allows for the reconstruction of the underlying hyper-Kähler manifold (under some further compatibility assumptions) by [20]. A natural question, due to Simpson [27], is whether all real sections are of the form (5). For ϱ-real sections this question can be answered in the affirmative, due to the Schwarzian reflexion principle, see [5]. For the real structure τ counterexamples have been constructed in [5, 11] revealing hidden structures, such as the existence of various connected components of real sections. The counterexamples constructed in [5] are of a different type than twistor lines, i.e., these never solve the self duality equation but correspond to harmonic maps into the deSitter space $\mathrm{SL}(2, \mathbb{C})/\mathrm{SL}(2, \mathbb{R})$, and the space of all these real sections does not lead to new hyper-Kähler spaces. Moreover, in [5] it is also shown that admissible τ-real sections are always twistor lines. Therefore, counterexamples of the same type as twistor lines are harder to construct. In [11] we found new solutions of the self-duality equations with singularities, which are only well-defined on an open and dense subset of the Riemann surface M. These so-called higher solutions of the self-duality equations are obtained by flowing certain Willmore tori using the generalized Whitham flow.

3 Higher Genus CMC Surfaces

While the long time existence of the heat flow for negatively curved spaces ensures existence and uniqueness of harmonic maps into $\mathbb{H}^3$, harmonic maps into positively curved spaces are more involving. Existence of compact minimal surfaces in S^3 and CMC surfaces in $\mathbb{R}^3$ of arbitrary genus was shown in [21–23] using gluing constructions and solutions of the Plateau problem. The integrable systems approach aims at a more detailed study of these surfaces and their moduli. For surfaces with non-abelian fundamental group, i.e., for surfaces with genus $g \geq 2$, there has been significant obstacles to generalize the very successful and well-developed theory for tori. The first glimpse that such an approach might work can be found in a series of papers by Heller [15–17], where necessary and sufficient conditions on (spectral) data for compact symmetric CMC surfaces are determined. Moreover, using the generalized Whitham flow [12], it was possible to construct spectral data satisfying these conditions.

The construction of higher genus CMC surfaces is tantamount to the construction of its associated family of flat connections $\tilde{\nabla}^\lambda$ (2) on $V \to M$. The map

$$\mathcal{D}\colon \lambda \mapsto [\tilde{\nabla}^\lambda]$$

induces a map into the moduli space of flat connections $\mathcal{M}_{dR}$ of V. The associated family and hence the surface can be constructed from $\mathcal{D}$ by loop group factorization [17]. In order to obtain a closed CMC surface of higher genus, $\mathcal{D}$ satisfies closing conditions:

- $D(\lambda)$ represents a unitary gauge class for all $\lambda \in S^1$;
- $D(\lambda_k)$ represents the trivial gauge class for $\lambda_k \in S^1$, $k = 1, 2$.

These closing conditions require the map $\mathcal{D}$ to intersect the singular set of $\mathcal{M}_{dR}$. The idea of the generalized Whitham flow is first to find appropriate coordinates of (a subset of) $\mathcal{M}_{dR}$ making the behavior of $\mathcal{M}_{dR}$ at the singularity transparent and the reality condition tractable. This is done by restricting to $\mathbb{Z}_{g+1}$-symmetric solutions and using the abelianization method [10]. Through these symmetries, the associated families $\tilde{\nabla}^\lambda_g$ of symmetric CMC surfaces of genus g are gauge equivalent to families of Fuchsian systems $\hat{\nabla}^\lambda_{\rho_g}$ over the same 4-punctured sphere with local monodromies ρ_g depending only on g. In a second step we vary ρ_g while preserving the closing conditions. This results in a unique variation of the spectral data leading to a (unique) deformation of the corresponding (symmetric) CMC surface. For rational ρ_g the

Fig. 2 The *stable* deformation of a 2-lobe Delaunay torus to a $\mathbb{Z}_3$-symmetric CMC surface of genus 2

Fig. 3 The *unstable* deformation of a 2-lobe Delaunay torus to a $\mathbb{Z}_3$-symmetric CMC surface of genus 2

surface is closed on a covering of the 4-punctured sphere. The local monodromies given by ρ_g might be interpreted as boundary conditions, see Figs. 1, 2, 3. The same idea also works for solutions to the self-duality equations by adjusting the reality condition accordingly.

3.1 Abelianization

Consider a compact Riemann surface M of genus g given by a $g+1$-fold covering of $\mathbb{C}P^1$ totally branched over 4 (distinct) points. Let f be a harmonic map from M into S^3 equivariant with respect to the $\mathbb{Z}_{g+1}$-symmetry. Then the associated family of flat connections $\tilde{\nabla}^\lambda$ (2) is also equivariant with respect to the symmetry. Biswas [4] showed that $\tilde{\nabla}^\lambda$ is then gauge equivalent to a family of invariant connections $\hat{\nabla}^\lambda$ by introducing removable first order poles at the fixed-points of the symmetry. Thus we can (generically) work with Fuchsian systems on the 4-punctured sphere

$$\hat{\nabla}^\lambda = d + \sum_{k=0}^{3} \frac{1}{z - p_k} A_k,$$

where $A_i \in \mathfrak{sl}(2, \mathbb{C})$. In [10] an abelianization procedure for flat connections on the 4-punctured sphere with prescribed local monodromies is carried out.

In view of the closing conditions, unitary Fuchsian systems are of particular importance. On a punctured Riemann surface, the Mehta-Seshadri theorem, a generalization of the Narasimhan-Seshadri theorem, associates to every stable parabolic structure a unique compatible unitary connection (with controlled singularities).

A parabolic structure on a holomorphic rank 2 vector bundle over a Riemann surface $E \to \Sigma$ is given by a finite number of points $p_k \in \Sigma$ together with lines $l_k \subset E_{p_k}$ and parabolic weights $\alpha_k \in [0; \frac{1}{2}[\subset \mathbb{R}$. The definition of parabolic stability is the same as the usual stability after replacing the degree of a sub line bundle by the parabolic degree

$$\text{par} - \deg(L) = \deg(L) + \sum_k \gamma_k,$$

where $\gamma_k = \alpha_k$ if $L_{p_k} = l_k$ and $\gamma_k = -\alpha_k$.

For Fuchsian systems the associated parabolic structure is given by the eigenlines with respect to the larger eigenvalues of the local monodromies A_k. Coming from an equivariant harmonic map on a genus g surface we can choose without loss of generality the four singular points on $\mathbb{C}P^1$ to be

$$p_0 = 0,\ p_1 = 1,\ p_2 = \infty,\ p_3 = m,$$

and all four weights to be $\alpha_k = \rho$. The generic stable parabolic structure is then given by the trivial rank 2 bundle, together with four different lines $l_k \subset \mathbb{C}^2$ interpreted as points in (a different) $\mathbb{C}P^1$, which can be normalized to

$$l_0 = 0, l_1 = 1, l_2 = \infty \text{ and } l_3 = u,$$

parametrized by their cross ratio u. If two of the four lines coalesce, i.e., if $u \to 0$, 1 or ∞, the corresponding parabolic structures are semi-stable but not stable. For $u = m$ there are three cases to consider (depending on ρ). Either the trivial $\mathbb{C}^2$ bundle with these parabolic lines is stable ($4\rho < 1$), it is strictly semi-stable $\rho = \frac{1}{4}$, or it is unstable. In the unstable case the trivial rank 2 bundle needs to be replaced by $\mathcal{O}(1) \oplus \mathcal{O}(-1)$ (with appropriate parabolic lines satisfying stability). Thus the moduli space of (semi-)stable holomorphic structures is then given by a $\mathbb{C}P^1$ with four distinguished points $u = 0, 1, \infty$ and $u = m$. The following theorem relates the moduli space $\mathcal{M}_\rho$ of Fuchsian systems over the 4-punctured sphere with local monodromy given by ρ to flat line bundle connections on a torus.

Theorem 1 ([10]) *Let $\rho \in]0, \frac{1}{2}[$ and let $T^2 \to \mathbb{C}P^1$ be the torus given by a twofold cover of $\mathbb{C}P^1$ branched over $p_0, \dots p_3$. Then, there exist a 2:1 covering of $\mathcal{M}_\rho$ by the moduli space of flat line bundles on the torus T^2 at stable Fuchsian Systems. This covering commutes with the projection of $\mathcal{M}_\rho$ to the space of parabolic bundles on the 4-punctured sphere and the projection to holomorphic line bundles on the torus.*

We make the above covering more explicit. Let w be the (global) holomorphic coordinate of the torus T^2. A flat line bundle connection on T^2 is given by

$$d + \alpha dw + \chi d\bar{w}$$

for complex constants α and χ. The pair (α, χ) gives rise to coordinates on $\mathcal{M}_\rho$ at stable Fuchsian systems via

$$(\alpha, \chi) \mapsto \nabla^{\alpha,\chi} = d + \alpha dw + \chi d\bar{w}\beta^- dw\beta^+ dw - \alpha dw - \chi d\bar{w}, \qquad (6)$$

where $\beta^\pm$ is uniquely determined by ρ and χ, see [11, Sect. 3.3.]. Moreover, these coordinates turn out to be Darboux coordinates. On the complement of the stable Fuchsian systems in $\mathcal{M}_\rho$, these coordinates have a particularly well-controlled behavior, see [10, Theorem 2].

3.2 *Deformation of Harmonic Maps*

For $\rho = \frac{1}{4}$, $\mathcal{M}_\rho$ is (an open and dense subset of) the moduli space of flat SL(2, $\mathbb{C}$)-connections on the torus. For generic $\lambda \in \mathbb{C}_*$ the connection $\tilde{\nabla}^\lambda$ is the direct sum of two line bundle connections, in particular $\beta^\pm = 0$.

Thus the coordinates $(\alpha^\lambda, \chi^\lambda)$ corresponding to $\tilde{\nabla}^\lambda$ give rise to holomorphic maps on a double cover Σ of the λ-plane. In particular, if the spectral curve Σ is of genus 0 or genus 1, the data (α, χ) are given very explicitly. Note that the spectral data for $\rho \neq \frac{1}{4}$ are not of an algebraic nature.

We take the well-known spectral data of a torus as initial data for the generalized Whitham flow. The flow parameter is given by $t = \rho - \frac{1}{4}$. In order to obtain closed surfaces for rational ρ, we require the closing conditions to be preserved (and some compatibility conditions to obtain a proper ϱ-section in $\mathcal{M}_{DH}$ in λ). The two closing conditions are dealt with successively.

The first closing condition is that $\tilde{\nabla}^\lambda$ is unitary for $\lambda \in S^1$. Through the Mehta-Seshadri theorem we obtain a unique α^λ for every given χ^λ with $\lambda \in S^1$. The short-time existence of the flow at Willmore stable homogenous tori and 2-lobed Delaunay tori are proven in [12], see Figs. 2, 3. The strategy is the following: for $t \sim 0$ we need to find a deformation of $\chi|_{t=0}$ such that the corresponding $\alpha(\chi)$ has a first order pole in $\lambda = 0$. This is achieved using the implicit function theorem. Let $M(\lambda\alpha)$ be the principal part of the holomorphic map $\lambda \cdot \alpha$ over $\lambda = 0$. Then we need to ensure

$$M(\lambda \cdot \alpha(\chi(t), t)) \equiv 0$$

for $t \sim 0$. Differentiating the equation yields:

$$M(\lambda \cdot \dot{\alpha}) + M(\lambda \cdot \partial_\chi \alpha|_{t=0})\dot{\chi} = 0.$$

Since $\dot{\alpha}$ is determined by the Mehta-Seshadri theorem, we can solve for $\dot{\chi}$, whenever $M(\lambda \cdot \partial_\chi \alpha|_{t=0})$ is invertible. The second closing condition requires the existence of trivial connections at two $\lambda_i \in S^1$ which is equivalent to $\chi(\lambda_i) = 0$ in this setup.

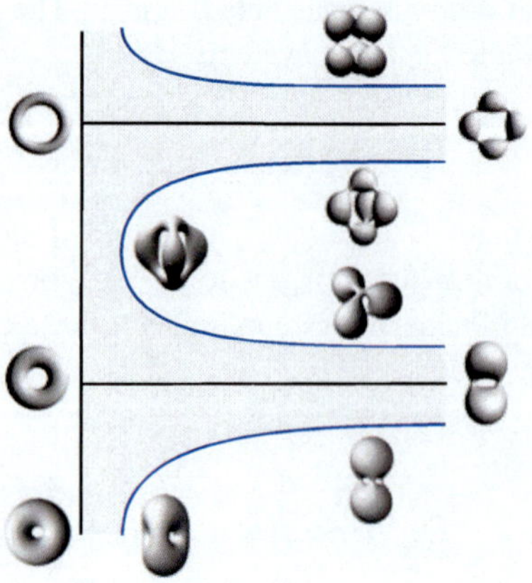

Fig. 4 Experimental moduli space of symmetric and embedded CMC surfaces of genus 2

The branch and umbilic order of the surface at the 4 singular points are completely determined by ρ. For rational ρ we thus obtain (possibly branched) compact CMC surfaces on a covering of $\mathbb{C}P^1$. If the flow exists till $t = \frac{g-1}{2g+2}$ we obtain a compact immersed CMC surface of genus 2. Therefore, the long-time existence of the flow gives a systematic way of constructing symmetric CMC surfaces of higher genus. Computer experiments indicate that such an approach might work, see Fig. 4, where we mapped out the moduli space of symmetric embedded CMC surfaces of genus 2 in S^3.

The study of long time existence of the generalized Whitham flow has further applications. The flow defines (after some choice) a map from the moduli space of harmonic tori into the moduli space of symmetric higher genus harmonic maps. The surjectivity of the map would give partial answers to a higher genus analogue of the Lawson conjecture. More generally, geometric properties, such as index and Willmore stability of the harmonic map can be studied along the flow. By restricting to appropriate boundary conditions strategies as used in [14, 24] to study the constrained Willmore problem along the family of homogenous tori might be made available to study higher genus analogues of the Willmore conjecture due to Kusner.

4 Higher Solutions and Constrained Willmore Surfaces

Constrained Willmore surfaces $f : M \to S^4$ are critical points of the Willmore functional

$$\int_M (|\mathbf{H}|^2 + 1)dA \qquad (7)$$

under variations preserving the conformal structure of the map. Here $\mathbf{H}$ denotes the mean curvature vector and dA is the induced area element. Willmore surfaces are characterized by the harmonicity of their associated mean curvature sphere congruence, a Moebius invariant Gauss map into the non-compact symmetric space of oriented 2-spheres in S^4, and therefore give rise to an associated family of flat connections $\tilde{\nabla}_4^{\mu}$, $\mu \in \mathbb{C}_*$, on a complex rank 4 bundle [7] of the form (1). When considering surfaces inside (a conformal) $S^3 \subset S^4$, the family of flat connections comes with an additional symmetry σ with $\sigma^*\tilde{\nabla}_4^{\mu} \cong \tilde{\nabla}_4^{\mu}$. Minimal and CMC surfaces in 3-dimensional space forms constitute important examples of constrained Willmore surfaces, where the rank 4 associated family of flat connections $\tilde{\nabla}_4^{\mu}$ globally splits into the direct sum of two equivalent rank 2 families, see [7, 13]. In the case of tori, the flat connections of the associated family are generically totally reducible and the spectral curve—the Riemann surface parametrizing the monodromies of $\tilde{\nabla}_4^{\mu}$—is a 4-fold covering of $\mathbb{C}P^1$ by [7]. For a CMC torus the quotient Σ of its spectral curve by σ is another $\mathbb{C}P^1$ and the associated family can be parametrized by $\Sigma \cong \mathbb{C}P^1$ instead to obtain its associated family as a CMC surface.

An important property is that by viewing these maps as constrained Willmore surfaces there is no distinction between the CMC surfaces in hyperbolic space $\mathbb{H}^3$ and in the other space forms, see [9]. The reason is that the intermediate spectral curve $\mathbb{C}P^1(\cong \Sigma)$ introduces a new spectral parameter λ which is a square root of the original parameter μ. Therefore, the involution on the associated family—parametrized by λ—is allowed to have either of the two signs on the λ-plane. Further, Babich and Bobenko [1] constructed smooth Willmore tori by gluing two minimal surfaces in $\mathbb{H}^3$ along the infinity boundary S^2. These minimal surfaces come from local solutions of the self-duality equations with singularities where the surface intersects the infinity boundary of the hyperbolic 3-space. By viewing the associated family of these Babich-Bobenko solutions as sections in the Deligne–Hitchin moduli, we observed that they satisfy the reality condition for self-duality solutions. On the other hand, such real sections cannot be of the form (5) as the corresponding self-duality solutions have singularities. More generally, for constrained Willmore tori in 3-space the quotient Σ of the spectral curve by the involution σ is always hyper-elliptic and naturally gives a rank 2 theory, where the associated family of flat connections is parametrized by Σ. We conjecture that all sections (for Riemann surfaces M of arbitrary genus) corresponding to Σ-families with a hyper-elliptic Σ satisfying the (self-duality) ρ-reality condition are local but not global self-duality solutions. It is interesting to note that as a solution of the rank 2 theory these examples have singularities, which are removed by putting them into the rank 4 framework.

In [11] the generalized Whitham flow was applied to deform the simplest Babich-Bobenko surfaces (equivariant cylinders) towards higher genus surfaces. Though the underlying idea remain the same as in the S^3 case, many technical details differ. As in the CMC in S^3 case we consider symmetric solutions, i.e., the family of flat connections ∇^{λ} is gauge equivalent to a family of Fuchsian systems $\tilde{\nabla}^{\lambda}$ on the 4-punctured sphere. Again, the local monodromies ρ_g are the same at each puncture and the eigenvalue depends only on the genus of the underlying surface and we can use the coordinates given by (α, χ) via (6).

Instead of a ϱ-real section of the Deligne–Hitchin moduli space, we are now dealing with a τ-real section. The τ-reality replaces the closing conditions for the harmonic map in S^3.

As τ covers the involution $\lambda \mapsto -\bar{\lambda}^{-1}$ on $\mathbb{C}P^1$, which has no fixed points, we need a two point approach to show that α^λ is still uniquely determined by χ^λ through a generalized Mehta-Seshadri theorem. As in the S^3 case we want to find a variation of χ_ρ such that the corresponding α_ρ has a simple pole over $\lambda = 0$ (together with some additional properties to define a τ-real section of $\mathcal{M}_{DH}$ in λ). The resulting surfaces are Willmore in S^3. Though the surface itself is not well-defined on M its curvatures are. Using the rotation of an elastic figure-8 curve in $\mathbb{H}^2$ as initial condition and applying the generalized Whitham flow in [12] would yield compact higher genus constrained Willmore surfaces for rational times.

The technical part in [11] is to show that these new solutions are in fact proper counterexamples of the question of Simpson, i.e., it solves the self-duality equation away codimension 1 singular sets. The first step is to show that there exist a family of **SL**(2, $\mathbb{C}$)-gauge transformation $g(\lambda)$ determined by

$$\overline{\nabla^{-\bar{\lambda}^{-1}}} = \nabla^\lambda.g(\lambda)$$

which up to normalization satisfies

$$\overline{g(-\bar{\lambda}-1)}g(\lambda) = \pm\text{id}.$$

The "+"sign gives real sections corresponding to harmonic maps into the deSitter space SL(2, $\mathbb{C}$)/SL(2, $\mathbb{R}$) while the "−"sign corresponds to harmonic maps into the hyperbolic space SL(2, $\mathbb{C}$)/$SU(2)$, i.e., solutions of the self-duality equations. For tori for which the connections are reducible, these two cases are not separated. For irreducible sections the sign is preserved when applying a continuous deformation. In the case of higher solutions constructed in [11], the sign was directly computed on the 4-punctured sphere, were the corresponding connections are generically irreducible even for the case of tori. Then, by applying a gauge transformation by the positive part of the Birkhoff factorization [26] of $g(\lambda)$ (on the open dense set where it exists) puts the lift ∇^λ of s into the self-duality form. Integrating the flat connection gives a equivariant conformal harmonic map into $\mathbb{H}^3$ which extends to a Moebius equivariant Willmore surface in S^3. The loop group factorization is only well-defined away from a 1-dimensional analytic subset of the Riemann surface M, [10], implying a non-trivial intersection of the surface with the boundary S^2 of $\mathbb{H}^3 \subset S^3$. In a recent preprint [2], an energy functional for sections of the Deligne–Hitchin moduli space is introduced to distinguish the higher solutions from twistor lines. While the energy is negative for all self-duality solutions, it is always positive for the higher solutions of [10].

The generalized Whitham flow applied to plain twistor lines, i.e. self-duality solutions, interpolates between the moduli spaces with respect to Riemann surfaces of different genera. The long time existence of the flow is guaranteed for an open and dense subset of solutions, as corresponding sections in $\mathcal{M}_{DH}$ remain stable and do not intersect the singular set of $\mathcal{M}_{DH}$.

The reduction of Willmore surfaces to higher solutions of the self-duality equations suggests that the study of Σ-curves in $\mathcal{M}_{DH}$ as done in [3], with a hyper-elliptic curve Σ, is equivalent to the study of a semi-stable higher rank systems parametrized by $\mathbb{C}P^1$. In particular, this gives a technique to study the singular strata of higher rank harmonic map and self-duality type equations.

References

1. Babich, M., Bobenko, A.: Willmore tori with umbilic lines and minimal surfaces in hyperbolic space. Duke Math. J. **72**(1), 151–185 (1993)
2. Beck, F., Heller, S., Roeser, M.: The energy functional on the space of sections of the Deligne-Hitchin moduli space. Preprint 2019
3. Bielawski, R.: Hyperkähler manifolds of curves in twistor spaces. SIGMA Symmetry Integrability Geom. Methods Appl. 10, Paper 033, 13 (2014)
4. Biswas, I.: Parabolic bundles as orbifold bundles. Duke Math. J. **88**(2), 305–325 (1997)
5. Biswas, I., Heller, S., Roeser, M.: Real holomorphic sections of the Deligne-Hitchin moduli space. Commun. Math. Phys. **366**(3), 1099–1133 (2019)
6. Bobenko, A.: All constant mean curvature tori in R3, S3, H3 in terms of theta-functions. Math. Ann. **290**(2), 209–245 (1991)
7. Bohle, Ch.: Constrained Willmore tori in the 4-sphere. J. Differential Geom. **86**(1), 71–131 (2010)
8. Donaldson, S.K.: Twisted harmonic maps and the self-duality equations. Proc. Lond. Math. Soc. (3) **55**(1), 127–131 (1987)
9. Heller, L.: Constrained Willmore tori and elastic curves in 2-dimensional space forms. Commun. Anal. Geom. **22**(2), 343–369 (2014)
10. Heller, L., Heller, S.: Abelianization of Fuchsian systems on a 4-punctured sphere and applications. J. Symplectic Geom. **14**(4), 1059–1088 (2016)
11. Heller, L., Heller, S.: Higher solutions of Hitchin's self-duality equations. Preprint 2018
12. Heller, L., Heller, S., Schmitt, N.: Navigating the space of symmetric CMC surfaces. J. Differential Geom. **110**(3), 413–455 (2018)
13. Heller, L., Heller, S., Nidaye, Ch.B.: Isothermic constrained Willmore Tori in 3-space. Preprint 2019
14. Heller, L., Ndiaye, C.B.: First explicit constrained willmore minimizers of non-rectangular conformal class. Preprint 2017
15. Heller, S.: Lawson's genus two surface and meromorphic connections. Math. Z. **274**(3–4), 745–760 (2013)
16. Heller, S.: Higher genus minimal surfaces in S^3 and stable bundles. J. Reine Angew. Math. **685**, 105–122 (2013)
17. Heller, S.: A spectral curve approach to Lawson symmetric CMC surfaces of genus 2. Math. Ann. **360**(3–4), 607–652 (2014)
18. Hitchin, N.J.: The self-duality equations on a Riemann surface. Proc. Lond. Math. Soc. (3) **55**(1), 59–126 (1987)
19. Hitchin, N.J.: Harmonic maps from a 2-torus to the 3-sphere. J. Differential Geom. **31**(3), 627–710 (1990)
20. Hitchin, N.J., Karlhede, A., Lindström, U., Roček, M.: Hyper-Kähler metrics and supersymmetry. Commun. Math. Phys. **108**(4), 535–589 (1987)
21. Kapouleas, N.: Compact constant mean curvature surfaces in Euclidean three-space. J. Differential Geom. **33**(3), 683–715 (1991)
22. Kapouleas, N.: Constant mean curvature surfaces constructed by fusing Wente tori. Invent. Math. **119**(3), 443–518 (1995)
23. Lawson Jr., H.B.: Complete minimal surfaces in S^3. Ann. Math. **2**(92), 335–374 (1970)
24. Ndiaye, ChB, Schätzle, R.: New examples of conformally constrained Willmore minimizers of explicit type. Adv. Calc. Var. **8**(4), 291–319 (2015)

25. Pinkall, U., Sterling, I.: On the classification of constant mean curvature tori. Ann. Math. **130**(2), 407– 451 (1989)
26. Pressley, A., Segal, G.: Loop Groups. Oxford Mathematical Monographs, The Clarendon Press, Oxford University Press, New York (1986). (Oxford Science Publications)
27. Simpson, C.: The Hodge filtration on nonabelian cohomology. Algebraic geometry–Santa Cruz 1995. In: Proceedings of Symposia in Pure Mathematics, vol. 62, pp. 217–281. American Mathematical Society, Providence (1997)
28. Simpson, C.: A weight two phenomenon for the moduli of rank one local systems on open varieties. From Hodge theory to integrability and TQFT tt*-geometry. In: Proceedings of Symposia in Pure Mathematics, vol. 78, pp. 175–214. American Mathematical Society, Providence (2008)

Notes on Translating Solitons for Mean Curvature Flow

David Hoffman, Tom Ilmanen, Francisco Martín, and Brian White

Abstract These notes provide an introduction to translating solitons for the mean curvature flow in $\mathbf{R}^3$. In particular, we describe a full classification of the translators that are complete graphs over domains in $\mathbf{R}^2$.

Keywords Mean curvature flow · Singularities · Monotonicity formula · Area estimates · Comparison principle

1 Introduction

Mean curvature flow is an exciting area of mathematical research. It is situated at the crossroads of several scientific disciplines: geometric analysis, geometric measure theory, partial differential equations, differential topology, mathematical physics, image processing, computer-aided design, among others. In these notes, we give a brief introduction to mean curvature flow and we describe recent progress on translators, an important class of solutions to mean curvature flow (Fig. 1).

F. Martín was partially supported by the MINECO/FEDER grant MTM2017-89677-P and by the Leverhulme Trust grant IN-2016-019.

B. White was partially supported by grants from the Simons Foundation (#396369) and from the National Science Foundation (DMS 1404282, DMS 1711293).

D. Hoffman (✉) · B. White
Department of Mathematics, Stanford University, Stanford, CA 94305, USA
e-mail: dhoffman@stanford.edu

B. White
e-mail: bcwhite@stanford.edu

T. Ilmanen
Department of Mathematics, E. T. H. Zürich, Rämistrasse 101, 8092 Zürich, Switzerland
e-mail: tom.ilmanen@math.ethz.ch

F. Martín
Departamento de Geometría y Topología, Universidad de Granada, 18071 Granada, Spain
e-mail: fmartin@ugr.es

T. Hoffmann et al. (eds.), *Minimal Surfaces: Integrable Systems and Visualisation*, Springer Proceedings in Mathematics & Statistics 349,
https://doi.org/10.1007/978-3-030-68541-6_9

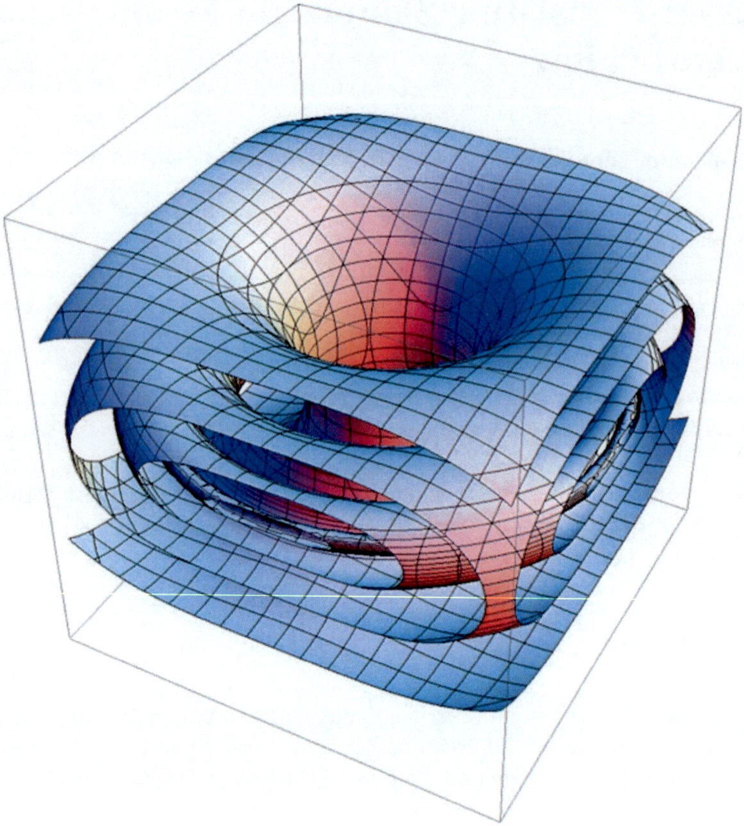

Fig. 1 A surface moving by mean curvature

In physics, diffusion is a process which equilibrates spatial variations in concentration. If we consider a initial concentration u_0 on a domain $\Omega \subseteq \mathbf{R}^2$ and seek solutions of the linear heat equation $\frac{\partial}{\partial t}u - \Delta u = 0$, with initial data u_0 and natural boundary conditions on $\partial\Omega$, we obtain smoothed concentrations $\{u_t\}_{t>0}$. If we are interested in the smoothing of perturbed surface geometries, it make sense to use analogous strategies. The geometrical counterpart of the Euclidean Laplace operator Δ on a smooth surface $M^2 \subset \mathbf{R}^3$ (or more generally, a hypersurface $M^n \subset \mathbf{R}^{n+1}$) is the Laplace-Beltrami operator, which we will denote as Δ_M. Thus, we obtain the geometric diffusion equation

$$\frac{\partial}{\partial t}\mathbf{x} = \Delta_{M_t}\mathbf{x}, \tag{1.1}$$

for the coordinates $\mathbf{x}$ of the corresponding family of surfaces $\{M_t\}_{t\in[0,T)}$.

A classical formula (see [9], for instance) says that, given a hypersurface in Euclidean space, one has:

$$\Delta_{M_t}\mathbf{x} = \mathbf{H},$$

where $\mathbf{H}$ represents the mean curvature vector. This means that (1.1) can be written as:

$$\frac{\partial}{\partial t}\mathbf{x}(p,t) = \mathbf{H}(p,t). \tag{1.2}$$

Using techniques of parabolic PDE's it is possible to deduce the existence and uniqueness of the mean curvature flow for a small time period in the case of compact manifolds (for details see [10, 23], among others).

Theorem 1 *Given a compact, immersed hypersurface M in $\mathbf{R}^{n+1}$ then there exists a unique mean curvature flow defined on an interval $[0, T)$ with initial surface M.*

The mean curvature is known to be the first variation of the area functional $M \mapsto \int_M d\mu$ (see [7, 26]). We will obtain for the $\text{Area}(\Omega(t))$ of a relatively compact $\Omega(t) \subset M_t$ that

$$\frac{d}{dt}(\text{Area}(\Omega(t))) = -\int_{\Omega(t)} |\mathbf{H}|^2 d\mu_t.$$

In other words, we get that the mean curvature flow is the corresponding gradient flow for the area functional:

Remark 1 The mean curvature flow is the flow of steepest descent of surface area.

Moreover, we also have a nice maximum principle for this particular diffusion equation.

Theorem 2 (Maximum/comparison principle) *If two compact immersed hypersurfaces of $\mathbf{R}^{n+1}$ are initially disjoint, they remain so. Furthermore, compact embedded hypersurfaces remain embedded.*

Similarly, convexity and mean convexity are preserved:

- If the initial hypersurface M is convex (i.e., all the geodesic curvatures are positive, or equivalently M bounds a convex region of $\mathbf{R}^{n+1}$), then M_t is convex, for any t.
- If M is mean convex ($H > 0$), then M_t is also mean convex, for any t.

Moreover, mean curvature flow has a property which is similar to the eventual simplicity for the solutions of the heat equation. This result was proved by Huisken and asserts:

Theorem 3 ([21]) *Convex, embedded, compact hypersurfaces converge to points $p \in \mathbf{R}^{n+1}$. After rescaling to keep the area constant, they converge smoothly to round spheres.*

As a consequence of the above theorems we have.

Corollary 1 (Existence of singularities in finite time) *Let M be a compact hypersurface in $\mathbf{R}^{n+1}$. If M_t represents its evolution by the mean curvature flow, then M_t*

must develop singularities in finite time. Moreover, if we denote this maximal time as $T_{\max}$, then

$$2\,n\,T_{\max} \leq (\mathrm{diam}_{\mathbf{R}^{n+1}}(M))^2\,.$$

Proof Since M is compact, it is contained an open ball $B(p, \rho)$. So, M must develop a singularity before the flow of $\mathbf{S}^n_p$ collapses at the point p, as otherwise we would contradict the previous theorem. The upper bound of $T_{\max}$ is just a consequence of the collapsing time of a sphere. □

A natural question is: What can we say when M is not compact? In this case, we can have long-time existence. A trivial example is the case of a complete, properly embedded minimal hypersurface M in $\mathbf{R}^{n+1}$. Under the mean curvature flow, M remains stationary, so the flow exists for all time. If we are looking for non-stationary examples, then we can consider the following example:

Example 1 (*grim reapers*) Consider the euclidean product $M = \Gamma \times \mathbf{R}^{n-1}$, where Γ is the *grim reaper* in $\mathbf{R}^2$ represented by the immersion

$$f : (-\pi/2, \pi/2) \to \mathbf{R}^2 \tag{1.3}$$

$$f(x) = (x, \log\cos x).$$

If, ignoring parametrization, we let M_t be the result of flowing M by mean curvature flow for time t, then $M_t = M - t\,e_{n+1}$, where again $\{\mathrm{e}_1, \ldots, \mathrm{e}_{n+1}\}$ represents the canonical basis of $\mathbf{R}^{n+1}$. In other words, M moves by vertical translations. By definition, we say that M is **a translating soliton** in the direction of $-\mathrm{e}_{n+1}$. More generally, any translator in the direction of $-\mathrm{e}_{n+1}$ which is a Riemannian product of a planar curve and an euclidean space $\mathbf{R}^{n-1}$ can be obtained from this example by a suitable combination of a rotation and a dilation (see [16] for further details.) We refer to these translating hypersurfaces as ***n*-dimensional grim reapers**, or simply grim reapers if the n is clear from the context (Fig. 2).

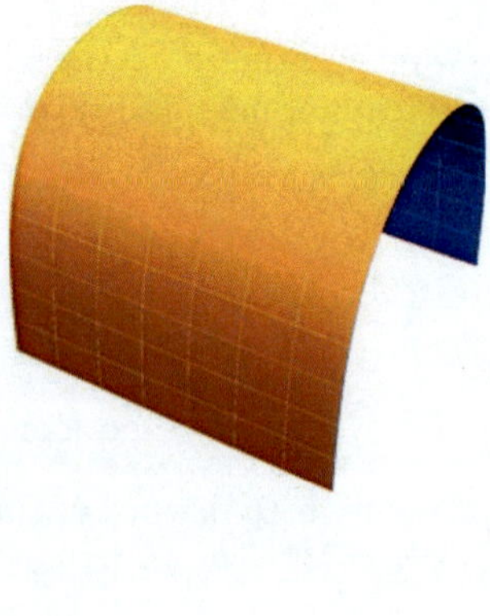

Fig. 2 A grim reaper

2 Some Remarks About Singularities

Throughout this section, we consider a fixed compact initial hypersurface M. Consider the maximal time $T = T_M$ such that a smooth solution of the MCF $F : M \times [0, T) \to \mathbf{R}^{n+1}$ as in Theorem 1 exists. Then the embedding vector F is uniformly bounded according to Theorem 2. It follows that some spatial derivatives of the embedding F_t have to become unbounded as $t \nearrow T$. Otherwise, we could apply Arzelà–Ascoli Theorem and obtain a smooth limit hypersurface, M_T, such that M_t converges smoothly to M_T as $t \nearrow T$. This is impossible because, in such a case, we could apply Theorem 1 to restart the flow. In this way, we could extend the flow smoothly all the way up to $T + \varepsilon$, for some $\varepsilon > 0$ small enough, contradicting the maximality of T. In fact, Huisken [20, Theorem 8.1] showed that the second spatial derivative (i.e., the norm of the second fundamental form) blows up as $t \to T$.

We would like to say more about the "blowing-up" of the norm of A, as $t \nearrow T$. The evolution equation for $|A|^2$ is

$$\frac{\partial}{\partial t}|A|^2 = \Delta|A|^2 - 2\left|\nabla A\right|^2 + 2|A|^4.$$

Define

$$|A|^2_{\max}(t) := \max_{M_t} |A|^2(\cdot, t).$$

Using Hamilton's trick (see [23]) we deduce that $|A|^2_{\max}$ is locally Lipschitz and that

$$\frac{d}{dt}\, |A|^2_{\max}(t_0) = \frac{\partial}{\partial t}\, |A|^2(p_0, t_0),$$

where p_0 is any point where $|A|^2(\cdot, t_0)$ reaches its maximum. Thus, using the above expression, we have

$$\begin{aligned} \frac{d}{dt}\, |A|^2_{\max}(t_0) &= \frac{\partial}{\partial t}\, |A|^2(p_0, t_0) \\ &= \Delta|A|^2(p_0, t_0) - 2\left|\nabla A(p_0, t_0)\right|^2 + 2|A|^4(p_0, t_0). \end{aligned}$$

It is well known that the Hessian of $|A|$ is negative semi-definite at any maximum. In particular the Laplacian of $|A|$ at these points is non-positive. Hence,

$$\frac{d}{dt}\, |A|^2_{\max}(t_0) \leq 2|A|^4(p_0, t_0) \leq 2|A|^4_{\max}(t_0).$$

Notice that $|A|^2_{\max}$ is always positive, since otherwise at some instant t we would have $|A(\cdot, t)| \equiv 0$ along M_t, which would imply that M_t is totally geodesic and therefore a hyperplane in $\mathbf{R}^{n+1}$, contrary to the fact that the initial surface was compact.

So, one can prove that $1/|A|^2_{\max}$ is locally Lipschitz. Then the previous inequality allows us to deduce that:

$$-\frac{d}{dt}\left(\frac{1}{|A|^2_{\max}}\right) \leq 2 \quad \text{a.e. in } t \in [0, T).$$

Integrating (respect to time) in any sub-interval $[t, s] \subset [0, T)$ we get

$$\frac{1}{|A|^2_{\max}(t)} - \frac{1}{|A|^2_{\max}(s)} \leq 2(s - t).$$

As $|A(\cdot, t)|$ is not bounded as to tends to T, there exists a time sequence $s_i \nearrow T$ such that

$$|A|^2_{\max}(s_i) \to +\infty.$$

Substituting $s = s_i$ in the above inequality and taking the limit as $i \to \infty$, we get

$$\frac{1}{|A|^2_{\max}(t)} \leq 2(T - t).$$

We collect all this information in the next proposition.

Proposition 1 *Consider the mean curvature flow for a compact initial hypersurface M. If T is the maximal time of existence, then the following lower bound holds*

$$|A|_{max}(t) \geq \frac{1}{\sqrt{2(T - t)}}$$

for all $t \in [0, T)$.

In particular,

$$\lim_{t \to T} |A|_{max}(t) = +\infty.$$

Definition 1 When this happens we say that T is *singular time* for the mean curvature flow.

So we have the following improved version of Theorem 1:

Theorem 4 *Given a compact, immersed hypersurface M in $\mathbf{R}^{n+1}$ then there exists a unique mean curvature flow defined on a maximal interval $[0, T_{max})$.*

Moreover, T_{max} is finite and

$$|A|_{max}(t) \geq \frac{1}{\sqrt{2(T_{max} - t)}}$$

for each $t \in [0, T_{max})$.

Remark 2 From the above proposition, we deduce the following estimate for the maximal time of existence of flow:

$$T_{\max} \geq \frac{1}{2|A|^2_{\max}(0)}.$$

Definition 2 Let T be the maximal time of existence of the mean curvature flow. If there is a constant $C > 1$ such that

$$|A|_{\max}(t) \leq \frac{C}{\sqrt{2(T-t)}},$$

then we say that the flow develops a *Type I singularity* at instant T. Otherwise, that is, if

$$\limsup_{t \to T} |A|_{\max}(t) \sqrt{(T-t)} = +\infty,$$

we say that is a *Type II singularity*.

We conclude this brief section by pointing out that there have been substantial breakthroughs in the study and understanding of the singularities of Type I, whereas Type II singularities have been much more difficult to study. This seems reasonable since, according to the above definition and the results we have seen, the singularities of Type I are those for which one has the best possible control of blow-up of the second fundamental form.

3 Translators

A standard example of Type II singularity is given by a loop pinching off to a cusp (see Fig. 3). Angenent [3] proved, in the case of convex planar curves, that singularities of this kind are asymptotic (after rescaling) to the above mentioned grim reaper curve (1.3), which moves set-wise by translation. In this case, up to inner diffeomorphisms of the curve, it can be seen as a solution of the curve shortening flow which evolves by translations and is defined for all time. In this paper we are interested in this type of solitons, which we will call **translating solitons (or translators)** from now on. Summarizing this information, we make the following definition:

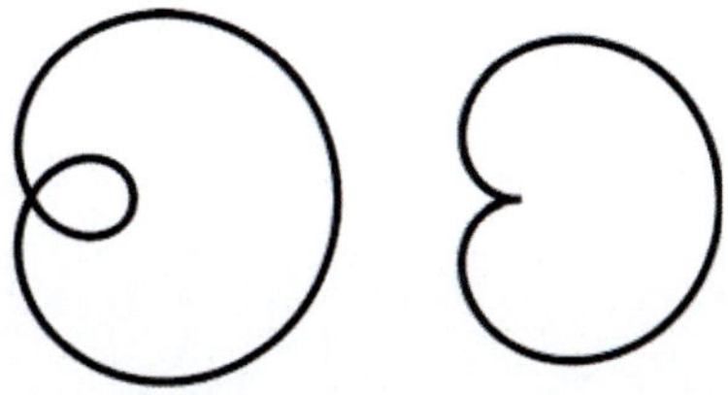

Fig. 3 A Type II singularity

Definition 3 (Translator) A translator is a hypersurface M in $\mathbf{R}^{n+1}$ such that

$$t \mapsto M - t\,\mathbf{e}_{n+1}$$

is a mean curvature flow, i.e., such that normal component of the velocity at each point is equal to the mean curvature at that point:

$$\mathbf{H} = -\mathbf{e}_{n+1}^{\perp}. \tag{3.1}$$

The cylinder over a grim-reaper curve, i.e. the hypersurface in $\mathbf{R}^{n+1}$ parametrized by $\mathscr{G} : \left(-\frac{\pi}{2}, \frac{\pi}{2}\right) \times \mathbf{R}^{n-1} \longrightarrow \mathbf{R}^{n+1}$ given by

$$\mathscr{G}(x_1, \ldots, x_n) = (x_1, \ldots, x_n, -\log\cos x_1),$$

is a translating soliton. It appears as limit of sequences of parabolic rescaled solutions of mean curvature flows of immersed mean convex hypersurfaces. For example, we can take product of the loop pinching off to a cusp times $\mathbf{R}^{n-1}$. We can produce others examples of solitons just by scaling and rotating the grim reaper. In this way, we obtain a 1-parameter family of translating solitons parametrized by $\mathscr{G}_\theta$: $\left(-\frac{\pi}{2\cos(\theta)}, \frac{\pi}{2\cos(\theta)}\right) \times \mathbf{R}^{n-1} \longrightarrow \mathbf{R}^{n+1}$

$$\mathscr{G}_\theta(x_1, \ldots, x_n) = (x_1, \ldots, x_n, \sec^2(\theta)\log\cos(x_1\cos(\theta)) - \tan(\theta)x_n), \tag{3.2}$$

where $\theta \in [0, \pi/2)$ (Fig. 4). Notice that the limit of the family F_θ, as θ tends to $\pi/2$, is a hyperplane parallel to $\mathbf{e}_{n+1}$.

Fig. 4 The regular grim reaper in $\mathbf{R}^3$ and the tilted grim reaper for $\theta = \pi/6$

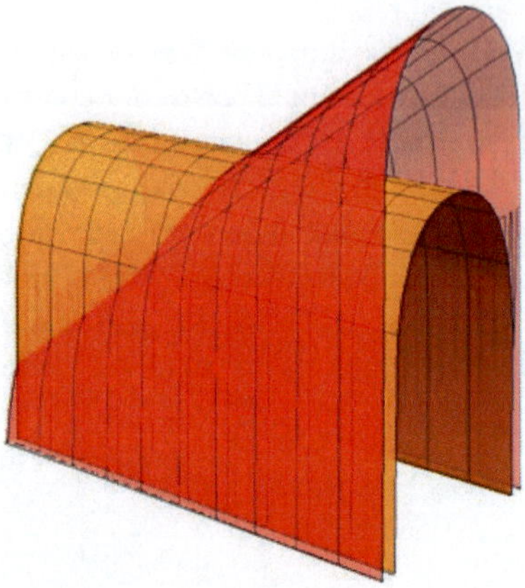

3.1 Variational Approach

Ilmanen [22] noticed that a translating soliton M in $\mathbf{R}^{n+1}$ can be seen as a minimal surface for the weighted volume functional

$$\mathscr{A}_f[M] = \int_M e^{-f}\, d\mu$$

where f represents the Euclidean height function, that is, the restriction of the last coordinate x_{n+1} to M. We have the following

Proposition 2 (Ilmanen) *Let M^n be a translating soliton in $\mathbf{R}^{n+1}$ and let N be its unit normal. Then the translator equation*

$$H = \langle \mathbf{e}_{n+1}, N\rangle \tag{3.3}$$

on the relatively compact domain $\Omega \subset M$ is the Euler–Lagrange equation of the functional

$$\mathscr{A}_f[\Omega] = \mathrm{vol}_f(\Omega) = \int_\Omega e^{-f}\, d\mu. \tag{3.4}$$

Moreover, the second variation formula for normal variations is given by

$$\delta^2 \mathscr{A}_f[\Omega]\cdot(u,u) = -\int_M e^{-f}\, u\, L_f u\, d\mu, \quad u \in C_0^\infty(\Omega), \tag{3.5}$$

where the stability operator L_f is defined by

$$L_f u = \Delta^f u + |A|^2 u \tag{3.6}$$

where Δ^f is the drift Laplacian given by

$$\Delta^f = \Delta - \langle \nabla f, \nabla\,\cdot\,\rangle = \Delta - \langle \mathbf{e}_{n+1}, \nabla\,\cdot\,\rangle. \tag{3.7}$$

Proof Given $\varepsilon > 0$, let $\Psi : (-\varepsilon,\varepsilon)\times M \to \mathbf{R}^{n+1}$ be a variation of M compactly supported in $\Omega \subset M$ with $\Psi(0,\,\cdot\,) = \mathrm{Id}$ and normal variational vector field

$$\left.\frac{\partial \Psi}{\partial s}\right|_{s=0} = uN + T$$

for some function $u \in C_0^\infty(\Omega)$ and a tangent vector field $T \in \Gamma(TM)$. Here, N denotes a local unit normal vector field along M. Then

$$\left.\frac{\mathrm{d}}{\mathrm{d}s}\right|_{s=0} \mathrm{vol}_f[\Psi_s(\Omega)] = \int_\Omega e^{-f}\,(\langle \mathbf{e}_{n+1}, N\rangle - H)u\, d\mu + \int_\Omega \mathrm{div}(e^{-f}T)\, d\mu.$$

Hence, stationary immersions for variations fixing the boundary of Ω are characterized by the scalar soliton equation

$$H - \langle \mathbf{e}_{n+1}, N\rangle = 0 \quad \text{on} \quad \Omega \subset\subset M,$$

which yields (3.3). Now we compute the second variation formula. At a stationary immersion we have

$$\frac{\mathrm{d}^2}{\mathrm{d}s^2}\bigg|_{s=0} \mathrm{vol}_f[\Psi_s(\Omega)] = \int_M e^{-f} \frac{d}{ds}\bigg|_{s=0} (\langle \mathbf{e}_{n+1}, N_s\rangle - H_s) u \, d\mu.$$

Using the fact that

$$\bar{\nabla}_{\partial_s} N = -\nabla u - \mathscr{W} T, \tag{3.8}$$

(where $\mathscr{W}$ means the Weingarten map) then we compute

$$\frac{\mathrm{d}}{\mathrm{d}s}\bigg|_{s=0} \langle \mathbf{e}_{n+1}, N\rangle = \langle \bar{\nabla}_{\partial_s} \mathbf{e}_{n+1}, N\rangle + \langle \mathbf{e}_{n+1}, \bar{\nabla}_{\partial_s} N\rangle = \langle \mathbf{e}_{n+1}, -\nabla u - \mathscr{W} T\rangle. \tag{3.9}$$

Since

$$\frac{\mathrm{d}}{\mathrm{d}s}\bigg|_{s=0} H = \Delta u + |A|^2 u + \mathscr{L}_T H \tag{3.10}$$

and $\nabla \eta = \mathbf{e}_{n+1}^{\top}$, we obtain for normal variations (when $T = 0$)

$$\begin{aligned} \frac{\mathrm{d}}{\mathrm{d}s}\big(H - \langle \mathbf{e}_{n+1}, N\rangle\big) &= \Delta u - \langle \mathbf{e}_{n+1}, \nabla u\rangle + |A|^2 u \\ &= \Delta^f u + |A|^2 u. \end{aligned}$$

This finishes the proof of the proposition. □

The previous result has important consequences. It means that a hypersurface $M \subset \mathbf{R}^{n+1}$ is a translator if and only if it is minimal with respect to the Riemannian metric

$$g_{ij}(x_1, \ldots, x_{n+1}) = \exp\left(-\frac{2}{n} x_{n+1}\right) \delta_{ij}.$$

Although the metric g_{ij} is not complete (notice that the length of vertical half-lines in the direction of $\mathbf{e}_{n+1}$ is finite), we can apply all the local results of the theory of minimal hypersurfaces in Riemannian manifolds. Thus we can freely use curvature estimates and compactness theorems from minimal surface theory; cf. [34, Chap. 3]. In particular, if M is a graphical translator, then (since vertical translates of it are also g-minimal) $\langle \mathbf{e}_3, \nu\rangle$ is a nowhere vanishing Jacobi field, so M is a stable g-minimal surface. It follows that any sequence M_i of complete translating graphs in $\mathbf{R}^3$ has a subsequence that converges smoothly to a translator M. Also, if a translator M is the graph of a function $u : \Omega \to \mathbf{R}$, then M and its vertical translates form a g-minimal foliation of $\Omega \times \mathbf{R}$, from which it follows that M is g-area minimizing in $\Omega \times \mathbf{R}$, and thus that if $K \subset \Omega \times \mathbf{R}$ is compact, then the g-area of $M \cap K$ is at most $1/2$ of the g-area of ∂K. Hence, if we consider sequences of translators that are manifolds-with-boundary, then the area bounds described above together with standard compactness theorems for minimal surfaces (such as those in [33, 35]) give smooth, subsequential convergence, including at the boundary. This has been a crucial tool in [16, 17, 19, 25] (The local area bounds and bounded topology mean that the only boundary singularities that could arise would be boundary branch points. In the situations that occur in these papers, obvious barriers preclude boundary branch points.)

The situation for higher dimensional translating graphs is more subtle; (see [16, Appendix A] and [12]).

4 Examples of Translators

Besides the grim reapers that we have already described, the last decades have witnessed the appearance of numerous examples of translators. Clutterbuck, Schnürer and Schulze [6] (see also [2]) proved that there exists an entire graphical translator in $\mathbf{R}^{n+1}$ which is rotationally symmetric, strictly convex with translating velocity $-\mathbf{e}_{n+1}$. This example is known as the **translating paraboloid** or **bowl soliton**. Moreover, they classified all the translating solitons of revolution, giving a one-parameter family $\{W^n_\lambda\}_{\lambda>0}$ of rotationally invariant cylinders called **translating catenoids**. The parameter λ controls the size of the neck of each translating soliton. The limit, as $\lambda \to 0$, of W^n_λ consists of two superimposed copies of the bowl soliton with a singular point at the axis of symmetry. Furthermore, all these hypersurfaces have the following asymptotic expansion as r approaches infinity:

$$\frac{r^2}{2(n-1)} - \log r + O(r^{-1}),$$

where r is the distance function in $\mathbf{R}^n$. These rotationally symmetric translating catenoids can be seen as the desingularization of two paraboloids connected by a small neck of some radius (Fig. 5).

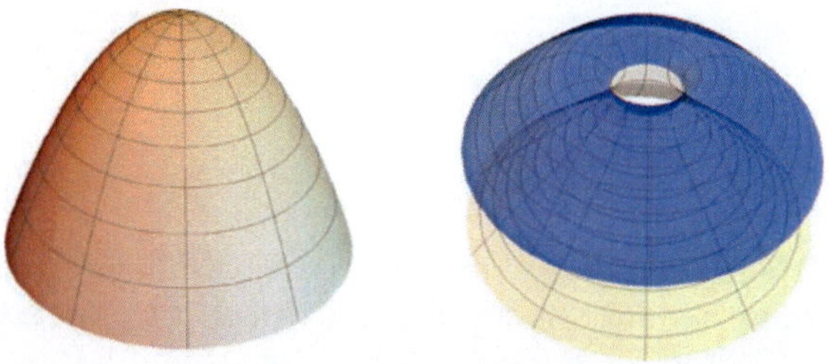

Fig. 5 The bowl soliton in $\mathbf{R}^3$ and the translating catenoid for $\lambda = 2$

Recall that the Costa–Hoffman–Meeks surfaces can be regarded as desingularizations of a plane and catenoid: a sequence of Costa–Hoffman–Meeks surfaces with genus tending to infinity converges (if suitably scaled) to the union of a catenoid and the plane through its waist. This suggests that one try to construct translators by desingularizing the union of a translating catenoid and a bowl solition. Dávila, del Pino, and Nguyen [8] were able to do that (for large genus) by glueing methods, replacing the circle of intersection by a surface similar to the singly periodic Scherk minimal surface. Previously, Nguyen in [27–29] had used similar techniques to desingularize the intersection of a grim reaper and a plane. In this way she obtained a complete periodic embedded translator of infinite genus, that she called a translating trident.

Once this abundance of translating solitons is guaranteed, there arises the need to classify them. One of the first classification results was given by Wang in [32]. He characterized the bowl soliton as the only convex translating soliton which is an entire graph.

Very recently, Spruck and Xiao [31] have proved that a complete translating soliton which is graph over a domain in $\mathbf{R}^2$ must be convex (see Sect. 6 below.) So, combining both results we have:

Theorem 5 *The bowl soliton is the only translator that is an entire graph over* $\mathbf{R}^2$.

Using the Alexandrov method of moving hyperplanes, Martín, Savas-Halilaj, and Smoczyk [24] showed that the bowl soliton is the only translator (not assumed to be graphical) that has one end and is C^∞-asymptotic to a bowl soliton. Hershkovits [15] improved this by showing uniqueness of the bowl soliton among (not necessarily graphical) translators that have one cylindrical end (and no other ends). Haslhofer [14] proved a related result in higher dimensions: he showed that any translator in $\mathbf{R}^{n+1}$ that is noncollapsed and uniformly 2-convex must be the n-dimensional bowl soliton. At this point, we would like to mention the recent classification result of Brendle and Choi [5]. They prove that the rotationally symmetric bowl soliton is the only noncompact ancient solution of mean curvature flow in $\mathbf{R}^3$ which is strictly convex and noncollapsed.

Martín, Savas-Halilaj and Smoczyk also obtained one of the first characterizations of the family of tilted grim reapers:

Theorem 6 ([24]) *Let* M *be a connected translating soliton in* $\mathbf{R}^{n+1}$, $n \geq 2$, *such that the function* $|A|^2 H^{-2}$ *has a local maximum in* $\{x \in M : H(x) \neq 0\}$. *Then* M *is a tilted grim reaper.*

5 Graphical Translators

If a translator M is the graph of function $u : \Omega \subset \mathbf{R}^n \to \mathbf{R}$, we will say that M is a **translating graph**; in that case, we also refer to the function u as a translator, and we say that u is complete if its graph is a complete submanifold of $\mathbf{R}^{n+1}$. Thus $u : \Omega \subset \mathbf{R}^n \to \mathbf{R}$ is a translator if and only if it solves the translator equation (the nonparametric form of (3.1)):

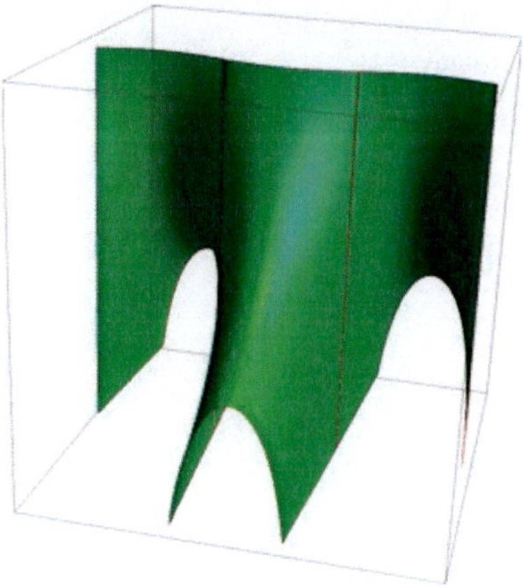

Fig. 6 A Scherkenoid is a singly periodic translator. As $z \to \infty$, it is asymptotic to a plane, and as $z \to -\infty$, it is asymptotic to an infinite family of parallel planes. There is a one-parameter family of such Scherkenoids, the parameter being the angle between the upper plane and the lower planes

$$D_i \left(\frac{D_i u}{\sqrt{1 + |Du|^2}} \right) = -\frac{1}{\sqrt{1 + |Du|^2}}. \tag{5.1}$$

The equation can also be written as

$$(1 + |Du|^2)\Delta u - D_i u \, D_j u \, D_{ij} u + |Du|^2 + 1 = 0. \tag{5.2}$$

In a recent preprint, we classify all complete translating graphs in $\mathbf{R}^3$. In two other papers [17, 19], we construct new families of complete, properly embedded (non-graphical) translators: a two-parameter family of translating annuli, examples that resemble Scherk's minimal surfaces, and examples that resemble helicoids. In [19], we also construct several new families of complete translators that are obtained as limits of the Scherk-type translators mentioned above. They include a 1-parameter family of single periodic surfaces called Scherkenoids (see Fig. 6) and a simply-connected translator called the pitchfork translator (see Fig. 7). The pitchfork translator resembles Nguyen's translating tridents [27] (see also [18]): like the tridents, it is asymptotic to a plane as $z \to \infty$ and to three parallel planes as $z \to -\infty$. However, the pitchfork has genus 0, whereas the tridents have infinite genus.

As a consequence of Theorem 6, we have that every translator $\mathbf{R}^3$ with zero Gauss curvature is a grim reaper surface, a tilted grim reaper surface, or a vertical plane.

In addition to the examples described in the previous section, Ilmanen (in unpublished work) proved that for each $0 < k < 1/2$, there is a translator $u : \Omega \to \mathbf{R}$ with the following properties: $u(x, y) \equiv u(-x, y) \equiv u(x, -y)$, u attains its maximum at $(0, 0) \in \Omega$, and

$$D^2 u(0,0) = \begin{bmatrix} -k & 0 \\ 0 & -(1-k) \end{bmatrix}.$$

The domain Ω is either a strip $\mathbf{R} \times (-b, b)$ or $\mathbf{R}^2$. He referred to these examples as **Δ-wings**. As $k \to 0$, he showed that the examples converge to the grim reaper

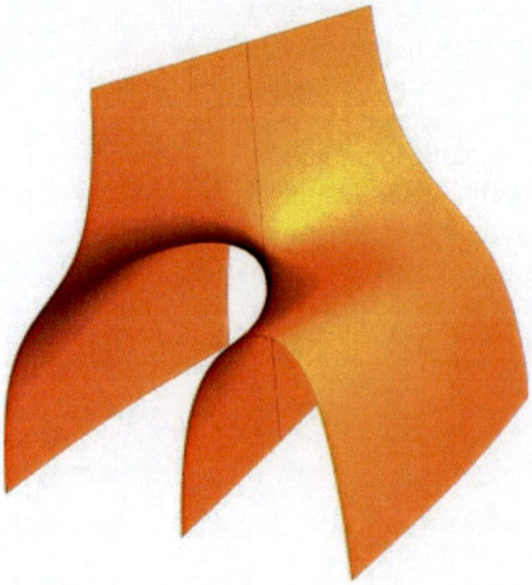

Fig. 7 The pitchfork translator is a simply connected translator that is asympotic to a plane as $z \to \infty$ and to three parallel planes as $z \to -\infty$

surface. Uniqueness (for a given k) was not known. It was also not known which strips $\mathbf{R} \times (-b, b)$ occur as domains of such examples. The main result in [16] is the following:

Theorem 7 *For every $b > \pi/2$, there is (up to translation) a unique complete, strictly convex translator $u^b : \mathbf{R} \times (-b, b) \to \mathbf{R}$. Up to isometries of $\mathbf{R}^2$, the only other complete translating graphs in $\mathbf{R}^3$ are the grim reaper surface, the tilted grim reaper surfaces, and the bowl soliton.*

Although the paper [16] is primarily about translators in $\mathbf{R}^3$, the last sections extend Ilmanen's original proof to get Δ-wings in $\mathbf{R}^{n+1}$ that have prescribed principal curvatures at the origin. For $n \geq 3$, the examples include entire graphs that are not rotationally invariant. At the end of the paper, we modify the construction to produce a family of Δ-wings in $\mathbf{R}^{n+2}$ over any given slab of width $> \pi$. See [32] for a different construction of some higher dimensional graphical translators (Fig. 8).

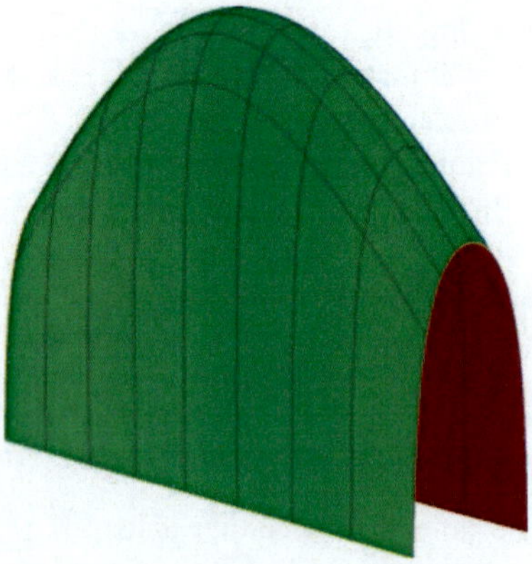

Fig. 8 The Δ-wing of width $\sqrt{2}\pi$. As $y \to \pm\infty$, this Δ-wing is asymptotic to the tilted grim reapers $\mathcal{G}_{-\frac{\pi}{4}}$ and $\mathcal{G}_{\frac{\pi}{4}}$, respectively

6 The Spruck–Xiao Convexity Theorem

One of the fundamental results in the recent development of soliton theory has been the paper by Spruck and Xiao [31], where they proved that complete graphical translators (or, more generally, complete translators of positive mean curvature) are convex. The ideas contained in this paper are really inspiring and we would like to provide a slightly simplified exposition of their proof.

At any non-umbilic point, we let $\kappa_1 > \kappa_2$ be the principal curvatures and $H = \kappa_1 + \kappa_2 > 0$ be the mean curvature. We let v_1 and v_2 be the principal direction unit vector fields, so

$$\kappa_i \equiv A(v_i, v_i) \text{ and } A(v_1, v_2) \equiv 0.$$

Note $\nabla_u v_1$ is perpendicular to v_1. Thus

$$\nabla_{v_1} v_1 = \alpha_1 v_2, \qquad \nabla_{v_2} v_1 = \alpha_2 v_2 \tag{6.1}$$

for some functions α_1 and α_2. Since $0 = \nabla_{v_i}(v_1 \cdot v_2) = (\nabla_{v_i} v_1) \cdot v_2 + v_1 \cdot (\nabla_{v_i} v_2)$, we see that

$$\nabla_{v_1} v_2 = -\alpha_1 v_1, \qquad \nabla_{v_2} v_2 = -\alpha_2 v_1. \tag{6.2}$$

Thus

$$\begin{aligned} \nabla_u \kappa_1 &= \nabla_u A(v_1, v_1) \\ &= (\nabla_u A)(v_1, v_1) + 2A(\nabla_u v_1, v_1). \end{aligned}$$

But $\nabla_u v_1$ is perpendicular to v_1, so $A(\nabla_u v_1, v_1) \equiv 0$. Thus

$$\nabla_u \kappa_1 = (\nabla_u A)(v_1, v_1). \tag{6.3}$$

In particular,

$$\nabla_i \kappa_j = h_{jj,i}, \tag{6.4}$$

where $h_{ij} = A(v_i, v_j)$ and $h_{ij,k} = (\nabla_{v_k} A)(v_i, v_j)$. From $A(v_1, v_2) \equiv 0$, we see that

$$\begin{aligned} h_{12,i} &= (\nabla_i) A(v_1, v_2) \\ &= \nabla_i (A(v_1, v_2)) - A(\nabla_i v_1, v_2) - A(v_1, \nabla_i v_2) \\ &= 0 - \alpha_i h_{22} + \alpha_i h_{11} \\ &= (\kappa_1 - \kappa_2)\alpha_i \end{aligned}$$

by (6.1) and (6.2). Thus

$$\alpha_i = \frac{h_{12,i}}{\kappa_1 - \kappa_2}. \tag{6.5}$$

Also, if we let $u_i = v_i$ at a particular point and extend by parallel transport on radial geodesics, we have

$$\begin{aligned}\Delta\kappa_1 &= \nabla_{u_i}\nabla_{u_i}(A(v_1, v_1)) \\ &= \nabla_{u_i}((\nabla_{u_i}A)(v_1, v_1)) \\ &= (\Delta A)(v_1, v_1) + 2(\nabla_{u_i}A)(\nabla_{u_i}v_1, v_1).\end{aligned}$$

Thus

$$\begin{aligned}\Delta\kappa_1 &= (\Delta A)(v_1, v_1) + 2(\nabla_i A)(\nabla_i v_1, v_1) \\ &= (\Delta A)(v_1, v_1) + 2(\nabla_i A)(\alpha_i v_2, v_1) \\ &= (\Delta A)(v_1, v_1) + 2\alpha_i h_{12,i} \\ &= (\Delta A)(v_1, v_1) + 2\frac{(h_{12,1})^2 + (h_{12,2})^2}{\kappa_1 - \kappa_2} \\ &= (\Delta A)(v_1, v_1) + \frac{2Q^2}{\kappa_1 - \kappa_2},\end{aligned}$$

where

$$Q^2 := (h_{12,1})^2 + (h_{12,2})^2 = (h_{11,2})^2 + (h_{22,1})^2. \tag{6.6}$$

(The second equality follows from the Codazzi equations.)

Now suppose the surface is a translator. Then, we have that (see [31, Lemma 2.1] or [24, Lemma 2.1]):

$$\Delta A - \nabla_{e_3^T} A + |A|^2 A = 0.$$

Hence,

$$\begin{aligned}(\Delta A)(v_1, v_1) &= -|A|^2 A(v_1, v_1) + (\nabla_{e_3^T} A)(v_1, v_1) \\ &= -|A|^2\kappa_1 + \nabla_{e_3^T}\kappa_1\end{aligned}$$

by (6.3). Thus

$$\Delta^f\kappa_1 = -|A|^2\kappa_1 + \frac{2Q^2}{\kappa_1 - \kappa_2}. \tag{6.7}$$

Recall that the *drift Laplacian* (see (3.7)) is given by

$$\Delta^f := \Delta - \langle\nabla f, \nabla\cdot\rangle = \Delta - \langle e_3, \nabla\cdot\rangle.$$

Likewise,

$$\Delta^f\kappa_2 = -|A|^2\kappa_2 - \frac{2Q^2}{\kappa_1 - \kappa_2}.$$

Adding these gives

$$\Delta^f H = -|A|^2 H. \tag{6.8}$$

From (6.7) and (6.8),

$$\kappa_1 \Delta^f H - H \Delta^f \kappa_1 = \frac{-2HQ^2}{\kappa_1 - \kappa_2}.$$

Thus

$$\begin{aligned}\Delta^f \left(\frac{H}{\kappa_1}\right) &= \frac{\kappa_1 \Delta^f H - H \Delta^f \kappa_1}{{\kappa_1}^2} - 2\frac{\nabla \kappa_1}{\kappa_1} \cdot \nabla \left(\frac{H}{\kappa_1}\right) \\ &= \frac{-2HQ^2}{(\kappa_1 - \kappa_2)(\kappa_1)^2} - 2\frac{\nabla \kappa_1}{\kappa_1} \cdot \nabla \left(\frac{H}{\kappa_1}\right),\end{aligned} \tag{6.9}$$

so

$$\Delta^f \left(\frac{H}{\kappa_1}\right) + 2\frac{\nabla \kappa_1}{\kappa_1} \cdot \nabla \left(\frac{H}{\kappa_1}\right) \leq 0. \tag{6.10}$$

Theorem 8 (Spruck–Xiao) *Let $M \subset \mathbf{R}^3$ be a complete translator with $H > 0$. Then M is convex.*

Proof Suppose the theorem is false. Then

$$\eta := \inf \frac{H}{\kappa_1} = \inf \left(1 + \frac{\kappa_2}{\kappa_1}\right) \in [0, 1). \tag{6.11}$$

Note that the set of points where $H/\kappa_1 < 2$ contains no umbilic points, which implies that H/κ_1 is smooth on that set and that we can apply the formulas in this section (§6), which were derived assuming that $\kappa_1 > \kappa_2$.

Step 1: The infimum η is not attained. For suppose it is attained at some point. Then by (6.10) and the strong minimum principle (see [13, Theorem 3.5] or [11, Chap. 6, Theorem 3]), H/κ_1 is constant on M. Therefore $\kappa_2/\kappa_1 = H/\kappa_1 - 1$ is constant on M. Since $H > 0$ and since H/κ_1 is constant, we see from (6.9) that $Q \equiv 0$, i.e., (see (6.6)) $h_{12,1} \equiv h_{12,2} \equiv 0$. Hence by (6.5), (6.1) and (6.2), the frame $\{v_1, v_2\}$ is parallel, so $\kappa_1 \kappa_2 \equiv 0$, contadicting (6.11). Thus the infimum is not attained.

Step 2: If $p_n \in M$ is a sequence with $H(p_n)/\kappa_1(p_n) \to \eta$, then (after passing to a subsequence) $M - p_n$ converges smoothly to a limit by the curvature estimates mentioned at the end of Sect. 3. We claim that the limit must be a vertical plane. For suppose not. Then (after passing to a subsequence) $M - p_n$ converges smoothly to a complete translator M' with $H > 0$. By the smooth convergence, H/κ_1 attains its minimum value on M' at the origin, and that minimum value is η, contradicting Step 1.

Step 3: Now we apply the Omori–Yau maximum principle (see Theorem 9 below) to get a sequence $p_n \in M$ such that

$$\frac{H}{\kappa_1} = 1 + \frac{\kappa_1}{\kappa_2} \to \eta, \tag{6.12}$$

$$\nabla\left(\frac{H}{\kappa_1}\right) \to 0, \tag{6.13}$$

$$\Delta\left(\frac{H}{\kappa_1}\right) \to \delta \in [0, \infty]. \tag{6.14}$$

From (6.13) and (6.14), we see that

$$\Delta^f\left(\frac{H}{\kappa_1}\right) \to \delta \in [0, \infty]. \tag{6.15}$$

By Step 2, we can assume that $M - p_n$ converges smoothly to a vertical plane. For the rest of the proof, any statement that some quantity tends to a limit refers only to the quantity at the points p_n.

Since A is a quadratic form with eigenvalues κ_1 and κ_2, A/κ_1 is a quadratic form with eigenvalues 1 and $\kappa_2/\kappa_1 = H/\kappa_1 - 1$. Thus (by passing to a subsequence) we can assume that A/κ_1 (at p_n) converges to a quadratic form with eigenvalues 1 and $\eta - 1$. (Note that the eigenvalue $\eta - 1$ is negative by Hypothesis (6.11).)

Recall that

$$\nabla H = A(\mathbf{e}_3^T, \cdot).$$

(See, for example, [24, Lemma 2.1].) Since $\mathbf{e}_3^T \to \mathbf{e}_3$, we see that $\nabla H/\kappa_1$ (at p_n) converges to a nonzero vector N:

$$\frac{\nabla H}{\kappa_1} \to N \neq 0. \tag{6.16}$$

Now

$$\nabla\left(\frac{H}{\kappa_1}\right) = \frac{\nabla H}{\kappa_1} - \frac{H}{\kappa_1}\frac{\nabla \kappa_1}{\kappa_1}. \tag{6.17}$$

By Omori–Yau (see (6.13)), this tends to 0, so

$$\frac{H}{\kappa_1}\frac{\nabla \kappa_1}{\kappa_1} \to N, \tag{6.18}$$

or, equivalently (see (6.4)),

$$\frac{H}{\kappa_1}\frac{h_{11,i}}{\kappa_1} \to N_i \qquad (i = 1, 2), \tag{6.19}$$

where $N_i = N \cdot v_i$.

We can rewrite (6.17) as

$$\nabla\left(\frac{H}{\kappa_1}\right) = \frac{\nabla\kappa_1 + \nabla\kappa_2}{\kappa_1} - \frac{H}{\kappa_1}\frac{\nabla\kappa_1}{\kappa_1}$$
$$= \left(1 - \frac{H}{\kappa_1}\right)\left(\frac{\nabla\kappa_1}{\kappa_1}\right) + \frac{\nabla\kappa_2}{\kappa_1}.$$

Multiply by H/κ_1:

$$\frac{H}{\kappa_1}\nabla\left(\frac{H}{\kappa_1}\right) = \left(1 - \frac{H}{\kappa_1}\right)\left(\frac{H}{\kappa_1}\frac{\nabla\kappa_1}{\kappa_1}\right) + \frac{H}{\kappa_1}\frac{\nabla\kappa_2}{\kappa_1}$$

By Omori–Yau (see (6.13)), this tends to 0, so (using (6.18)),

$$\frac{H}{\kappa_1}\frac{\nabla\kappa_2}{\kappa_1} \to (\eta - 1)N, \tag{6.20}$$

or, equivalently (by (6.4)),

$$\frac{H}{\kappa_1}\frac{h_{22,i}}{\kappa_1} \to (\eta - 1)N_i \qquad (i = 1, 2). \tag{6.21}$$

Combining (6.19) with $i = 2$ and (6.21) with $i = 1$ gives

$$\left(\frac{H}{\kappa_1}\right)^2\left(\frac{Q}{\kappa_1}\right)^2 \to (N_2)^2 + (\eta - 1)^2(N_1)^2 := \lambda^2 > 0. \tag{6.22}$$

Note that $\lambda^2 > 0$ because $N \neq 0$ and $\eta < 1$ by Hypothesis (6.11).

Now multiply (6.9) by H/κ_1:

$$\left(\frac{H}{\kappa_1}\right)\Delta^f\left(\frac{H}{\kappa_1}\right) = \frac{-2}{1 - \frac{\kappa_2}{\kappa_1}}\left(\frac{H}{\kappa_1}\right)^2\left(\frac{Q}{\kappa_1}\right)^2 - 2\left(\frac{H}{\kappa_1}\frac{\nabla\kappa_1}{\kappa_1}\right)\cdot\nabla\left(\frac{H}{\kappa_1}\right).$$

Using (6.12) and (6.15) for the left side, (6.12) and (6.22) for the first term on the right, and (6.18) and (6.13) for the second term, we can let $n \to \infty$ to get:

$$\eta\delta \leq \frac{-2}{2 - \eta}\lambda^2 + 0,$$

a contradiction (since η and δ are nonnegative). □

We used the Omori–Yau Theorem (see, for example, [1]):

Theorem 9 (Omori–Yau Theorem) *Let M be a complete Riemannian manifold with Ricci curvature bounded below. Let $f : M \to \mathbf{R}$ be a smooth function that is bounded below. Then there is a sequence p_n in M such that*

$$f(p_n) \to \inf_M f, \tag{6.23}$$

$$\nabla f(p_n) \to 0, \tag{6.24}$$

$$\liminf \Delta f(p_n) \geq 0. \tag{6.25}$$

The theorem remains true if we replace the assumption that f is smooth by the assumption that f is smooth on $\{f < \alpha\}$ for some $\alpha > \inf_M f$.

To see the last assertion, let $\phi : \mathbf{R} \to \mathbf{R}$ be a smooth, monotonic function such that $\phi(t) = 0$ for $t \geq \alpha$ and such that $\phi \equiv 1$ on an open interval containing $\inf_M f$. Then $\phi \circ f$ is smooth, so the Omori–Yau Theorem holds for $\phi \circ f$, from which it follows immediately that the Omori–Yau Theorem also holds for f.

In our case, the function H/κ_1 is smooth except at umbilic points. At such points, $H/\kappa_1 = 2$. Since we assumed that the infimum was <1, we could invoke the Omori–Yau Theorem.

7 Characterization of Translating Graphs in $\mathbf{R}^3$

As we mentioned before, the authors of these notes have obtained the complete classification of the complete graphical translators in Euclidean 3-space.

Recall that by translator we mean a smooth function $u : \Omega \to \mathbf{R}$ such that $M = \text{Graph}(u)$ is a translator. Then u must be solution of the equation:

$$(1 + u_y^2)\, u_{xx} - 2u_x\, u_y\, u_{xy} + (1 + u_x^2)\, u_{yy} + u_x^2 + u_y^2 + 1 = 0. \tag{7.1}$$

If we impose that M is complete, then we will say that u is a **complete translator**. In this setting Shahriyari [30] proved in her thesis the following

Theorem 10 (Shahriyari) *If M is complete, then Ω must be a strip, a halfspace, or all of $\mathbf{R}^2$.*

In [32], X. J. Wang proved that the only entire convex translating graph is the bowl soliton, and that there are no complete translating graphs defined over halfplanes. Thus by the Spruck–Xiao Convexity Theorem, the bowl soliton is the only complete translating graph defined over a plane or halfplane.

Hence, it remained to classify the translators $u : \Omega \to \mathbf{R}$ whose domains are strips. Our main new contributions in this line are:

1. For each $b > \pi/2$, we prove [16, Theorem 5.7] existence and uniqueness (up to translation) of a complete translator $u^b : \mathbf{R} \times (-b, b) \to \mathbf{R}$ that is not a tilted grim reaper. We call u^b the **Δ-wing** of width $2b$.
2. We give a simpler proof (see [16, Theorem 6.7]) that there are no complete graphical translators in $\mathbf{R}^3$ defined over halfplanes in $\mathbf{R}^2$.

Furthermore, there are no complete translating graphs defined over strips of width $< \pi$ (see [4, 31]), and the grim reaper surface is the only translating graph over a strip of width π (see [16]). Consequently, we have a classification: every complete, translating graph in $\mathbf{R}^3$ is one of the following: a grim reaper surface or tilted grim reaper surface, a Δ-wing, or the bowl soliton.

We remark that Bourni, Langford, and Tinaglia gave a different proof of the existence (but not uniqueness) of the Δ-wings in (1) [4].

References

1. Alías, L.J., Mastrolia, P., Rigoli, M.: Maximum Principles and Geometric Applications. Springer Monographs in Mathematics. Springer, Cham (2016). MR3445380
2. Altschuler, S.J., Wu, L.F.: Translating surfaces of the non-parametric mean curvature flow with prescribed contact angle. Calc. Var. Partial Differential Equations **2**(1), 101–111 (1994)
3. Angenent, S.: On the formation of singularities in the curve shortening flow. J. Differential Geom. **33**, 601–633 (1991)
4. Bourni, T., Langford, M., Tinaglia, G.: On the existence of translating solutions of mean curvature flow in slab regions. Anal. PDE **13**(4), 1051–1072 (2020). arXiv:1805.05173
5. Brendle, S., Choi, K.: Uniqueness of convex ancient solutions to mean curvature flow in $\mathbf{R}^3$. Invent. Math. **217**(1), 35–76 (2019)
6. Clutterbuck, J., Schnürer, O., Schulze, F.: Stability of translating solutions to mean curvature flow. Calc. Var. Part. Differ. Equ. **29**, 281–293 (2007)
7. Colding, T.H., Minicozzi, W.: A course on minimal surfaces. Graduate Studies in Mathematics, vol. 121. AMS, New York (2011)
8. Dávila, J., del Pino, M., Nguyen, X.H.: Finite topology self-translating surfaces for the mean curvature flow in $\mathbb{R}^3$. Adv. Math. **320**, 674–729 (2017), MR3709119. https://doi.org/10.1016/j.aim.2017.09.014
9. Dierkes, U., Hildebrandt, S., Küster, A., Wohlrab, O.: Minimal surfaces I, Boundary Value Problems, Grundlehren der mathematischen Wissenschaften, vol. 295. Springer, Berlin (1992)
10. Ecker, K.: Regularity Theory for Mean Curvature Flow. Birkhauser, Boston (2004)
11. Evans, L.C.: Partial Differential Equations, 2nd ed. Graduate Studies in Mathematics, vol. 19. American Mathematical Society, Providence (2010), MR2597943
12. Gama, E.S., Martín, F.: Translating solitons of the mean curvature flow asymptotic to hyperplanes in $\mathbf{R}^{n+1}$. Int. Math. Res. Not. (to appear)
13. Gilbarg, D., Trudinger, N.S.: Elliptic partial differential equations of second. Classics in Mathematics. Reprint of the 1998 edition. Springer, Berlin (2001), MR1814364
14. Haslhofer, R.: Uniqueness of the bowl soliton. Geom. Top. **19**(4), 2393–2406 (2015)
15. Hershkovits, O.: Translators asymptotic to cylinders. J. reine angew. Math. **766**, 61–71 (2020). arXiv:1805.10553
16. Hoffman, D., Ilmanen, T., Martín, F., White, B.: Graphical translators for mean curvature flow. Calc. Var. Part. Differ. Equ. **58**, 117 (2019). https://doi.org/10.1007/s00526-019-1560-x
17. Hoffman, D., Martín, F., White, B.: Scherk-like translators for mean curvature flow. J. Differ. Geom. (to appear) (2019). arXiv:1903.04617
18. Hoffman, D., Martín, F., White, B.: Nguyen's Tridents and the Classification of Semigraphical Translators for Mean Curvature Flow (2019). arXiv:1909.09241
19. Hoffman, D., Martín, F., White, B.: Translating Annuli for Mean Curvature Flow, in preparation
20. Huisken, G.: Flow by Mean Curvature of Convex Surfaces into Spheres. J. Differ. Geom. **20**, 237–266 (1984)
21. Huisken, G.: Asymptotic behaviour for singularities of the mean curvature flow. J. Differ. Geom. **31**, 285–299 (1990)
22. Ilmanen, T.: Elliptic regularization and partial regularity for motion by mean curvature. Memoirs of the American Mathematical Society, vol. 108, no. 520, pp. x+90 (1994)

23. Mantegazza, C.: Lectures Notes on Mean Curvature Flow. Birkhäuser, Boston (2011)
24. Martín, F., Savas-Halilaj, A., Smoczyk, K.: On the topology of translating solitons of the mean curvature flow. Calc. Var. Part. Differ. Equ. **54**(3), 2853–2882 (2015)
25. Martín, F., Pérez-García, J., Savas-Halilaj, A., Smoczyk, K.: A characterization the of the grim reaper cylinder. J. Reine Angew. Math. **2019**(746), 209–234 (2019)
26. Meeks, W.H., Pérez, J.: A Survey on Classical Minimal Surface Theory. University Lecture Series (AMS) vol. 60 (2012)
27. Nguyen, X.H.: Translating tridents. Commun. Part. Differ. Equ. **34**(3), 257–280 (2009)
28. Nguyen, X.H.: Complete embedded self-translating surfaces under mean curvature flow. J. Geom. Anal. **23**(3), 1379–1426 (2013)
29. Nguyen, X.H.: Doubly periodic self-translating surfaces for the mean curvature flow. Geom. Dedicata **174**(1), 177–185 (2015)
30. Shahriyari, L.: Translating graphs by mean curvature flow. Geom. Dedicata **175**, 57–64 (2015)
31. Spruck, J., Xiao, L.: Complete translating solitons to the mean curvature flow in $\mathbf{R}^3$ with nonnegative mean curvature. Amer. J. of Math. **142**(3), 993–1015 (June, 2020). arXiv:1703.01003v2
32. Wang, X.J.: Convex solutions of the mean curvature flow. Ann. Math. **173**, 1185–1239 (2011)
33. White, B.: Curvature estimates and compactness theorems in 3-manifolds for surfaces that are stationary for parametric elliptic functionals. Invent. Math. **88**(2), 243–256 (1987), MR880951. https://doi.org/10.1007/BF01388908
34. White, B.: Introduction to Minimal Surface Theory, Geometric Analysis. IAS/Park City Mathematics Series, vol. 22, pp. 387–438. American Mathematical Society, Providence (2016), MR3524221,
35. White, B.: Controlling area blow-up in minimal or bounded mean curvature varieties. J. Differ. Geom. **102**(3), 501–535 (2016), MR3466806

Uniqueness Problem for Closed Non-smooth Hypersurfaces with Constant Anisotropic Mean Curvature and Self-similar Solutions of Anisotropic Mean Curvature Flow

Miyuki Koiso

Abstract An anisotropic surface energy is the integral of an energy density that depends on the surface normal over the considered surface, and it is a generalization of surface area. Equilibrium surfaces with volume constraint are called CAMC (constant anisotropic mean curvature) surfaces and they are not smooth in general. We show that, if the energy density function is two times continuously differentiable and convex, then, like isotropic (constant mean curvature) case, the uniqueness for closed stable CAMC surfaces holds under the assumption of the integrability of the anisotropic principal curvatures. Moreover, we show that, unlike the isotropic case, uniqueness of closed embedded CAMC surfaces with genus zero in the three-dimensional euclidean space does not hold in general. We also give nontrivial self-similar shrinking solutions of anisotropic mean curvature flow. These results are generalized to hypersurfaces in the Euclidean space with general dimension. This article is an announcement of two forthcoming papers by the author.

Keywords Anisotropic mean curvature · Anisotropic surface energy · Wulff shape · Anisotropic mean curvature flow · Crystalline variational problem

1 Introduction

We study hypersurfaces with constant anisotropic mean curvature (CAMC hypersurfaces) in the $(n+1)$-dimensional euclidean space $\mathbb{R}^{n+1}$ which are a generalization of constant mean curvature (CMC) hypersurfaces. CAMC hypersurfaces are critical points of an anisotropic surface energy for variations which preserve the enclosed $(n+1)$-dimensional volume (we will call such a variation a volume-preserving variation). Anisotropic surface energy is the integral of an energy density that depends on

M. Koiso (✉)
Institute of Mathematics for Industry, Kyushu University, Motooka Nishi -ku, Fukuoka 819 -0395, Japan
e-mail: koiso@math.kyushu-u.ac.jp
URL: https://www.imi.kyushu-u.ac.jp/eng/academic_staffs/view/85

T. Hoffmann et al. (eds.), *Minimal Surfaces: Integrable Systems and Visualisation*, Springer Proceedings in Mathematics & Statistics 349,
https://doi.org/10.1007/978-3-030-68541-6_10

the surface normal over the considered hypersurface, and it was introduced to model the surface tension of a small crystal [19, 20]. From this origin, one can easily expect that equilibrium hypersurfaces may have singular points such as edges and vertexes. In the special case where the energy density is constant (which is sometimes referred to as the isotropic case), the anisotropic surface energy is the constant times the usual n-dimensional volume, and critical points for volume-preserving variations are CMC hypersurfaces. A CMC (resp. CAMC) hypersurface is said to be stable if the second variation of the n-dimensional volume (resp. the anisotropic surface energy) is nonnegative for all volume-preserving variations.

It is known that if a closed CMC hypersurface M in $\mathbb{R}^{n+1}$ satisfies a "good" property such as one of the following (I)–(III), then M is a round sphere.

(I) M does not have self-intersection.

(II) M is stable.

(III) $n = 2$ and the genus of M is 0.

We show that if the anisotropic energy density function is convex, then closed stable CAMC hypersurface with integrable anisotropic principal curvatures is unique (up to translation and homothety). On the other hand we show that, the uniqueness under the assumption (I) or (III) does not hold in general. Moreover, we give a non-uniqueness result for closed self-similar shrinking solutions with genus 0 of anisotropic mean curvature flow, which is also different from the isotropic case.

Let us give the presice statements of our study. Let $\gamma : S^n \to \mathbb{R}_{>0}$ be a positive continuous function on the unit sphere $S^n = \{\nu \in \mathbb{R}^{n+1} \mid \|\nu\| = 1\}$ in $\mathbb{R}^{n+1}$, which will be the energy density function for hypersurfaces in $\mathbb{R}^{n+1}$. γ is said to be convex if the homogeneous extension $\overline{\gamma} : \mathbb{R}^{n+1} \to \mathbb{R}_{\geq 0}$ of γ that is defined by $\overline{\gamma}(rX) := r\gamma(X)$, $(\forall X \in S^n, \forall r \geq 0)$, is a convex function (cf. [17]). Let X be a closed piecewise-C^2 weakly immersed hypersurface in $\mathbb{R}^{n+1}$ (the definition of piecewise-C^2 weakly immersed hypersurface will be given in Sect. 2). X will be represented as a piecewise-C^2 mapping $X : M \to \mathbb{R}^{n+1}$ from an n-dimensional oriented connected compact C^∞ manifold M into $\mathbb{R}^{n+1}$. And the unit normal vector field ν along X is defined on M except a set with measure zero. The anisotropic energy $\mathcal{F}_\gamma(X)$ of X is defined by $\mathcal{F}_\gamma(X) = \int_M \gamma(\nu)\, dA$, where dA is the n-dimensional volume form of M induced by X.

It is known that, for any positive number $V > 0$, among all closed hypersurfaces in $\mathbb{R}^{n+1}$ enclosing the same $(n+1)$-dimensional volume V, there exists a unique (up to translation in $\mathbb{R}^{n+1}$) minimizer $W_\gamma(V)$ of $\mathcal{F}_\gamma$ [17]. The minimizer $W_\gamma(V_0)$ for a specific volume V_0 is called the Wulff shape for γ, and we denote it by W_γ (see Sect. 2.2 for details). If $\gamma \equiv 1$, $\mathcal{F}_\gamma(X)$ is the usual n-dimensional volume of the hypersurface X, and W_γ is the unit sphere S^n. All $W_\gamma(V)$ are homothetic to W_γ. The Wulff shape W_γ is convex, but, unlike the isotropic case, it is not necessarily smooth.

Each equilibrium hypersurface X of $\mathcal{F}_\gamma$ for volume-preserving variations has constant anisotropic mean curvature. Here the anisotropic mean curvature Λ of a piecewise-C^2 hypersurface X is defined at each regular point of X as (cf. [11, 16]) $\Lambda := (1/n)(-\mathrm{div}_M D\gamma + nH\gamma)$, where $D\gamma$ is the gradient of γ on S^n and H is the mean curvature of X (see Sect. 3 for details). If $\gamma \equiv 1$, then $\Lambda = H$ holds.

As for the condition (II) above, we can prove the following uniqueness result.

Theorem 1 ([10]) *Assume that $\gamma : S^n \to \mathbb{R}_{>0}$ is of class C^2 and convex. Then, the image of any closed stable piecewise-C^2 CAMC hypersurface for γ whose rth anisotropic mean curvature for γ is integrable for $r = 1, \ldots, n$ is (up to translation and homothety) the Wulff shape W_γ.*

The assumption on the rth anisotropic mean curvatures (Definition 1) in Theorem 1 is a technical one. In fact, one of the key tools we use in the proof of Theorem 1 is a "Steiner-type integral formula" ([10]. He and Li [6] for smooth case) and it includes integrals of the rth anisotropic mean curvatures. Note that, because each rth anisotropic mean curvature of the Wulff shape W_γ at each regular point is $(-1)^r$ and hence it is integrable (Fact 1). However, we emphasize that the principal curvatures of our hypersurfaces can be unbounded (see Example 1 in Appendix 1).

As for the conditions (I) and (III) above, we have the following non-uniqueness results.

Theorem 2 ([9]) *There exist C^∞ functions $\gamma : S^n \to \mathbb{R}_{>0}$ such that there exist closed embedded CAMC hypersurfaces in $\mathbb{R}^{n+1}$ for γ each of which is not (any homothety or translation of) the Wulff shape.*

Theorem 3 ([9]) *There exist C^∞ functions $\gamma : S^2 \to \mathbb{R}_{>0}$ such that there exist closed embedded CAMC surfaces in $\mathbb{R}^3$ with genus zero for γ each of which is not (any homothety or translation of) the Wulff shape.*

Theorems 2 and 3 are proved by giving examples (Sect. 5). In those examples, the energy density functions γ are not convex. Therefore, there is possibility that the uniqueness under the assumption (I) or (III) also holds for any convex γ. The same examples can be applied to the anisotropic mean curvature flow to obtain the following non-uniqueness result (Sect. 6).

Theorem 4 ([9]) *There exist C^∞ functions $\gamma : S^n \to \mathbb{R}_{>0}$ such that there exist closed embedded self-similar shrinking solutions for anisotropic mean curvature flow in $\mathbb{R}^{n+1}$ for γ each of which is homeomorphic to S^n and is not (any homothety or translation of) the Wulff shape.*

In contrast with this result, the round sphere is the only closed embedded self-similar shrinking solution of mean curvature flow in $\mathbb{R}^3$ with genus zero [5].

We should mention preceding studies about the uniqueness for closed CAMC hypersurfaces. If we assume that the Wulff shape W_γ is a smooth strictly convex hypersurface, then any closed CAMC hypersurface X is also smooth and the uniqueness was already proved under the assumption that the hypersurface satisfies one of the conditions (I)–(III) above in the following papers. (I): [1] for $\gamma \equiv 1$, [7] for general γ. (II): [3] for $\gamma \equiv 1$, [14] for general γ. (III): [8] for $\gamma \equiv 1$, [2, 12] for general γ. As for non-smooth critical points of anisotropic energy, the following results are known. As for planer curves, [13] proved that, if $\gamma : S^1 \to \mathbb{R}_{>0}$ is continuous and convex, then any closed equilibrium rectifiable curve for $\mathcal{F}_\gamma$ in $\mathbb{R}^2$ with area constraint is (up to translation and homothety) a covering of the Wulff shape. About uniqueness of closed stable equilibria in $\mathbb{R}^3$, [15] proved the same conclusion as Theorem 1 but under the assumptions that $\gamma : S^2 \to \mathbb{R}_{>0}$ is of class C^3 and considered surfaces and the Wulff shape satisfy some extra assumptions.

This article is organized as follows. Section 2 is the section of preliminaries. In Sect. 3, the definition of anisotropic curvatures and the first variation formula of the anisotropic surface energy are given. In Sect. 4, we give an outline of the proof of Theorem 1. In Sect. 5 we give examples, and by using them, we prove Theorems 2 and 3. In Sect. 6, the anisotropic mean curvature flow will be introduced briefly and an outline of the proof of Theorem 4 will be given. Finally, in Appendix 1 we give a simple 1-dimensional example of a convex energy density function whose Wulff shape has singular points and its curvature is unbounded, and in Appendix 2 we prove that the anisotropic mean curvature flow diminishes the energy.

2 Preliminaries

2.1 *Definitions of Piecewise-C^r Weak Immersion and Its Anisotropic Energy*

First we recall the definition of a *piecewise-C^r weakly immersed hypersurface*, ($r \in \mathbb{N}$), defined in [10], which is the base of our study. Let $M = \cup_{i=1}^k M_i$ be an n-dimensional oriented compact connected C^∞ manifold, where each M_i is an n-dimensional connected compact submanifold of M with piecewise-C^∞ boundary, and $M_i \cap M_j = \partial M_i \cap \partial M_j$, ($i, j \in \{1, \ldots, k\}$, $i \neq j$). We call a map $X : M \to \mathbb{R}^{n+1}$ a piecewise-C^r weak immersion (or a piecewise-C^r weakly immersed hypersurface) if X satisfies the following conditions (A1), (A2), and (A3) ($i \in \{1, \ldots, k\}$).

(A1) X is continuous, and each $X_i := X|_{M_i} : M_i \to \mathbb{R}^{n+1}$ is of class C^r.

(A2) The restriction $X|_{M_i^o}$ of X to the interior M_i^o of M_i is a C^r-immersion.

(A3) The unit normal vector field $\nu_i : M_i^o \to S^n$ along $X_i|_{M_i^o}$ can be extended to a C^{r-1}-mapping $\nu_i : M_i \to S^n$. Here, if $(u^1, \ldots, u^n)$ is a local coordinate system in M_i, then $\{\partial/\partial u^1, \ldots, \partial/\partial u^n, \nu_i\}$ gives the canonical orientation in $\mathbb{R}^{n+1}$.

The anisotropic energy of a piecewise-C^1 weak immersion $X : M \to \mathbb{R}^{n+1}$ is defined as follows. Assume that $\gamma : S^n \to \mathbb{R}_{\geq 0}$ is a nonnegative continuous function. Denote by $S(X)$ the set of all singular points of X, here a singular point of X is a point in M at which X is not an immersion. Let $\nu : M \setminus S(X) \to S^n$ be the unit normal vector field along $X|_{M \setminus S(X)}$. The anisotropic energy $\mathcal{F}_\gamma(X)$ of X is defined as

$$\mathcal{F}_\gamma(X) := \int_M \gamma(\nu)\, dA := \sum_{i=1}^k \int_{M_i} \gamma(\nu_i)\, dA. \tag{1}$$

Note that, since the n-dimensional Hausdorff measure of $X(S(X))$ is zero [10], the improper integral in the right hand side of (1) converges.

2.2 Wulff Shape and Convexity of Integrands

Assume that $\gamma : S^n \to \mathbb{R}_{>0}$ is a positive continuous function. The boundary W_γ of the convex set $\tilde{W}[\gamma] := \cap_{\nu \in S^n} \{X \in \mathbb{R}^{n+1} | \langle X, \nu \rangle \le \gamma(\nu)\}$ is called the Wulff shape for γ, where $\langle\ ,\ \rangle$ stands for the standard inner product in $\mathbb{R}^{n+1}$. We should remark that originally $\tilde{W}[\gamma]$ itself was called the Wulff shape.

The Wulff shape W_γ is not smooth in general. It is smooth and strictly convex if and only if γ is of class C^2 and the $n \times n$ matrix $D^2\gamma + \gamma \cdot I_n$ is positive definite at any point in S^n, where $D^2\gamma$ is the Hessian of γ on S^n and I_n is the identity matrix of size n. If γ is of class C^2, γ is convex if and only if $D^2\gamma + \gamma \cdot I_n$ is positive semi-definite. For such γ, the Wuff shape can have singular points as Example 1 (see Appendix 1) shows. For simplicity, we say that γ is strictly convex if $D^2\gamma + \gamma \cdot I_n$ is positive definite.

Assume now that γ is of class C^2. The Cahn–Hoffman map $\xi_\gamma : S^n \to \mathbb{R}^{n+1}$ (for γ) is defined as

$$\xi(\nu) := \xi_\gamma(\nu) := D\gamma|_\gamma + \gamma(\nu)\nu, \quad \nu \in S^n, \tag{2}$$

here $T_\nu(S^n)$ for each $\nu \in S^n$ is naturally identified with a hyperplane in $\mathbb{R}^{n+1}$. If $\gamma \equiv 1$, then ξ_γ is the inclusion: $S^n \to \mathbb{R}^{n+1}$.

The Wulff shape W_γ is a subset of the image $\hat{W}_\gamma := \xi_\gamma(S^n)$ of the Cahn–Hoffman map ξ_γ. The equality $\hat{W}_\gamma = W_\gamma$ holds if and only if γ is convex. Hence, in this case, the map $\xi_\gamma(\nu) = D\gamma|_\nu + \gamma(\nu)\nu$ gives the representation of W_γ.

A simple calculation shows that ξ_γ is represented by using the homogeneous extension $\overline{\gamma} : \mathbb{R}^{n+1} \to \mathbb{R}_{\ge 0}$ of γ as

$$\xi_\gamma(\nu) = \overline{D}\overline{\gamma}|_\nu, \quad \nu \in S^n, \tag{3}$$

where $\overline{D}$ is the gradient in $\mathbb{R}^{n+1}$.

3 First Variation Formula, Anisotropic Curvatures, and Anisotropic Gauss Map

From now on, we assume that $\gamma : S^n \to \mathbb{R}_{>0}$ is of class C^2. Let $X : M = \cup_{i=1}^k M_i \to \mathbb{R}^{n+1}$ be a piecewise-C^2 weak immersion. We use the same notation as in Sect. 2. The Cahn–Hoffman field $\tilde{\xi}_i$ along $X_i = X|_{M_i}$ for γ (or the anisotropic Gauss map of X for γ) is defined as $\tilde{\xi}_i := \xi_\gamma \circ \nu_i : M_i \to \mathbb{R}^{n+1}$. If $\gamma \equiv 1$, then $\tilde{\xi}_i$ coincides with the Gauss map ν_i. For any point $p \in M_i^o$, the linear map $S_p^\gamma : T_pM_i \to T_pM_i$ given by the $n \times n$ matrix $S^\gamma := -d\tilde{\xi}_i$ is called the anisotropic shape operatior of X_i.

Definition 1 (*anisotropic curvatures, cf.* [7, 16]) (i) The eigenvalues of S^γ are called the anisotropic principal curvatures of X. We denote them by $k_1^\gamma, \ldots, k_n^\gamma$.

(ii) Let σ_r^γ be the elementary symmetric functions of $k_1^\gamma, \ldots, k_n^\gamma$:

$$\sigma_r^\gamma := \sum_{1 \le l_1 < \cdots < l_r \le n} k_{l_1}^\gamma \cdots k_{l_r}^\gamma, \quad r = 1, \ldots, n. \tag{4}$$

Set $\sigma_0^\gamma := 1$. $H_r^\gamma := \sigma_r^\gamma / {}_nC_r$ is called the rth anisotropic mean curvature of X.

(iii) $\Lambda := H_1^\gamma$ is called the anisotropic mean curvature of X.

Remark 1 S^γ is not symmetric in general. However, we have the following good properties of the anisotropic curvatures.

(i) [6] If $A := d\xi_\gamma = D^2\gamma + \gamma \cdot I_n$ is positive definite at a point $\nu(p)$, $(p \in M_i^o)$, then all of the anisotropic principal curvatures of X at p are real.

(ii) [10] k_i^γ is not a real value in general. However, each H_r^γ is always a real valued function on M_i^o.

The following formula for the anisotropic mean curvature is often useful. At any regular point of X, it holds that (cf. [11])

$$\Lambda = -\frac{1}{n}\mathrm{trace}_M(D^2\gamma + \gamma \cdot I_n) \circ d\nu = -\frac{1}{n}\mathrm{trace}_M\ d(\tilde{\xi}_\gamma). \tag{5}$$

From the first equality in (5), the equation '$\Lambda =$ constant' is elliptic or hyperbolic or any other type depends on $A = D^2\gamma + \gamma \cdot I_n$.

We have the following first variation formula for the anisotropic surface energy $\mathcal{F}_\gamma$.

Proposition 1 ([10]) *Assume that the map $X : M_0 \to \mathbb{R}^{n+1}$ satisfies (A1), (A2), and (A3) in Sect. 2 with $r = 2$, $X_i = X$, $M_i = M_0$, and $\nu_i = \nu$. Let $X_\varepsilon : M_0 \to \mathbb{R}^{n+1}$, $(\varepsilon \in J := [-\varepsilon_0, \varepsilon_0])$, be a variation of X, that is, $\varepsilon_0 > 0$ and $X_0 = X$. Assume for simplicity that X_ε is of class C^∞ in ε. We also assume that, for each $\varepsilon \in J$, the anisotropic mean curvature Λ_ε of X_ε is bounded on M_0^o. Set*

$$\delta X := \left.\frac{\partial X_\varepsilon}{\partial \varepsilon}\right|_{\varepsilon=0}, \quad \psi := \langle \delta X, \nu \rangle.$$

Then the first variation of the anisotropic energy $\mathcal{F}_\gamma$ is given as follows.

$$\left.\frac{d\mathcal{F}_\gamma(X_\varepsilon)}{d\varepsilon}\right|_{\varepsilon=0} = -\int_{M_0} n\Lambda\psi\, dA - \oint_{\partial M_0} \langle \delta X, R(p(\tilde{\xi}))\rangle\, d\tilde{s}, \tag{6}$$

here N is the outward-pointing unit conormal along ∂M_0, $d\tilde{s}$ is the $(n-1)$-dimensional volume form of ∂M_0 induced by X, R is the $\pi/2$-rotation on the (N, ν)-plane, and p is the projection from $\mathbb{R}^{n+1}$ to the (N, ν)-plane.

Note that, since the principal curvatures of the map X in Proposition 1 are not defined on ∂M_0 and they can be unbounded on M_0, the first integral in the right hand side of (6) is an improper integral (see Example 1). However, we can prove that it converges [10].

On the other hand the first variation of the $(n+1)$-dimensional volume enclosed by X_ε is $\delta V = \int_{M_0} \psi \, dA$ (cf. [4]). This with (6) gives the following Euler–Lagrange equations.

Proposition 2 (Euler–Lagrange equations [10]. For $n = 2$, see [15]) *A piecewise-C^2 weak immersion $X : M = \cup_{i=1}^k M_i \to \mathbb{R}^{n+1}$ is a critical point of the anisotropic energy $\mathcal{F}_\gamma$ for volume-preserving variations if and only if the following (i) and (ii) hold.*

(i) The anisotropic mean curvature of X is constant on $M \setminus S(X)$.

(ii) $\tilde{\xi}_i(\zeta) - \tilde{\xi}_j(\zeta) \in T_\zeta M_i \cap T_\zeta M_j = T_\zeta(\partial M_i \cap \partial M_j)$ holds at any $\zeta \in \partial M_i \cap \partial M_j$, where a tangent space of a submanifold of $\mathbb{R}^{n+1}$ is naturally identified with a linear subspace of $\mathbb{R}^{n+1}$.

Definition 2 ([10]) A piecewise-C^2 weak immersion $X : M = \cup_{i=1}^k M_i \to \mathbb{R}^{n+1}$ is called a hypersurface with constant anisotropic mean curvature (CAMC) if both of (i) and (ii) in Proposition 2 hold.

Let us give a remark on the anisotropic mean curvature of the Cahn–Hoffman map ξ_γ and the Wulff shape W_γ.

Fact 1 ([10, 11]) Since ξ_γ^{-1} gives the unit normal vector field ν_{ξ_γ} for the Cahn–Hoffman map ξ_γ at the regular points of ξ_γ, the anisotropic shape operator of ξ_γ is $S^\gamma = -d(\xi_\gamma \circ \nu_{\xi_\gamma}) = -d(\mathrm{id}_{S^n}) = -I_n$. Hence, at any regular point of ξ_γ, the anisotropic principal curvatures of ξ_γ are -1, and hence each rth anisotropic mean curvature of ξ_γ is $(-1)^r$. Particularly the anisotropic mean curvature of ξ_γ for the normal ν and that of W_γ for the outward-pointing unit normal is -1 at any regular point.

4 Outline of the Proof of Theorem 1

The idea of the proof of Theorem 1 is to generalize the proof of the uniqueness of stable closed CMC hypersurface in $\mathbb{R}^{n+1}$ given by [18], which was used also in [14, 15].

In order to prove Theorem 1, we first recall the striking result that says that the Cahn–Hoffman field $\tilde{\xi}$ along a closed CAMC hypersurface $X : M = \cup_{i=1}^k M_i \to \mathbb{R}^{n+1}$ is well-defined on the whole of M:

Theorem 5 ([10]) *Assume $\gamma : S^n \to \mathbb{R}_{>0}$ is of class C^2 and convex. Let $X : M = \cup_{i=1}^k M_i \to \mathbb{R}^{n+1}$ be a closed piecewise-C^2 CAMC hypersurface with unit normal $\nu_i : M_i \to S^n$. Then, the Cahn–Hoffman fields $\tilde{\xi}_i := \xi \circ \nu_i : M_i \to \mathbb{R}^{n+1}$, $(i = 1, \dots k)$, defines a C^0 map $\tilde{\xi} : M \to \mathbb{R}^{n+1}$.*

Because of Theorem 5, we can consider the so-called anisotropic parallel hypersurfaces $X_t := X + t\tilde{\xi}$, $(t \in \mathbb{R}, |t| << 1)$ of X. By taking homotheties of X_t if necessary, we have a volume-preserving variation $Y_t = \mu(t)X_t = \mu(t)(X + t\tilde{\xi})$, $(\mu(t) > 0$, $\mu(0) = 1)$ of X.

By using the Steiner-type formula and the Minkowski-type formula ([10]. He and Li [6] for smooth case), we prove that

$$\frac{d^2\mathcal{F}_\gamma(Y_t)}{dt^2}\Big|_{t=0} = \frac{-1}{n}\int_M \gamma(\nu) \sum_{1\le i<j\le n} (k_i^\gamma - k_j^\gamma)^2 \, dA \tag{7}$$

holds, where k_i^γ are the anisotropic principal curvatures of X. Since γ is convex, all k_i^γ are real values on $M \setminus S(X)$ (Remark 1). Hence, if X has constant anisotropic mean curvature Λ and stable, the formula (7) implies that $k_1^\gamma = \cdots = k_n^\gamma = \Lambda/n \neq 0$ holds on $M \setminus S(X)$. Therefore, from Corollary 1 in [16],

$$X(M \setminus S(X)) \subset rW_\gamma \tag{8}$$

holds for some $r > 0$. Because M is closed and the Wulff shape W_γ has anisotropic mean curvature -1 (Fact 1), (8) implies that

$$X(M) = (1/|\Lambda|)W_\gamma$$

holds. □

5 Examples of Non-convex Integrand and Proofs of Theorems 2 and 3

In this section, we give examples which will be used to prove Theorems 2, 3, and 4. Detailed explanations and computations are given in [9]. Throughout this section $\gamma : S^1 \to \mathbb{R}_{>0}$ is the function defined by

$$\gamma((\cos\theta, \sin\theta)) := \cos^6\theta + \sin^6\theta. \tag{9}$$

5.1 A One-Dimensional Example

By simple computation using (3), the Cahn–Hoffman map $\xi_\gamma : S^1 \to \mathbb{R}^2$ for γ is represented as follows.

$$\begin{aligned}\xi_\gamma((\cos\theta, \sin\theta)) = \ &\big((\cos\theta)(\cos^6\theta + 6\cos^4\theta\sin^2\theta - 5\sin^6\theta),\\ &(\sin\theta)(-5\cos^6\theta + 6\cos^4\theta\sin^2\theta + \sin^6\theta)\big).\end{aligned} \tag{10}$$

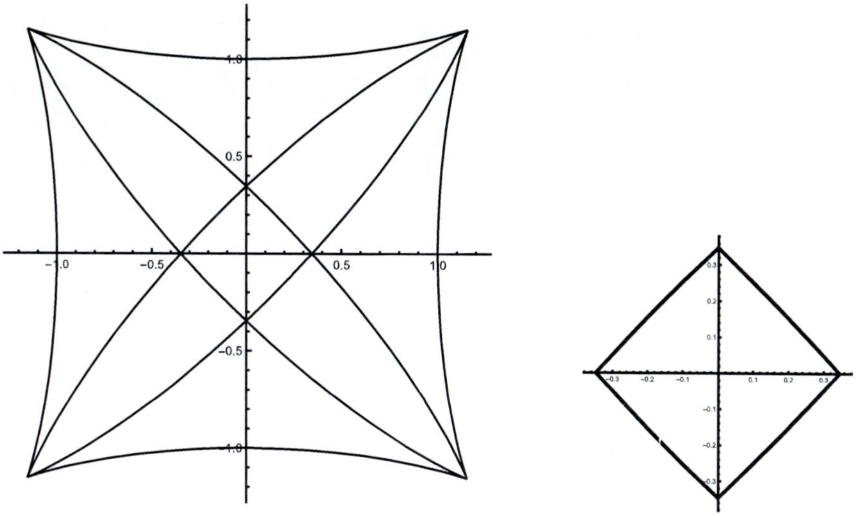

Fig. 1 The image $\xi_\gamma(S^1)$ of the Cahn–Hoffman map ξ_γ (left) and the Wulff shape W_γ (right) for γ defined by (9). W_γ is a subset of $\xi_\gamma(S^1)$

The image $\xi_\gamma(S^1)$ of ξ_γ and the Wulff shape W_γ are shown in Fig. 1.

In Fig. 2, five closed piecewise-C^∞ curves which are subsets of $\xi_\gamma(S^1)$ are shown. Because of Fact 1, the anisotropic (mean) curvature of them for the outward-pointing normal is -1. The curve at the upper left end is the Wulff shape. There are more piecewise-C^∞ CAMC closed curves which are subsets of $\xi_\gamma(S^1)$. The complete list is given in [9].

On the other hand, the closed curve in Fig. 3 is also a subset of $\xi_\gamma(S^1)$. However it is not CAMC. For, again by Fact 1, for the outward-pointing normal, the anisotropic (mean) curvature is -1 at each point in the solid arc, while it is 1 at each point in the dashed arcs.

5.2 *Higher Dimensional Examples*

Higher dimensional examples are obtained by rotating γ which was defined by (9).

First, by rotating γ around the second axis, we obtain the function $\gamma_1 : S^2 \to \mathbb{R}_{>0}$ defined by

$$\gamma_1(\nu_1, \nu_2, \nu_3) = (\nu_1^2 + \nu_2^2)^3 + \nu_3^6, \quad (\nu_1, \nu_2, \nu_3) \in S^2. \tag{11}$$

The corresponding Cahn–Hoffman map $\xi_{\gamma_1} : S^2 \to \mathbb{R}^3$ is given by (Fig. 4)

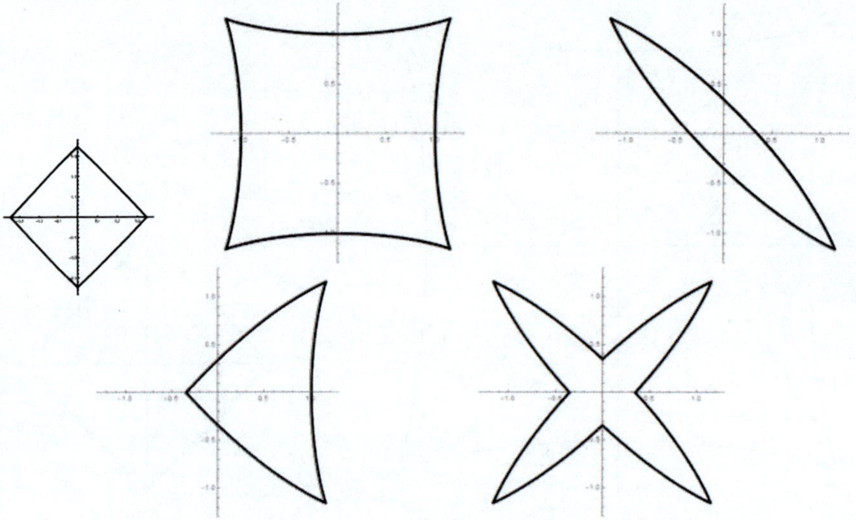

Fig. 2 Some of the closed CAMC curves which are subsets of $\xi_\gamma(S^1)$ for γ defined by (9) (cf. Fig. 1, left). The curve at the upper left end is the Wulff shape. From the upper left to lower right, we call W_γ, $(C_\gamma)_1$, $(C_\gamma)_2$, $(C_\gamma)_3$, $(C_\gamma)_4$. The anisotropic (mean) curvature for the outward-pointing normal is -1 for all of them

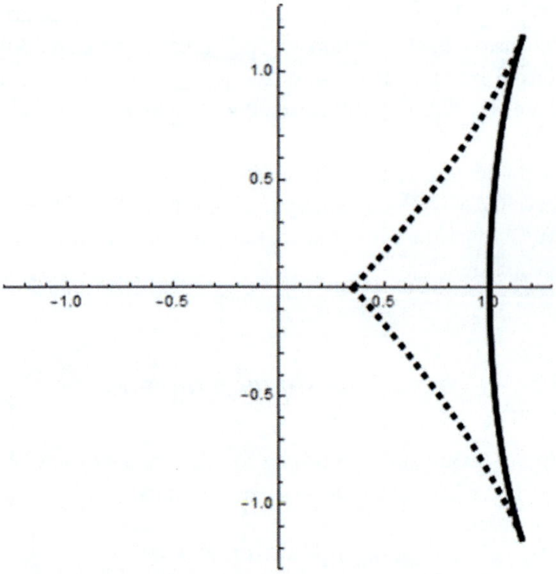

Fig. 3 Non-CAMC closed curve which is a subset of $\xi_\gamma(S^1)$ for γ defined by (9). We call it $(C_\gamma)_5$. For the outward-pointing normal, the anisotropic (mean) curvature is -1 at each point in the solid arc, while it is 1 at each point in the dashed arcs

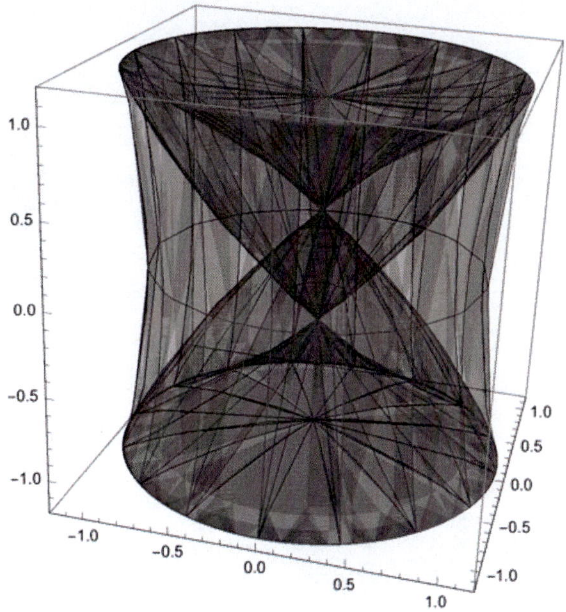

Fig. 4 The image of the Cahn–Hoffman map $\xi_{\gamma_1} : S^2 \to \mathbb{R}^3$ for $\gamma_1 : S^2 \to \mathbb{R}_{>0}$ defined by (11)

$$\begin{aligned}\xi_{\gamma_1}(\nu) = \big(&(\cos\theta)(\cos^6\theta + 6\cos^4\theta\sin^2\theta - 5\sin^6\theta)(\cos\varrho),\\ &(\cos\theta)(\cos^6\theta + 6\cos^4\theta\sin^2\theta - 5\sin^6\theta)(\sin\varrho),\\ &(\sin\theta)(-5\cos^6\theta + 6\cos^4\theta\sin^2\theta + \sin^6\theta)\big),\end{aligned} \tag{12}$$

$(\nu = (\cos\theta\cos\varrho, \cos\theta\sin\varrho, \sin\theta) \in S^2)$.

The Wulff shape W_{γ_1} is the surface of revolution (Fig. 5, upper left) given by rotating W_γ (Fig. 2, upper left) around the vertical axis. The two piecewise-C^∞ closed surfaces (Fig. 5, upper right and lower left) given by rotating the closed curves $(C_\gamma)_1$, $(C_\gamma)_4$ (Fig. 2) around the vertical axis are embedded and CAMC. The piecewise-C^∞ closed surface (Fig. 5, lower right) given by rotating the closed curve $(C_\gamma)_5$ (Fig. 3) is not CAMC.

We can get another example by rotating γ around the straight line $\theta = \pi/4$. For convenience, first we rotate γ around the origin by $\pi/4$. We call the new energy density function $\gamma_2^0 : S^1 \to \mathbb{R}_{>0}$ (Fig. 6, left). Then we rotate γ_2^0 around the vertical axis to obtain

$$\gamma_2(\nu_1, \nu_2, \nu_3) = (\nu_1^2 + \nu_2^2)^3 + 15(\nu_1^2 + \nu_2^2)^2\nu_3^2 + 15(\nu_1^2 + \nu_2^2)\nu_3^4 + \nu_3^6, \tag{13}$$

$((\nu_1, \nu_2, \nu_3) \in S^2)$. The corresponding Cahn–Hoffman map $\xi_{\gamma_2} : S^2 \to \mathbb{R}^3$ is given as follows (Fig. 6, right).

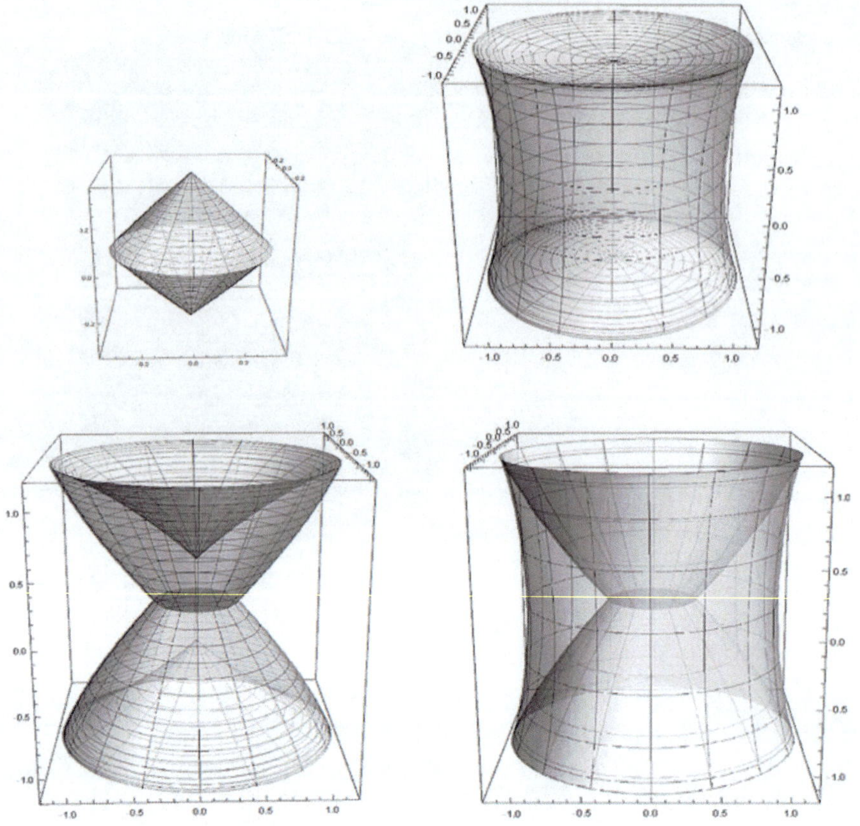

Fig. 5 Some of the closed surfaces which are subsets of $\xi_{\gamma_1}(S^2)$ for γ_1 defined by (11) (Fig. 4). They are surfaces given by rotating the curves W_γ, $(C_\gamma)_1$, $(C_\gamma)_4$, $(C_\gamma)_5$. Upper left, upper right, and lower left: The anisotropic mean curvature for the outward-pointing normal is -1. Lower right: The anisotropic mean curvature is -1 on the 'outer part', while it is 1 on the 'inner part'. Hence, this surface is not CAMC

$$\begin{aligned}\xi_{\gamma_2}(\nu) = \frac{1}{4}\big(&(\cos\theta)(\cos^6\theta - 9\cos^4\theta\sin^2\theta + 15\cos^2\theta\sin^4\theta + 25\sin^6\theta)(\cos\varrho),\\ &(\cos\theta)(\cos^6\theta - 9\cos^4\theta\sin^2\theta + 15\cos^2\theta\sin^4\theta + 25\sin^6\theta)(\sin\varrho),\\ &(\sin\theta)(25\cos^6\theta + 15\cos^4\theta\sin^2\theta - 9\cos^2\theta\sin^4\theta + \sin^6\theta)\big), \qquad (14)\end{aligned}$$

$(\nu = (\cos\theta\cos\varrho, \cos\theta\sin\varrho, \sin\theta) \in S^2)$.

By the same way as above, we get several closed embedded piecewise-C^∞ CAMC surfaces (Fig. 7) for γ_2 which are subsets of $\xi_{\gamma_2}(S^2)$.

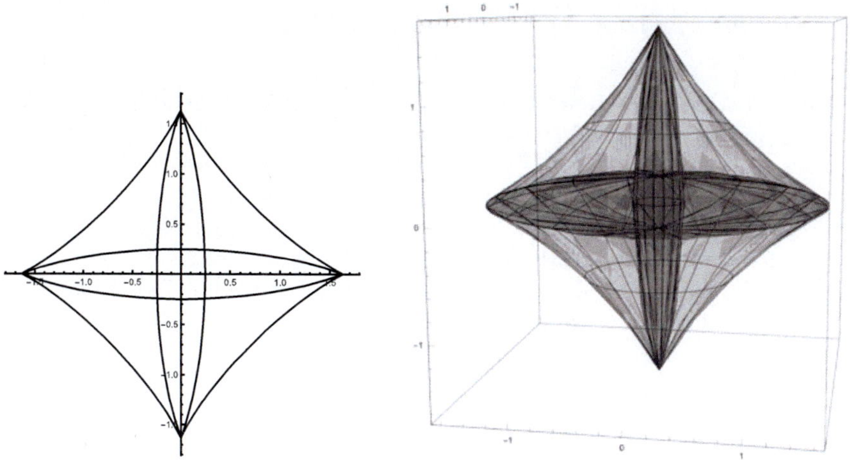

Fig. 6 Left: The image of the Cahn–Hoffman map $\xi_{\gamma_2^0} : S^1 \to \mathbb{R}^2$ for $\gamma_2^0 : S^1 \to \mathbb{R}_{>0}$. Right: The image of the Cahn–Hoffman map $\xi_{\gamma_2} : S^2 \to \mathbb{R}^3$ for $\gamma_2 : S^2 \to \mathbb{R}_{>0}$ defined by (13)

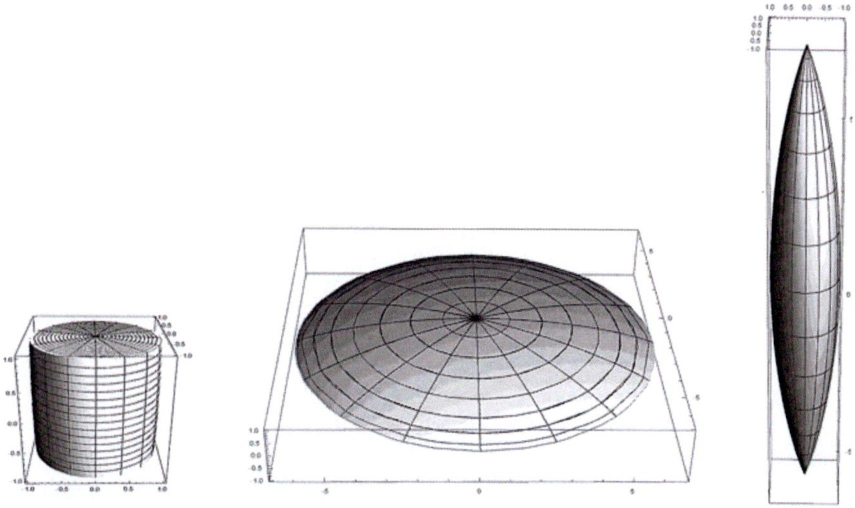

Fig. 7 The Wulff shape W_{γ_2} (left) and some of the closed embedded surfaces which are subsets of $\xi_{\gamma_2}(S^2)$ for γ_2 defined by (13) (Fig. 6, right). The anisotropic mean curvatures for the outward-pointing normal are -1

5.3 *Outline of Proofs of Theorems 2 and 3*

The function $\gamma : S^1 \to \mathbb{R}_{>0}$ defined by (9) and the closed embedded CAMC curves $(C_\gamma)_1$, $(C_\gamma)_2$, $(C_\gamma)_3$ and $(C_\gamma)_4$ shown in Fig. 2 prove Theorem 2 for $n = 1$. The function $\gamma_1 : S^2 \to \mathbb{R}_{>0}$ defined by (11) and the closed embedded CAMC surfaces of revolution with genus 0 obtained by rotating the curves $(C_\gamma)_1$, $(C_\gamma)_4$ (Fig. 5) prove Theorems 2 and 3 for $n = 2$. Also, the function $\gamma_2 : S^2 \to \mathbb{R}_{>0}$ defined by (13) and the two closed embedded CAMC surfaces of revolution with genus 0 shown in the center and the right in Fig. 7 give an example which proves Theorems 2 and 3 for $n = 2$. Higher dimensional examples which prove Theorem 2 can be obtained by suitable rotations of $\gamma : S^1 \to \mathbb{R}_{>0}$ in a similar way.

6 Anisotropic Mean Curvature Flow

Let $X_t : M \to \mathbb{R}^{n+1}$ be a one-parameter family of piecewise-C^2 weak immersion with anisotropic mean curvature Λ_t and unit normal ν_t. Assume that the Cahn–Hoffman field $\tilde{\xi}_t$ along X_t is well-defined on M. If X_t satisfies $\partial X_t / \partial t = \Lambda_t \tilde{\xi}_t$, it is called an anisotropic mean curvature flow, which diminishes the anisotropic energy if $\Lambda_t \not\equiv 0$ (Appendix 2). By a simple observation, one can show the following.

Theorem 6 ([9]) *Let c be a positive constant. Set*

$$X_t := \sqrt{2(c-t)}\,\xi_\gamma, \quad t \le c.$$

Then X_t is a self-similar shrinking solution, that is
(i) $\partial X_t / \partial t = \Lambda_t \tilde{\xi}_t$, and
(ii) X_t is homothetic to ξ_γ and it shrinks as t increases.

Outline of the proof of Theorem **6** Since the anisotropic mean curvature of ξ_γ is -1 (Fact 1), $\Lambda_t = \dfrac{-1}{\sqrt{2(c-t)}}$ holds. On the other hand, since, by the definition of the Cahn–Hoffman map, one can show that the unit normal vector field of ξ_γ is given by ν, the equality $\tilde{\xi}_t = \xi_\gamma$ holds. These two facts imply that (i) and (ii) hold. □

Proof of Theorem **4** Each of the examples of closed embedded CAMC hypersurfaces appeared in Sect. 5 is a subset of the image of the Cahn–Hoffman map and it is homeomorphic to S^n. Hence, by Theorem 6, each of them gives a self-similar shrinking solution for anisotropic mean curvature flow. Therefore, they are examples which prove Theorem 4. □

Acknowledgements The author would like to thank the referee for the valuable comments which helped to improve the manuscript. This work was partially supported by JSPS KAKENHI Grant Number JP18H04487.

Appendix 1: An Example

Here we give a simple example which shows that if the energy density function γ is not strictly convex, even if it is convex, the Wulff shape can have singular points and its principal curvatures can be unbounded.

Example 1 Set $n = 1$. For $m \in \mathbb{N}$, define $\gamma_m(\nu) := (\nu_1^{2m} + \nu_2^{2m})^{1/(2m)}$. Then γ_m is of class C^∞ and convex (Fig. 8).

The Cahn–Hoffman map ξ_m for γ_m is represented as follows.

$$\begin{aligned}\xi_m(\cos\theta, \sin\theta) &= (\cos^{2m}\theta + \sin^{2m}\theta)^{(1/(2m))-1}(\cos^{2m-1}\theta, \sin^{2m-1}\theta) \\ &=: (f_m(\theta), g_m(\theta)).\end{aligned}$$

Moreover we have

$$\begin{aligned}A_m &:= D^2\gamma_m + \gamma_m \cdot 1 \\ &= (2m-1)\cos^{2m-2}\theta \sin^{2m-2}\theta(\cos^{2m}\theta + \sin^{2m}\theta)^{(1/(2m))-2}.\end{aligned}$$

Hence,

(i) If $m \geq 2$, $A_m = 0$ at $\theta = (1/2)\ell\pi$, $(\ell \in \mathbb{Z})$.

(ii) A_m is positive definite on $S^1 \setminus \{(\cos\theta, \sin\theta) \mid \theta = (1/2)\ell\pi, (\ell \in \mathbb{Z})\}$.

The curvature κ_m of ξ_m with respect to the outward-pointing normal ν is represented as

$$\begin{aligned}\kappa_m(\theta) &= \frac{-f_m' g_m'' + f_m'' g_m'}{((f_m')^2 + (g_m')^2)^{3/2}} \\ &= \frac{-1}{2m-1}\cos^{-2m+2}\theta \sin^{-2m+2}\theta(\cos^{2m}\theta + \sin^{2m}\theta)^{2-\frac{1}{2m}}.\end{aligned}$$

Hence, near each point where A_m is degenerate, κ_m is unbounded, that is,

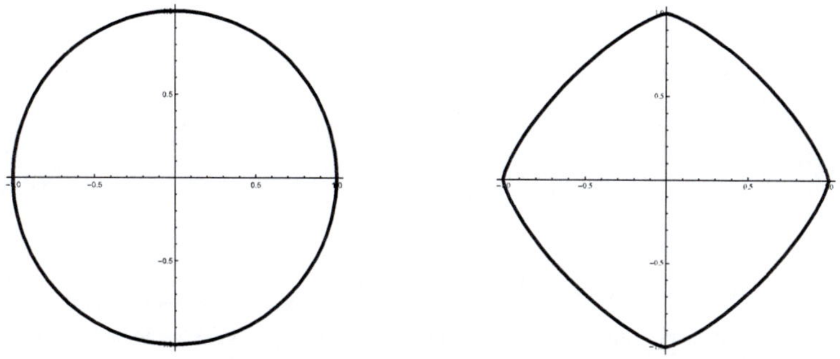

Fig. 8 Left: W_{γ_m} for $m = 1$, right: W_{γ_m} for $m = 2$, for $\gamma_m(\nu) := (\nu_1^{2m} + \nu_2^{2m})^{1/(2m)}$ in Example 1

$$\lim_{\theta \to \frac{\ell}{2}\pi} \kappa_m(\theta) = -\infty, \quad \ell \in \mathbb{Z},\ m \geq 2$$

holds. However, since

$$\kappa_m\, ds = -d\theta$$

holds, we have

$$\int_{\theta=0}^{\theta=2\pi} \kappa_m\, ds = -\int_0^{2\pi} d\theta = -2\pi.$$

Appendix 2: Remark on the Anisotropic Mean Curvature Flow

Here we prove that an anisotropic mean curvature flow X_t diminishes the anisotropic energy if the anisotropic mean curvature Λ_t of X_t does not vanish identically. Let $\gamma : S^n \to \mathbb{R}_{>0}$ be of class C^2 with Cahn–Hoffman map ξ_γ. Let $X_t : M \to \mathbb{R}^{n+1}$ be one-parameter family of embedded piecewise-C^2 weakly immersed hypersurfaces with anisotropic mean curvature Λ_t. Assume that the Cahn–Hoffman field $\tilde{\xi}_t$ along X_t is defined on M. If X_t satisfies $\partial X_t/\partial t = \Lambda_t \tilde{\xi}_t$, it is called an anisotropic mean curvature flow. Then

$$\begin{aligned} \frac{d\mathcal{F}_\gamma(X_t)}{dt} &= -\int_M n\Lambda_t \Big\langle \frac{\partial X_t}{\partial t}, \nu_t \Big\rangle\, dA_t = -\int_M n\Lambda_t^2 \langle D\gamma + \gamma(\nu_t)\nu_t, \nu_t\rangle\, dA_t \\ &= -\int_M n\Lambda_t^2 \gamma(\nu_t)\, dA_t \leq 0 \end{aligned} \tag{15}$$

holds.

References

1. Alexandrov, A.D.: A characteristic property of spheres. Ann. Mat. Pura Appl. **58**, 303–315 (1962)
2. Ando, N.: Hartman-Wintner's theorem and its applications. Calc. Var. Partial Differ. Eq. **43**, 389–402 (2012)
3. Barbosa, J.L., do Carmo, M.: Stability of hypersurfaces with constant mean curvature. Math. Z. **185**, 339–353 (1984)
4. Barbosa, J.L., do Carmo, M., Eschenburg, J.: Stability of hypersurfaces of constant mean curvature in Riemannian manifolds. Math. Z. **197**, 123–138 (1988)
5. Brendle, S.: Embedded self-similar shrinkers of genus 0. Ann. Math. **183**, 715–728 (2016)
6. He, Y., Li, H.: Integral formula of Minkowski type and new characterization of the Wulff shape. Acta Math. Sinica (English Series) **24**, 697–704 (2008)
7. He, Y., Li, H., Ma, H., Ge, J.: Compact embedded hypersurfaces with constant higher order anisotropic mean curvatures. Indiana Univ. Math. J. **58**, 853–868 (2009)

8. Hopf, H.: Differential Geometry in the Large. Lecture Notes in Mathematics, vol. 1000. Springer, Berlin (1989)
9. Jikumaru, Y., Koiso, M.: Non-uniqueness of closed embedded non-smooth hypersurfaces with constant anisotropic mean curvature. preprint. arXiv:1903.03958 [math.DG]
10. Koiso, M.: Uniqueness of stable closed non-smooth hypersurfaces with constant anisotropic mean curvature. preprint. arXiv:1903.03951 [math.DG]
11. Koiso, M., Palmer, B.: Geometry and stability of surfaces with constant anisotropic mean curvature. Indiana Univ. Math. J. **54**, 1817–1852 (2005)
12. Koiso, M., Palmer, B.: Anisotropic umbilic points and Hopf's theorem for surfaces with constant anisotropic mean curvature. Indiana Univ. Math. J. **59**, 79–90 (2010)
13. Morgan, F.: Planar Wulff shape is unique equilibrium. Proc. Amer. Math. Soc. **133**, 809–813 (2005)
14. Palmer, B.: Stability of the Wulff shape. Proc. Amer. Math. Soc. **126**, 3661–3667 (1998)
15. Palmer, B.: Stable closed equilibria for anisotropic surface energies: surfaces with edges. J. Geom. Mech. **4**, 89–97 (2012)
16. Reilly, R.C.: The relative differential geometry of nonparametric hypersurfaces. Duke Math. J. **43**, 705–721 (1976)
17. Taylor, J.E.: Crystalline variational problems. Bull. Amer. Math. Soc. **84**, 568–588 (1978)
18. Wente, H.C.: A note on the stability theorem of J. L. Barbosa and M. do Carmo for closed surfaces of constant mean curvature. Pac. J. Math. **147**, 375–379 (1991)
19. Winterbottom, W.L.: Equilibrium shape of a small particle in contact with a foreign substrate. Acta Metall. **15**, 303–310 (1967)
20. Wulff, G.: Zur Frage der Geschwindigkeit des Wachsthums und der Auflösung der Krystallflächen. Z. für Kryst. und Mineral. **34**, 449–530 (1901)

The Translating Soliton Equation

Rafael López

Abstract We give an analytic approach to the translating soliton equation with a special emphasis in the study of the Dirichlet problem in convex domains of the plane.

Keywords Translating soliton · Maximum principle · Dirichlet problem · Perron method

1 Historical Introduction and Motivation

In this paper we consider the equation of mean curvature type

$$\operatorname{div}\left(\frac{Du}{\sqrt{1+|Du|^2}}\right)=\frac{1}{\sqrt{1+|Du|^2}} \tag{1}$$

in a smooth domain $\Omega \subset \mathbb{R}^2$, where $u \in C^2(\Omega) \cap C^0(\overline{\Omega})$. We call (1) the *translating soliton equation*. The geometry behind this equation is the following. Let (x, y, z) be the canonical coordinates in Euclidean space $(\mathbb{R}^3, \langle , \rangle)$ and denote by $\Sigma_u = \{(x, y, u(x, y)) : (x, y) \in \Omega\}$ the graph of a function u. The left-hand side of (1) is twice the mean curvature H of Σ_u at each point $(x, y, u(x, y))$. Here H is the average of the principal curvatures calculated with respect to the unit normal vector field

$$N=\frac{1}{\sqrt{1+|Du|^2}}(-Du, 1).$$

Partially supported by the grant no. MTM2017-89677-P, MINECO/AEI/FEDER, UE.

R. López (✉)
Departamento de Geometría y Topología, Universidad de Granada, 18071 Granada, Spain
e-mail: rcamino@ugr.es
URL: http://www.ugr.es/local/rcamino

T. Hoffmann et al. (eds.), *Minimal Surfaces: Integrable Systems and Visualisation*, Springer Proceedings in Mathematics & Statistics 349,
https://doi.org/10.1007/978-3-030-68541-6_11

Hence the right-hand side of (1) is the z-coordinate of N. Consequently, a surface in Euclidean space satisfies locally the translating soliton equation if and only if the mean curvature at each point is the half of the cosine of the angle that makes N with the vertical direction $\mathbf{a} = (0, 0, 1)$.

As far as the author knows, it was S. Bernstein in 1910 the first who studied Eq. (1) in a couple of papers [5, 6] in the context of the solvability of the Dirichlet problem for elliptic equations. In [5, p. 240], the translating soliton equation appears numbering as (6) and Bernstein names *l'équation des surfaces, dont la courbure en chaque point is proportionnelle (égale) au cosinus de l'angle de la normale en ce point avec l'axe des z*. On the other hand, in [6, p. 515] Bernstein considers a family of equations numbered as (2') in classical notation

$$(1+q^2)r - 2pqs + (1+p^2)t = (1+p^2+q^2)^{n/2}, \tag{2}$$

where n is an integer number. In particular, for $n = 2$ this equation coincides with (1). The Dirichlet problem consists of finding a smooth solution of (2) with prescribed boundary data

$$u = \varphi \quad \text{on } \partial\Omega, \tag{3}$$

where $\varphi \in C^0(\partial\Omega)$. Bernstein proved that (2)–(3) is solvable for arbitrary analytic functions φ when Ω is an analytic convex domain and $n \leq 2$. In particular, this result holds for the translating soliton equation as well.

Sixty years later, the second approach to Eq. (1) is due to J. Serrin. In the eighty-pages article [40], Serrin gave a systematic treatment of the Dirichlet problem for a large class of quasilinear non-uniformly second order elliptic equations. Following the Leray-Schauder fixed point theorem and Hölder estimates theory of Ladyzenskaja and Ural'ceva, Serrin establishes the necessary and sufficient conditions for the solvability of the Dirichlet problem for arbitrary boundary data. Possibly, the most known result of [40] is the case of the constant mean curvature equation, that is, when the right-hand side of (1) is replaced by a constant $2H$. In such a case, the Dirichlet problem has a solution for arbitrary smooth boundary data φ if and only if the curvature κ of $\partial\Omega$ with respect to the inward normal direction satisfies $\kappa \geq 2|H|$. If the solution exists, it is unique. See [41] when the boundary $\partial\Omega$ is not necessarily smooth.

However, the article [40] covers many other types of quasilinear elliptic equations and this is the situation of the translating soliton equation. Exactly in pp. 477–478, Serrin considers two families of quasilinear elliptic equations and one of them coincides with (2). The Eq. (96) of [40] is

$$(1+q^2)r - 2pqs + (1+p^2)t = 2H(1+p^2+q^2)^{n/2}, \tag{4}$$

where now H and n are two real constants: recall that in (2), n is an integer number. Notice that if $n = 3$, the expression (4) is the constant mean curvature equation. As a consequence of the results previously obtained, Serrin proves the following existence result [40, p. 478].

Theorem 1 *Let $\Omega \subset \mathbb{R}^2$ be a bounded C^2-domain. Then (2)–(3) is solvable for an arbitrarily given C^2 function φ for*

1. *$n \leq 2$, if and only if $\kappa \geq 0$,*
2. *$2 < n < 3$, if and only if $\kappa > 0$.*

When $n > 3$, the Dirichlet problem is not generally solvable, whatever the domain.

Definitively, for the translating soliton equation, we conclude:

Corollary 1 *Let $\Omega \subset \mathbb{R}^2$ be a bounded C^2-domain. Then (1)–(3) is solvable for arbitrarily given C^2 function φ if and only if the curvature κ of the boundary Ω with respect to the inward normal direction satisfies $\kappa \geq 0$.*

Remark 1 1. The case $n = 3$ in (4), which does not appear in Theorem 1, is the constant mean curvature equation, where the solvability occurs if and only if $\kappa \geq 2|H|$.
2. Serrin generalizes the result of Bernstein in [5] changing analyticity by smoothness of Ω.
3. The results of [40] for the Eq. (3) are established in arbitrary dimension.

Possibly due to the length of the paper [40], Eq. (1) seems to be forgotten in the literature. It is in 80s when the translating soliton equation appears in two different contexts at the same time.

Firstly in the singularity theory of the mean curvature flow of Huisken and Ilmanen [19, 21]. A *translating soliton* is a surface $\Sigma \subset \mathbb{R}^3$ that is a solution of the mean curvature flow when Σ evolves purely by translations along some direction $\mathbf{a} \in \mathbb{R}^3 \setminus \{0\}$. In other words, Σ is a translating soliton if $\Sigma + t\mathbf{a}$, $t \in \mathbb{R}$, satisfies that fixed t, the normal component of the velocity vector $\mathbf{a}$ at each point is equal to the mean curvature at that point. For the initial surface Σ, this implies that $2H = \langle N, \mathbf{a}\rangle$. After a change of coordinates, if $\mathbf{a} = (0, 0, 1)$, then $2H = \langle N, \mathbf{a}\rangle$ coincides locally with (1). Translating solitons appear in the singularity theory of the mean curvature flow. After scaling, near type II-singularity points on the surfaces evolved by mean curvature vector, Huisken, Sinestrari and White demonstrated that the limit flow with initial convex surface is a convex translating soliton [19, 20, 44]. On the other hand, Ilmanen observed that Σ translates with velocity $\mathbf{a}$ if and only if it is stationary for the weighted area $\int_\Sigma e^{\langle p, \mathbf{a}\rangle} dA$. In fact, $2H = \langle N, \mathbf{a}\rangle$ is the Euler–Lagrange equation for this functional and thus Σ is a minimal surface with respect to the Riemannian metric $e^{\langle p, \mathbf{a}\rangle}\langle , \rangle$.

From the last viewpoint, Eq. (1) links with the theory of manifolds with density of Gromov [16]. Namely, let e^ϕ be a positive density function in $\mathbb{R}^3$, $\phi \in C^\infty(\mathbb{R}^3)$, which serves as a weight for the volume and the surface area. Note that this is not equivalent to scaling the metric conformally by e^ϕ because the area and the volume change with different scaling factors. For a given compactly supported variation of Σ_t of Σ that fixes the boundary $\partial\Sigma$ of Σ, let $A_\phi(t)$ and $V_\phi(t)$ denote the weighted area and the enclosed weighted volume of Σ_t, respectively. Then the first variations of $A_\phi(t)$ and $V_\phi(t)$ are

$$A'_\phi(0) = -2\int_\Sigma H_\phi \langle N, \xi\rangle \, dA_\phi, \quad V'_\phi(0) = \int_\Sigma \langle N, \xi\rangle \, dA_\phi,$$

where ξ is the variational vector field of Σ_t and $H_\phi = H - \langle N, \nabla\phi\rangle/2$ is called the *weighted mean curvature*. If we choose $\phi(p) = \langle p, \mathbf{a}\rangle$, $p \in \mathbb{R}^3$, then

$$H_\phi = H - \frac{\langle N, \mathbf{a}\rangle}{2}. \tag{5}$$

We say that $\mathbf{a}$ is the *density vector*. Thus we have the next characterizations of a translating soliton.

Proposition 1 *Let Σ be a surface in $\mathbb{R}^3$. The following statements are equivalent:*

1. *Σ satisfies locally (1).*
2. *Σ translates with velocity $\mathbf{a}$ by means of the mean curvature flow.*
3. *Σ is a critical point of the area A_ϕ for the density $\phi(p) = \langle p, \mathbf{a}\rangle$*

In view of both approaches, we point out that similar results of Theorem 1 and Corollary 1 have been recently treated in the literature. We indicate some of them.

1. Corollary 1 appears in [3, Theorem 2] assuming Ω is contained in a disc of radius 1 and satisfying an enclosing sphere condition. But in a Remark, Bergner asserts that the assumption to be contained in a ball can drop if there exist C^0 estimates, such as it occurs for (1): see Proposition 10 below.
2. Corollary 1 appears in [34, Remark 3]. Initially, it is assumed that $|\Omega| < 4\pi$ in a general result, but for (1) this hypothesis drops. Using the same proof than in [34], the existence holds for $n \leq 2$ in Eq. (2) under the assumption that $|\Omega| < 4\pi$.
3. Theorem 1 appears in [22, Lemma 2.2] for $0 < n < 3$ assuming $\kappa > 0$ and $|\Omega| < 4\pi$.
4. Corollary 1 appears in [43, Theorem 1.1] assuming that $\mathrm{diam}(\Omega) < 2$.

We finish this section giving two generalizations of the translating soliton equation. First, consider the flow of surfaces by powers of mean curvature according to [38, 39, 42]. If $\alpha > 0$ is a constant, then the surface $z = u(x, y)$ evolves by translations of the H^α-power of mean curvature flow if

$$\mathrm{div}\left(\frac{Du}{\sqrt{1+|Du|^2}}\right) = \left(\frac{1}{\sqrt{1+|Du|^2}}\right)^\alpha.$$

Notice that this equation coincides with (2) of Bernstein and Serrin with the relation $n = 3 - \alpha$.

The second generalization is by considering critical points of the area A_ϕ for a fixed weighted volume. As a consequence of the Lagrange multipliers, Σ satisfies that H_ϕ is a constant function and thus, in nonparametric form, we have

$$\mathrm{div}\frac{Du}{\sqrt{1+|Du|^2}} = \frac{1}{\sqrt{1+|Du|^2}} + \mu, \tag{6}$$

where μ is a constant. This equation has received a recent interest: [8, 14, 29]. Even more general, we may study the mean curvature flow with a forcing term, so the constant μ in (6) is replaced by a function $f = f(u, Du)$ [23, 37]. For example, the mean curvature type equation

$$\mathrm{div}\frac{Du}{\sqrt{1+|Du|^2}} = H_1(x,u,Du) + H_2(x,u,Du)\frac{1}{\sqrt{1+|Du|^2}}$$

has been studied in [3, 24, 34].

Convention. After a change of coordinates, we will assume that $\mathbf{a} = (0, 0, 1)$.

This paper is organized as follows. In Sect. 2 we recall the translating solitons that are invariant by a uniparametric group of translations and of rotations. Section 3 is devoted to the tangency principle and some consequences derived by its applications to control the shape of a compact translating soliton. Sections 4 and 5 solve the Dirichlet problem on bounded convex domains for the translating soliton Eq. (1) and the constant weighted mean curvature Eq. (6), respectively. Here the boundary gradient estimates are obtained by means of the classical maximum principle to suitable choices of barrier functions. Finally, in Sect. 6 we study the Dirichlet problem for (1) in unbounded domains. We will consider two cases, namely, the domain is a strip and the boundary data are two copies of a convex function or the domain is an unbounded convex domain contained in a strip and the boundary data are constant.

2 Examples of Translating Solitons

Let $\{e_1, e_2, e_3\} = \{(1, 0, 0), (0, 1, 0), (0, 0, 1)\}$ denote the canonical basis of $\mathbb{R}^3$. The Euclidean plane $\mathbb{R}^2$ will identified with plane of equation $z = 0$. We also use the terminology horizontal and vertical to indicate an orthogonal direction to $\mathbf{a}$ or a parallel direction to $\mathbf{a}$, respectively.

Notice that any translation of $\mathbb{R}^3$ preserves solutions of the translating soliton equation. The same occurs for a rotation about an axis parallel to e_3. Also, Eq. (1) is preserved by reversing the orientation on the surface.

In this section, we are interested by examples of translating solitons that are invariant by a uniparametric group of motions, more precisely, surfaces invariant along one direction and surfaces of revolution. In both cases, the Eq. (1) converts into an ODE and one may apply the standard theory.

2.1 Cylindrical Surfaces

A cylindrical surface Σ is a surface invariant along a direction $\mathbf{v} \in \mathbb{R}^3$, or in other words, Σ is a ruled surface where all the rulings are parallel to $\mathbf{v}$. We ask for those translating solitons of cylindrical type. Notice that there is not an a priori relation between the direction $\mathbf{v}$ and the density vector $\mathbf{a}$.

A parametrization of Σ is $X(t,s)=\gamma(s)+t\mathbf{v}$, $t\in\mathbb{R}$ and $\gamma: I\subset\mathbb{R}\to\mathbb{R}^3$ is a planar curve orthogonal to $\mathbf{v}$. We parametrize $\gamma=\gamma(s)$ by the arc length s such that $\gamma'(s)\times\mathbf{n}(s)=\mathbf{v}$, being $\mathbf{n}$ the unit principal normal vector of γ. The Gauss map of Σ is $N(X(s,t))=\mathbf{n}(s)$ and $2H=\kappa(s)$, being κ the inward curvature of γ as a planar curve. Thus Σ is a translating soliton if $\kappa(s)=\langle\mathbf{n}(s),\mathbf{a}\rangle$, hence we conclude that there is not a relation between the vectors $\mathbf{v}$ and $\mathbf{a}$. For example, if $\mathbf{v}$ is parallel to $\mathbf{a}$, then $\langle\mathbf{n}(s),\mathbf{a}\rangle=0$ for every $s\in I$, so γ is a straight line and Σ is a plane parallel to $\mathbf{a}$.

In a first step, we investigate the case that $\mathbf{v}$ is orthogonal to $\mathbf{a}$ which, after a rotation about $\mathbf{a}$, we suppose $\mathbf{v}=e_1$. Let us observe that a vertical plane parallel to e_1, that is, a plane parallel to the yz-plane, is a translating soliton of cylindrical type. If we write γ as $z=w(y)$, then (1) converts to

$$w''=1+w'^2.$$

By simple quadratures, the solution of this equation is

$$w(y)=-\log(\cos(y+b))+a,\quad a,b\in\mathbb{R}, \tag{7}$$

and this solution is called the *grim reaper*. Although this holds for graphs $z=w(y)$, it is true in general: if there is a vertical tangent vector at some point of γ, then γ is a vertical line by uniqueness of ODE. This can be also obtained as follows. We parametrize γ by the arc length. Then $\gamma(s)=(0,y(s),z(s))$, with $y'(s)=\cos\psi(s)$, $z'(s)=\sin\psi(s)$ for some function ψ. Then $X(t,s)=(t,y(s),z(s))$ and (1) becomes $2\psi'(s)=\cos\psi(s)$. If γ is not a graph on the y-axis, there is $s=s_0$ such that $\cos\theta(s_0)=0$. By uniqueness, the solution is $\theta(s)=\pm\pi/2$, $\gamma(s)=(0,a,\pm s+b)$, $a,b\in\mathbb{R}$, γ is a vertical line and Σ is the vertical plane of equation $y=a$.

Once obtained the translating solitons of cylindrical type when the vector $\mathbf{v}$ is orthogonal to $\mathbf{a}$, the rest of cylindrical surfaces are obtained by rotating the about surfaces about a horizontal axis. The resulting surfaces are all translating solitons of cylindrical type (after translations and rotations about a vertical axis). We present these surfaces, which will be called *grim reapers* again (Fig. 1).

Definition 1 The uniparametric family of grim reapers $w_\theta=w_\theta(x,y)$ are defined as

$$w_\theta(x,y)=-\frac{1}{(\cos\theta)^2}\log(\cos(\cos\theta y))+(\tan\theta)x+a, \tag{8}$$

where $\theta\in(-\pi/2,\pi/2)$, $a\in\mathbb{R}$.

Here we recall that planes parallel to the xz-plane are cylindrical translating solitons, which would correspond with the critical values $\theta=\pm\pi/2$.

Proposition 2 *All translating solitons of cylindrical type are planes parallel to the xz-plane or the grim reapers w_θ.*

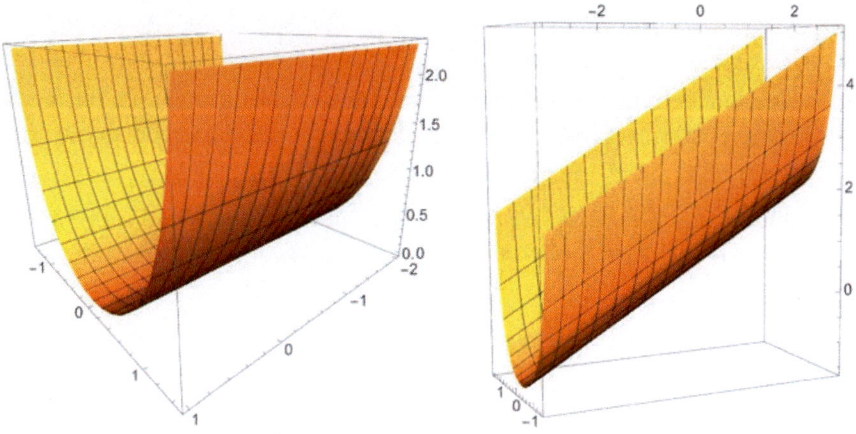

Fig. 1 The grim reapers w_θ. Left: $\theta = 0$; Right: $\theta = \pi/6$

Proof If $\mathbf{v} = \mathbf{a}$, we know that the surface is (7), which coincides, up to a reparametrization, with (8) for the choice $\theta = 0$.

Suppose $\mathbf{v}$ be a vector which is not orthogonal to $\mathbf{a}$. After a rotation with respect to the z-axis, we assume that $\mathbf{v} = \cos\theta e_1 + \sin\theta e_3$, $|\theta| < \pi/2$. Let $\mathbf{e} = -\sin\theta e_1 + \cos\theta e_3$. We write the generating curve as a graph $g = g(s)$ on the y-axis. Then parametrization of the surface is $X(t, s) = se_2 + g(s)\mathbf{e} + t\mathbf{v}$. A computation shows that (1) writes as $g'' = \cos\theta(1 + g'^2)$ and its integration gives $g(s) = -\log(\cos(\cos\theta s + b))/\cos\theta + a$, $a, b \in \mathbb{R}$. Then

$$X(t, s) = (-\sin\theta g(s) + t\cos\theta, s, t\sin\theta + \cos\theta g(s)).$$

Writing $X(t, s) = (x, y, u(x, y))$, we deduce easily that u coincides with the function w_θ in (8).

The maximal domain of w_θ is the strip

$$\Omega^\theta = \left\{(x, y) \in \mathbb{R}^2 : -\frac{\pi}{2\cos\theta} < y < \frac{\pi}{2\cos\theta}\right\}.$$

In particular, if $0 \leq \theta_1 < \theta_2$, it follows that $\Omega^{\theta_1} \subset \Omega^{\theta_2}$ and thus the domain Ω^0, namely,

$$\Omega^0 = \left\{(x, y) \in \mathbb{R}^2 : -\frac{\pi}{2} < y < \frac{\pi}{2}\right\}, \tag{9}$$

is contained in all Ω^θ for any $\theta \in (-\pi/2, \pi/2)$.

2.2 Rotational Surfaces

The second family of translating solitons of our interest are those ones of rotational type. If Σ is a surface of revolution about a rotation axis $\mathbf{v}$, we ask about the relation between the vector $\mathbf{v}$ and the density vector $\mathbf{a}$.

Proposition 3 *Let Σ be a surface of revolution with respect to the vector $\mathbf{v}$. If Σ is a translating soliton, then $\mathbf{v}$ is parallel to $\mathbf{a}$ or Σ is a plane parallel to $\mathbf{a}$ where $\mathbf{v}$ is orthogonal to $\mathbf{a}$.*

Proof The value of the mean curvature H is constant along a parallel of Σ. On the other hand, the Gauss map N makes a constant angle with $\mathbf{v}$ along a parallel of the surface. Since $2H = \langle N, \mathbf{a}\rangle$, the function $\langle N, \mathbf{a}\rangle$ is constant along every parallel of Σ. Hence, we have only two possibilities, namely, $\mathbf{v}$ is parallel to $\mathbf{a}$ or $\langle \mathbf{v}, \mathbf{a}\rangle = 0$ with $\langle N, \mathbf{a}\rangle = 0$ on Σ. In the latter case, Σ is a plane parallel to $\mathbf{a}$.

After a translation of $\mathbb{R}^3$, we will assume that the rotation axis is the z-axis. If we parametrize Σ as $z = u(r)$, $r^2 = x^2 + y^2$, Eq. (1) becomes

$$u'' + \frac{u(1+u'^2)}{r} = 1 + u'^2. \tag{10}$$

Therefore, by standard theory of ODE, there are solutions of (10) of initial conditions $u(r_0) = u_0$, $u'(r_0) = u_0'$, with $r_0 > 0$. The classification of the translating solitons of rotational type was done in [2, 13]: see Fig. 2.

Definition 2 There are two types of rotational translating solitons depending if the surface meets or does not meet the rotation axis:

1. Bowl solitons. They are strictly convex entire graphs with a global minimum in the z-axis and intersect orthogonally the rotation axis. The surfaces are asymptotic to a paraboloid.
2. Surfaces of winglike shape. These surfaces do not intersect the rotation axis.

The bowl soliton corresponds with the solution of (10) with initial condition $u(0) = u'(0) = 0$ where the existence is not a direct consequence of the standard

Fig. 2 Rotational translating solitons. Left: the bowl soliton; Right: surface with winglike-shape

theory because (10) presents a singularity at $r = 0$. On the other hand, the winglike-shape solutions corresponds with solutions of (10) with $r_0 > 0$ and $u'(r_0) = 0$, whose existence is immediate.

The existence of the bowl soliton was done in [2, Corollary 3.3]. The authors solve (1) in a round disk with Neumann boundary condition $\partial u/\partial n = \cos\alpha/\sqrt{1+|Du|^2}$ and, after an argument of continuity varying the parameter α, they obtain the desired rotational solution. In this paper, we give two alternative proofs of the existence of the bowl solitons. One will appear in Remark 2 using Corollary 1 and an argument by means of the Alexandrov reflection method. We now present the other proof, which follows standard techniques of radial solutions for some equations of mean curvature type [7, 11]. This technique was introduced for the first time in the context of manifolds with density by the author in [32]. We write (10) as

$$\frac{u''(r)}{(1+u'(r)^2)^{3/2}} + \frac{u'(r)}{r\sqrt{1+u'(r)^2}} = \frac{1}{\sqrt{1+u'(r)^2}}. \tag{11}$$

Multiplying (11) by r, and integration by parts, we wish to establish the existence of a classical solution of

$$\begin{cases} \left(\dfrac{ru'(r)}{\sqrt{1+u'(r)^2}}\right)' = \dfrac{r}{\sqrt{1+u'(r)^2}}, & \text{in } (0,\delta) \\ u(0) = 0, \quad u'(0) = 0. \end{cases} \tag{12}$$

Let us observe that Eq. (12) is singular at $r = 0$. We notice that the condition $u(0) = 0$ is not restrictive because the above problem is invariant by vertical translations. The condition on $u'(0) = 0$ means that the profile curve meets orthogonally the z-axis. One may consider other initial condition $u'(0)$. However, we have:

Proposition 4 *Any solution of (11) intersecting the rotation axis must do it perpendicularly.*

Proof Viewing (11) as (12) and integrating between 0 and r, with $r < \delta$, we have

$$\frac{ru'(r)}{\sqrt{1+u'(r)^2}} = \int_0^r \frac{t}{\sqrt{1+u'(t)^2}}dt,$$

or equivalently,

$$\frac{u'(r)}{\sqrt{1+u'(r)^2}} = \frac{1}{r}\int_0^r \frac{t}{\sqrt{1+u'(t)^2}}dt.$$

Letting $r \to 0$ and by the L'Hôpital rule, we find

$$\lim_{r\to 0} \frac{u'(r)}{\sqrt{1+u'(r)^2}} = \lim_{t\to 0} \frac{1}{\sqrt{1+u'(t)^2}},$$

concluding $\lim_{r\to 0} u'(r) = 0$.

We now prove the existence of solutions of (12).

Proposition 5 *The initial value problem (12) has a solution $u \in C^2([0, R])$ for some $R > 0$ which depends continuously on the initial data.*

Proof Define the functions $g : \mathbb{R} \times \mathbb{R} \to \mathbb{R}$ and $f : \mathbb{R} \to \mathbb{R}$ by

$$g(x, y) = \frac{1}{\sqrt{1+y^2}}, \quad f(y) = \frac{y}{\sqrt{1+y^2}}.$$

It is clear that a function $u \in C^2([0, \delta])$, for some $\delta > 0$, is a solution of (12) if and only if $(rf(u'))' = rg(u, u')$ and $u(0) = 0$, $u'(0) = 0$.

Fix $\delta > 0$ to be determined later, and define the operator T by

$$(\mathsf{T}u)(r) = \int_0^r f^{-1}\left(\int_0^s \frac{t}{s} g(u')dt\right) ds.$$

Note that a fixed point of the operator T is a solution of the initial value problem (12). We claim now that T is a contraction in the space $C^1([0, \delta])$ endowed with the usual norm $\|u\| = \|u\|_\infty + \|u'\|_\infty$. To see this, the functions g and f^{-1} are Lipschitz continuous of constant $L > 0$ in $[-\epsilon, \epsilon] \times [-\epsilon, \epsilon]$ and $[-\epsilon, \epsilon]$ respectively, provided $\epsilon < 1$. Then for all $u, v \in \overline{B(0, \epsilon)}$ and for all $r \in [0, \delta]$,

$$|(\mathsf{T}u)(r) - (\mathsf{T}v)(r)| \leq \frac{L^2}{4} r^2 \left(\|u - v\|_\infty + \|u' - v'\|_\infty\right)$$

$$|(\mathsf{T}u)'(r) - (\mathsf{T}v)'(r)| \leq \frac{L^2}{2} r \left(\|u - v\|_\infty + \|u' - v'\|_\infty\right)$$

By choosing $\delta > 0$ small enough, we deduce that T is a contraction in the closed ball $\overline{B(0, \delta)}$ in $C^1([0, \delta])$. Thus the Schauder Point Fixed theorem proves the existence of one fixed point of T, so the existence of a local solution of the initial value problem (12). This solution lies in $C^1([0, \delta]) \cap C^2((0, \delta])$. The C^2-regularity up to 0 is verified directly by using the L'Hôpital rule because (11) leads to

$$u''(0) + \lim_{r\to 0} \frac{u'(r)}{r} = 1,$$

that is,

$$\lim_{r\to 0} u''(r) = \frac{1}{2}.$$

The continuous dependence of local solutions on the initial data is a consequence of the continuous dependence of the fixed points of T.

From the classification of the rotational translating solitons, we observe that there do not exist closed surfaces (compact without boundary). Even more, we prove that there are not closed translating solitons. Usually the proof that appears in the literature of this result uses the touching principle (see Proposition 8 below). However, it is easier the following argument that we present, which only utilizes the divergence theorem [29].

Proposition 6 *There do not exist closed translating solitons.*

Proof The proof is by contradiction. Suppose that Σ is a closed translating soliton. Since the Laplacian Δ of the height function $\langle p, \mathbf{a}\rangle$ is $\Delta\langle p, \mathbf{a}\rangle = 2H\langle N, \mathbf{a}\rangle$, and $2H = \langle N, \mathbf{a}\rangle$, then

$$\Delta\langle p, \mathbf{a}\rangle = \langle N, \mathbf{a}\rangle^2. \tag{13}$$

Integrating in Σ and using the divergence theorem, we deduce

$$0 = \int_\Sigma \langle N, \mathbf{a}\rangle^2 \, d\Sigma, \tag{14}$$

because $\partial\Sigma = \emptyset$. Hence $\langle N, \mathbf{a}\rangle = 0$ in Σ. This is a contradiction because on a closed surface, the Gauss map N is surjective on the unit sphere $\mathbb{S}^2$.

3 Properties of the Solutions of the Translating Soliton Equation

This section establishes some properties of the solutions u of the translating soliton equation, with a special interest in the control of $|u|$ and $|Du|$ when Ω is a bounded domain.

It is easily seen that the difference of two solutions of Eq. (1) satisfies the maximum principle. As a consequence, we give a statement of the comparison principle in our context. First, Eq. (1) can be expressed as $Q[u] = 0$, where Q is the operator

$$Q[u] = (1 + |Du|^2)\Delta u - u_i u_j u_{i;j} - (1 + |Du|^2), \tag{15}$$

being $u_i = \partial u/\partial x_i$, $i = 1, 2$, and we assume the summation convention of repeated indices. The comparison principle asserts [15, Theorem 10.1]:

Proposition 7 (Comparison principle) *If $u, v \in C^2(\Omega) \cap C^0(\overline{\Omega})$ satisfy $Q[u] \geq Q[v]$ in Ω and $u \leq v$ on $\partial\Omega$, then $u \leq v$ in Ω. If we replace $Q[u] \geq Q[v]$ by $Q[u] > Q[v]$, then $u < v$ in Ω.*

As a consequence of the comparison principle and since the translating solitons are real analytic, we deduce:

Proposition 8 (Touching principle) *Let Σ_1 and Σ_2 be two translating solitons with possibly non-empty boundaries $\partial\Sigma_1$, $\partial\Sigma_2$. If Σ_1 and Σ_2 have a common tangent interior point and Σ_1 lies above Σ_2 around p, then Σ_1 and Σ_2 coincide everywhere. The same statement is also valid if p is a common boundary point and the tangent lines to $\partial\Sigma_i$ coincide at p.*

The tangency principle allows to control the shape of a given translating soliton by comparing, if possible, with other known surfaces [28, 30]. For instance, it is easy to deduce that there do not exist closed translating solitons (Proposition 6). For this purpose, let Σ be a such surface. Take a vertical plane Π, which is a translating soliton, far from Σ so $\Sigma\cap\Pi=\emptyset$ since Σ is a compact set. Let us move Π towards Σ until the first touching point, which occurs necessarily at some interior point because $\partial\Sigma=\emptyset$. Then the tangency principle implies that Σ is included in Π, which is impossible.

In the following proposition, we use the tangency principle for compact translating solitons. By virtue of Proposition 6, the boundary of a compact translating soliton is not an empty set. We will see that the boundary of the surface determines, in some sense, the shape of the whole surface that spans. For instance, we characterize the compact translating solitons with circular boundary.

Proposition 9 *Let Σ be a compact translating soliton with boundary $\partial\Sigma$.*

1. *Let $D\subset\mathbb{R}^2$ be the domain bounded by convex hull of the orthogonal projection of $\partial\Sigma$ on $\mathbb{R}^2$. Then int(Σ) is contained in the solid cylinder $D\times\mathbb{R}$.*
2. *If $\partial\Sigma$ is a graph on $\partial\Omega$, $\Omega\subset\mathbb{R}^2$ a bounded convex domain, then Σ is a graph on Ω.*
3. *The maximum of the height of Σ is attained at some boundary point, that is, $\max_{p\in\Sigma} z(p)=\max_{p\in\partial\Sigma} z(p)$.*

As a consequence, if $\partial\Sigma$ is a circle contained in a horizontal plane, then Σ is a rotational surface contained in a bowl soliton [35, 36].

Proof 1. Let $v\in\mathbb{R}^3$ be a fixed arbitrary horizontal direction. Consider a vertical plane Π and orthogonal to v. Take Π sufficiently far so $\Sigma\cap\Pi=\emptyset$. We move down Π along the direction v towards Σ until the first touching point. By the tangency principle, the intersection must occur at some boundary point of Σ. By repeating this argument for all horizontal vectors, we conclude the proof.

2. By the above item and the convexity of Ω, we deduce that int$(\Sigma)\subset\Omega\times\mathbb{R}$. Suppose, contrary to our claim, that Σ is not a graph on Σ, in particular, there are two distinct points $p, q\in$ int(Σ) such that their orthogonal projections coincide on $\mathbb{R}^2$. Let $\Sigma^t=\Sigma+t\mathbf{a}$ be a vertical translation of Σ by the vector $t\mathbf{a}$. Move up Σ sufficiently far so $\Sigma^t\cap\Sigma=\emptyset$ for t sufficiently large. Now we come back Σ^t by letting $t\searrow 0$ until the first time t_1 such that $\Sigma^{t_1}\cap\Sigma\neq\emptyset$. The existence of the points p and q ensures that $t_1>0$. Furthermore, this intersection occurs at some common interior point of both surfaces because the boundaries of both surfaces are included in $\partial\Omega\times\mathbb{R}$. By the tangency principle, $\Sigma^{t_1}=\Sigma$, a contradiction because their boundaries, namely, $\partial\Sigma^{t_1}=\partial\Sigma+t_1\mathbf{a}$ and $\partial\Sigma$, do not coincide because $t_1\neq 0$.

3. Taking $\mathbf{a} = (0, 0, 1)$ in (13), we find $\Delta z \geq 0$ and the result follows as a consequence of the maximum principle.
   The proof of the last statement is as follows. By items (2) and (3), Σ is a graph on the round disc Ω bounded by $\partial\Sigma$ and the interior of Σ lies below the plane P containing $\partial\Sigma$. Then $\Sigma \cup \Omega$ bounds a 3-domain. By using the technique of the Alexandrov reflection by vertical planes [1], it is straightforward to see that Σ is invariant by any rotation whose axis is the vertical line through the center of Ω. Accordingly, Σ a surface of revolution, and since its boundary is a circle, then Σ is contained in a bowl soliton.

Remark 2 The last statement of the above proposition gives other argument for the existence of the bowl soliton. Indeed, let $\Omega = D_r = \{(x, y) \in \mathbb{R}^2 : x^2 + y^2 < r^2\}$ in (1) and take the boundary data $\varphi = 0$ in (3). By Corollary 1, the existence and uniqueness of (1)–(3) is assured and Proposition 9 asserts that the solution is a radial function. Because the rotation axis meets orthogonally the domain D_r, then Σ is a surface of revolution intersecting orthogonally the z-axis.

Remark 3 (*Tangency principle*) An inspection of the comparison argument in the proof of item (3) in Proposition 9 allows to extend the tangency principle as follows. Let Σ_1 and Σ_2 be two surfaces with weighted mean curvature H^1_ϕ and H^2_ϕ, respectively. Suppose that Σ_1 and Σ_2 have a common tangent interior point p and the orientations in both surfaces coincide at p. If $H^1_\phi \leq H^2_\phi$ around p, and Σ_2 lies above Σ_1 around p with respect to $N(p)$, then Σ_1 and Σ_2 coincide at an open set around p. The same statement holds if p is a common boundary point and the tangent lines to $\partial\Sigma_i$ coincide at p.

We derive height and interior gradient estimates for a solution of the translating soliton equation.

Proposition 10 *Let $\Omega \subset \mathbb{R}^2$ be a bounded domain.*

1. *The solution of (1)–(3), if exists, is unique.*
2. *There is a constant $C_1 = C_1(\varphi, \Omega)$ such that if u is a solution of (1)–(3), then*

$$C_1 \leq u \leq \max_{\partial\Omega} \varphi \quad \text{in } \Omega. \tag{16}$$

3. *If u is a solution of (1)–(3), then*

$$\sup_{\Omega} |Du| = \max_{\partial\Omega} |Du|.$$

Proof 1. The uniqueness of solutions of (1)–(3) is a consequence of the maximun principle.
2. The inequality in the right-hand side of (16) is immediate from the item (3) of Proposition 9.
   The lower estimate for u in (16) is obtained by means of bowl solitons as comparison surfaces. Let $R > 0$ be sufficiently large so $\overline{\Omega} \subset D_R$. Let $\mathcal{B}$ be a bowl

soliton defined by a radial function $\mathbf{b} = \mathbf{b}(r)$ such that $\partial D_R \subset \mathcal{B}$, that is, $\mathbf{b}$ is a solution of (1) in D_R with $\mathbf{b} = 0$ on ∂D_R. Let $\mathcal{B}_R$ denote the compact portion of $\mathcal{B}$ below the plane $z = 0$. Move vertically down $\mathcal{B}_R$ sufficiently far so Σ_u lies above $\mathcal{B}_R$, that is, if $(x, y, z) \in \Sigma_u$, $(x, y, z') \in \mathcal{B}_R$, then $z > z'$. Then move up $\mathcal{B}_R$ until the first touching point with Σ_u. If the first contact occurs at some interior point, then the touching principle implies $\Sigma_u \subset \mathcal{B}_R$. The other possibility is that the first contact point occurs when $\mathcal{B}_R$ touches a boundary point of Σ_u. In both cases, we conclude $\mathbf{b}(0) \leq u - \min_{\partial\Omega} \varphi$ and consequently, the constant $C_1 = \mathbf{b}(0) + \min_{\partial\Omega} \varphi$ satisfies $C_1 \leq u$.

3. Define the function $v^i = u_i$, $i = 1, 2$, and differentiate (15) with respect to the variable x_k, obtaining

$$\left((1 + |Du|^2)\delta_{ij} - u_i u_j\right) v^k_{i;j} + 2\left(u_i \Delta u - u_j u_{i;j} - u_i\right) v^k_i = 0, \tag{17}$$

for each $k = 1, 2$. Hence v^k satisfies a linear elliptic equation with no zeroth order terms and by the maximum principle, $|v^k|$ has not a maximum at some interior point. Consequently, the maximum of $|Du|$ on the compact set $\overline{\Omega}$ is attained at some boundary point.

4 The Dirichlet Problem in Bounded Convex Domains

In this section we prove Corollary 1. Recall that the existence result of Serrin is also valid for the general family of Eq. (4). In order to make the paper self-contained, let us provide a proof following ideas of [40]. We apply the continuity method which requires the existence of C^0 and C^1 a priori estimates for a solution in order to provide the necessary compactness properties. These will be derived by proving that u admits barriers from above and from below along $\partial\Omega$. Higher order regularities of solutions hold under smoothness hypothesis: [15, Theorems 6.17, 6.19, 13.8].

Theorem 2 *Let $\Omega \subset \mathbb{R}^2$ be a bounded $C^{2,\alpha}$-domain whose curvature κ of the boundary Ω with respect to the inward normal direction satisfies $\kappa \geq 0$. If $\varphi \in C^0(\partial\Omega)$, then there is a unique solution of (1)–(3).*

In Proposition 10, we found height estimates for u and we proved that the interior gradient estimates are obtained once we have gradient estimates of u along $\partial\Omega$. Thus, we now establish these estimates on the boundary.

Proposition 11 (Boundary gradient estimates) *Let $\Omega \subset \mathbb{R}^2$ be a bounded domain with C^2-boundary, $\kappa \geq 0$ and let $\varphi \in C^2(\partial\Omega)$. If $u \in C^2(\Omega) \cap C^1(\overline{\Omega})$ is a solution of (1)–(3), then there is a constant $C_2 = C_2(\Omega, C_1, \|\varphi\|_{2;\overline{\mathcal{N}_\epsilon}})$ such that*

$$\max_{\partial\Omega} |Du| \leq C_2,$$

where φ is extended to some tubular neighborhood $\mathcal{N}_\epsilon$ of $\partial\Omega$.

Proof We consider the operator $Q[u]$ defined in (15), which we write now as

$$Q[u] = a_{ij}u_{i;j} - (1 + |Du|^2), \quad a_{ij} = (1 + |Du|^2)\delta_{ij} - u_i u_j. \tag{18}$$

An upper barrier for u is obtained by considering the solution v^0 of the Dirichlet problem for the minimal surface equation in Ω with the same boundary data φ: the existence of v^0 is assured in [40]. Because $Q[v^0] < 0 = Q[u]$ and $v^0 = u$ on $\partial\Omega$, we conclude $v^0 > u$ in Ω by the comparison principle.

We now find a lower barrier for u. Here we use the distance function in a small tubular neighborhood of $\partial\Omega$ in Ω. Consider on $\overline{\Omega}$ the distance function to $\partial\Omega$, $d(x) = \mathrm{dist}(x, \partial\Omega)$ and let $\epsilon > 0$ be sufficiently small so $\mathcal{N}_\epsilon = \{x \in \overline{\Omega} : d(x) < \epsilon\}$ is a tubular neighborhood of $\partial\Omega$. We parametrize $\mathcal{N}_\epsilon$ using normal coordinates $x \equiv (t, \pi(x)) \in \mathcal{N}_\epsilon$, where we write $x = \pi(x) + t\nu(\pi(x))$ for some $t \in [0, \epsilon)$, $\pi : \mathcal{N}_\epsilon \to \partial\Omega$ is the orthogonal projection and ν is the unit normal vector to $\partial\Omega$ pointing to Ω. Among the properties of the function d, we know that d is C^2, $|Dd|(x) = 1$, and $\Delta d \leq -\kappa(\pi(x))$ for all $x \in \mathcal{N}_\epsilon$.

We extend φ on $\mathcal{N}_\epsilon$ by $\varphi(x) = \varphi(\pi(x))$. Define in $\mathcal{N}_\epsilon$ the function

$$w = -h \circ d + \varphi,$$

where

$$h(t) = a\log(1 + bt), \quad a = \frac{c}{\log(1 + b)},$$

where $b > 0$ will be chosen later. Here c is any constant with

$$c > 2\left(\|\varphi\|_0 - C_1\right), \tag{19}$$

and C_1 is the constant of (16). Here and subsequently, $\|\cdot\|$ denotes the norm computed in $\overline{\mathcal{N}_\epsilon}$. It is immediate that $h \in C^\infty([0, \infty))$, $h' > 0$ and $h'' = -h'^2/a$. The first and second derivatives of w are $w_i = -h'd_i + \varphi_i$ and $w_{i;j} = -h''d_i d_j - h'd_{i;j} + \varphi_{i;j}$. The computation of $Q[w]$ leads to

$$Q[w] = -h''a_{ij}d_i d_j - h'a_{ij}d_{i;j} + a_{ij}\varphi_{i;j} - (1 + |Dw|^2). \tag{20}$$

From $|Dd| = 1$, it follows that $\langle D(Dd)_x\xi, Dd(x)\rangle = 0$ for all $\xi \in \mathbb{R}^2$. If $\{v_1, v_2\}$ is the canonical basis of $\mathbb{R}^2$, by taking $\xi = v_i$, we find $d_{i;j}d_j = 0$. Thus

$$\begin{aligned} w_i w_j d_{i;j} &= (-h'd_i + \varphi_i)(-h'd_j + \varphi_j)d_{i;j} = (h'^2 d_i - 2h'\varphi_i)d_j d_{i;j} + \varphi_i\varphi_j d_{i;j} \\ &= \varphi_i\varphi_j d_{i;j} \geq |D\varphi|^2\Delta d, \end{aligned}$$

where the last inequality is due to $D^2 d$ is negative. Using this inequality and the definition of a_{ij} in (18), we derive

$$a_{ij}d_{i;j} = (1 + |Dw|^2)\Delta d - w_i w_j d_{i;j} \leq (1 + |Dw|^2 - |D\varphi|^2)\Delta d. \tag{21}$$

Notice that

$$|Dw|^2 = h'^2 + |D\varphi|^2 - 2h'\langle Dd, D\varphi\rangle. \tag{22}$$

Then

$$\begin{aligned}1 + |Dw|^2 - |D\varphi|^2 &= 1 + h'^2 - 2h'\langle D\varphi, Dd\rangle \geq 1 + h'^2 - 2h'|D\varphi| \\ &= 1 + \frac{c^2b^2}{\log(1+b)^2(1+bt)^2} - \frac{2cb}{\log(1+b)(1+bt)}|D\varphi| > 0\end{aligned}$$

if b is sufficiently large, with b a constant depending on $\partial\Omega$, c and $|D\varphi|$. Since $\Delta d \leq 0$ because D^2d is negative, we deduce from (21) that $a_{ij}d_{i;j} \leq 0$.

The ellipticity of $A = (a_{ij})$ can be written as $|\xi|^2 \leq a_{ij}\xi_i\xi_j \leq (1 + |Dw|^2)|\xi|^2$ for all $\xi \in \mathbb{R}^2$. Taking $\xi = Dd$, then $1 = |Dd|^2 \leq a_{ij}d_id_j \leq (1 + |Dw|^2)$. Since $h'' < 0$, we have

$$h''(a_{ij}d_id_j) \leq h''. \tag{23}$$

On the other hand, if $\cdot$ is the usual scalar product in the set of the square matrix,

$$|A|^2 = A \cdot A = 1 + (1 + |Dw|^2)^2 \leq 2(1 + |Dw|^2)^2,$$

hence

$$a_{ij}\varphi_{i;j} = A \cdot D^2\varphi \geq -|A||D^2\varphi| \geq -\sqrt{2}|D^2\varphi|(1 + |Dw|^2).$$

By combining this inequality with $\Delta d \leq 0$, and inserting (21) and (23) in (20), we deduce

$$\begin{aligned}Q[w] &\geq -h'' - h'(1 + |Dw|^2 - |D\varphi|^2)\Delta d - (1 + \sqrt{2}|D^2\varphi|)(1 + |Dw|^2) \\ &\geq -h'' - (1 + \sqrt{2}|D^2\varphi|)(1 + |Dw|^2) \geq -h'' - \beta(1 + |Dw|^2),\end{aligned}$$

where $\beta = 1 + \sqrt{2}\|D^2\varphi\|_0$. Take b sufficiently large if necessary, to ensure that $1/a - \beta > 0$, so β depends on $\|D^2\varphi\|_0$, C_1 and $\|\varphi\|_0$. Using that $h'' = -h'^2/a$ and (22), we obtain

$$\begin{aligned}Q[w] &\geq \frac{h'^2}{a} - \beta(1 + |Dw|^2) = \left(\frac{1}{a} - \beta\right)h'^2 + 2h'\beta\langle Dd, D\varphi\rangle - \beta(1 + \|D\varphi\|_0^2) \\ &\geq \left(\frac{1}{a} - \beta\right)h'^2 - 2h'\beta\|D\varphi\|_0 - \beta(1 + \|D\varphi\|_0^2) \\ &= \left(\frac{1}{a} - \beta\right)\frac{a^2b^2}{(1+bt)^2} - 2\beta\frac{ab}{1+bt}\|D\varphi\|_0 - \beta(1 + \|D\varphi\|_0^2).\end{aligned} \tag{24}$$

We write the last term as a function on $\mathcal{N}_\epsilon$, namely, $g(x) = g(t, \pi(x))$. At $t = 0$,

$$
\begin{aligned}
g(0) &= \left(\frac{1}{a} - \beta\right) \frac{c^2 b^2}{\log(1+b)^2} - 2\beta \frac{cb}{\log(1+b)} \|D\varphi\|_0 - \beta(1 + \|D\varphi\|_0^2) \\
&= \frac{cb}{\log(1+b)} \left(\left(\frac{1}{a} - \beta\right) \frac{cb}{\log(1+b)} - 2\beta \|D\varphi\|_0 \right) - \beta(1 + \|D\varphi\|_0^2)
\end{aligned}
$$

Therefore, if b is sufficiently large, $g(0) > 0$. Since $\partial\Omega$ is compact, by an argument of continuity, b can be chosen sufficiently large to ensure that $g(t) > 0$ in $\mathcal{N}_\epsilon$. For this choice of b, we find $Q[w] > 0$.

In order to assure that w is a local lower barrier in $\mathcal{N}_\epsilon$, we have to see that

$$
w \leq u \quad \text{in } \partial\mathcal{N}_\epsilon. \tag{25}
$$

In $\partial\mathcal{N}_\epsilon \cap \partial\Omega$, the distance function is $d = 0$, so $w = \varphi = u$. On the other hand, let us further require b large enough so $\log(1 + b\epsilon)/\log(1 + b) \geq 1/2$. Then in $\partial\mathcal{N}_\epsilon \setminus \partial\Omega$, we find from (19) that

$$
\begin{aligned}
w &= -c \frac{\log(1+b\epsilon)}{\log(1+b)} + \varphi \leq -\frac{\|\varphi\|_0 - C_1}{2} \frac{\log(1+b)}{\log(1+b\epsilon)} + \varphi \\
&\leq C_1 - \|\varphi\|_0 + \varphi \leq C_1 \leq u
\end{aligned}
$$

in $\partial\mathcal{N}_\epsilon \setminus \partial\Omega$. Definitively, (25) holds in $\partial\mathcal{N}_\epsilon \setminus \partial\Omega$. Because $Q[w] > 0 = Q[u]$, we conclude $w \leq w$ in $\mathcal{N}_\epsilon$ by the comparison principle.

Consequently, we have proved the existence of lower and upper barriers for u in $\mathcal{N}_\epsilon$, namely, $w \leq u \leq v^0$. Hence

$$
\max_{\partial\Omega} |Du| \leq C_2 := \max\{\|Dw\|_{0;\partial\Omega}, \|Dv^0\|_{0;\partial\Omega}\}
$$

and both values $\|Dw\|_{0;\partial\Omega}$, $\|Dv^0\|_{0;\partial\Omega}$ depend only on Ω, C_1 and φ. This completes the proof of proposition.

Remark 4 1. It is possible, instead the function v^0, to use $w = h \circ d + \varphi$ for an upper barrier of u.

2. The use of the auxiliary function $h(d) = a \log(1 + bd)$ for obtaining boundary gradient estimates is standard in the theory of elliptic equations (see [15, Chap. 14] as a general reference). It should also be mentioned that Bernstein was the first author whose employed this function to construct barriers for solutions in elliptic equations in two variables, assuming analytic hypothesis: [4, pp. 265–6].

Proof (of Theorem 2)
In a first step, we demonstrate the theorem when $\varphi \in C^{2,\alpha}(\overline{\Omega})$. We establish the solvability of the Dirichlet problem (1)–(3) by applying a slightly modification of the method of continuity [15, Sect. 17.2]. Define the family of Dirichlet problems parametrized by $t \in [0, 1]$ by

$$\begin{cases} Q_t[u] = 0 \text{ in } \Omega \\ u \quad = \varphi \text{ on } \partial\Omega, \end{cases}$$

where

$$Q_t[u] = (1+|Du|^2)\Delta u - u_i u_j u_{i;j} - t(1+|Du|^2).$$

As usual, let

$$\mathcal{A} = \{t \in [0,1] : \exists u_t \in C^{2,\alpha}(\overline{\Omega}),\ Q_t[u_t] = 0,\ u_{t|\partial\Omega} = \varphi\}.$$

The theorem is established if $1 \in \mathcal{A}$. For this purpose, we prove that $\mathcal{A}$ is a non-empty open and closed subset of $[0, 1]$.

1. The set $\mathcal{A}$ is not empty. Let us observe that $0 \in \mathcal{A}$ because the minimal solution v^0 defined in Proposition 11 corresponds with $t = 0$.
2. The set $\mathcal{A}$ is open in $[0, 1]$. Given $t_0 \in \mathcal{A}$, we need to prove that there exists $\epsilon > 0$ such that $(t_0 - \epsilon, t_0 + \epsilon) \cap [0, 1] \subset \mathcal{A}$. Define the map $T(t, u) = Q_t[u]$ for $t \in \mathbb{R}$ and $u \in C^{2,\alpha}(\overline{\Omega})$. Then $t_0 \in \mathcal{A}$ if and only if $T(t_0, u_{t_0}) = 0$. If we show that the derivative of Q_t with respect to u, say $(DQ_t)_u$, at the point u_{t_0} is an isomorphism, it follows from the Implicit Function Theorem the existence of an open set $\mathcal{V} \subset C^{2,\alpha}(\overline{\Omega})$, with $u_{t_0} \in \mathcal{V}$ and a C^1 function $\psi : (t_0 - \epsilon, t_0 + \epsilon) \to \mathcal{V}$ for some $\epsilon > 0$, such that $\psi(t_0) = u_{t_0} > 0$ and $T(t, \psi(t)) = 0$ for all $t \in (t_0 - \epsilon, t_0 + \epsilon)$: this guarantees that $\mathcal{A}$ is an open set of $[0, 1]$.
   To show that $(DQ_t)_u$ is one-to-one is equivalent that say that for any $f \in C^{\alpha}(\overline{\Omega})$, there is a unique solution $v \in C^{2,\alpha}(\overline{\Omega})$ of the linear equation $Lv := (DQ_t)_u(v) = f$ in Ω and $v = \varphi$ on $\partial\Omega$. The computation of L was done in Proposition 10, namely,
   $$Lv = (DQ_t)_u v = a_{ij}(Du)v_{i;j} + \mathcal{B}_i(Du, D^2u)v_i,$$
   where a_{ij} is as in (15) and $\mathcal{B}_i = 2(u_i\Delta u - u_j u_{i;j} - tu_i)$. The existence and uniqueness is assured by standard theory [15, Theorem 6.14].
3. The set $\mathcal{A}$ is closed in $[0, 1]$. Let $\{t_k\} \subset \mathcal{A}$ with $t_k \to t \in [0, 1]$. For each $k \in \mathbb{N}$, there is $u_k \in C^{2,\alpha}(\overline{\Omega})$ such that $Q_{t_k}[u_k] = 0$ in Ω and $u_k = \varphi$ in $\partial\Omega$. Define the set
   $$\mathcal{S} = \{u \in C^{2,\alpha}(\overline{\Omega}) : \exists t \in [0,1] \text{ such that } Q_t[u] = 0 \text{ in } \Omega,\ u_{|\partial\Omega} = \varphi\}.$$
   Then $\{u_k\} \subset \mathcal{S}$. If we see that the set $\mathcal{S}$ is bounded in $C^{1,\beta}(\overline{\Omega})$ for some $\beta \in [0, \alpha]$, and since $a_{ij} = a_{ij}(Du)$ in (15), the Schauder theory proves that $\mathcal{S}$ is bounded in $C^{2,\beta}(\overline{\Omega})$, in particular, $\mathcal{S}$ is precompact in $C^2(\overline{\Omega})$ (Theorem 6.6 and Lemma 6.36 in [15]). Hence there is a subsequence $\{u_{k_l}\} \subset \{u_k\}$ converging to some $u \in C^2(\overline{\Omega})$ in $C^2(\overline{\Omega})$. Since $T : [0, 1] \times C^2(\overline{\Omega}) \to C^0(\overline{\Omega})$ is continuous, we obtain $Q_t[u] = T(t, u) = \lim_{l\to\infty} T(t_{k_l}, u_{k_l}) = 0$ in Ω. Moreover, $u_{|\partial\Omega} = \lim_{l\to\infty} u_{k_l|\partial\Omega} = \varphi$ on $\partial\Omega$, so $u \in C^{2,\alpha}(\overline{\Omega})$ and consequently, $t \in \mathcal{A}$.

Definitively, $\mathcal{A}$ is closed in [0, 1] provided we find a constant M independent on $t \in \mathcal{A}$, such that

$$\|u_t\|_{C^1(\overline{\Omega})} = \sup_{\Omega} |u_t| + \sup_{\Omega} |Du_t| \leq M.$$

Let $t_1 < t_2$, $t_i \in [0, 1]$, $i = 1, 2$. Then $Q_{t_1}[u_{t_1}] = 0$ and

$$Q_{t_1}[u_{t_2}] = (t_2 - t_1)(1 + |Du_{t_2}|^2) > 0 = Q_{t_1}[u_{t_1}].$$

Since $u_{t_1} = u_{t_2}$ on $\partial\Omega$, the comparison principle yields $u_{t_2} < u_{t_1}$ in Ω. This proves that the solutions u_{t_i} are ordered in decreasing sense according the parameter t. It turns out that $u_1 \leq u_t < v^0$ for all t, where u_1 is the solution of (1)–(3). According to (29), we have $C_1 \leq u_t \leq \sup_\Omega u_0 \leq \max_{\partial\Omega} \varphi$ and we conclude

$$\|u_t\|_{0;\overline{\Omega}} \leq C_3, \quad C_3 = \max\{|C_1|, \|\varphi\|_{0;\partial\Omega}\}. \tag{26}$$

In order to find the desired gradient estimates for the solution u_t, by Proposition 10, we have to find estimates of $|Du_t|$ along $\partial\Omega$. On the other hand, the same computations given in Proposition 11 conclude that $\sup_{\partial\Omega} |Du_t|$ is bounded by a constant depending on Ω, φ and $\|u_t\|_{0;\overline{\Omega}}$. However, and by using (26), the value $\|u_t\|_{0;\overline{\Omega}}$ is bounded by C_3, which depends only on φ and Ω.

Until here, we have proved the part of existence in Theorem 2. The uniqueness is a consequence of Proposition 10 and this completes the proof of theorem if $\varphi \in C^2(\partial\Omega)$.

Finally we suppose $\varphi \in C^0(\partial\Omega)$. Let $\{\varphi_k^+\}, \{\varphi_k^-\} \in C^{2,\alpha}(\partial\Omega)$ be a monotonic sequence of functions converging from above and from below to φ in the C^0 norm. By virtue of the first part of this proof, there are solutions $u_k^+, u_k^- \in C^{2,\alpha}(\overline{\Omega})$ of the translating soliton equation (1) such that $u^+_{k\,|\partial\Omega} = \varphi_k^+$ and $u^-_{k\,|\partial\Omega} = \varphi_k^-$. By the comparison principle, we find

$$u_1^- \leq \ldots \leq u_k^- \leq u_{k+1}^- \leq \ldots \leq u_{k+1}^+ \leq u_k^+ \leq \ldots \leq u_1^+$$

for every k, hence the sequences $\{u_k^\pm\}$ are uniformly bounded in the C^0 norm. By the proof of Theorem 2, the sequences $\{u_k^\pm\}$ have a priori C^1 estimates depending only on Ω and φ. Using classical Schauder theory again [15, Theorem 6.6], the sequence $\{u_k^\pm\}$ contains a subsequence $\{v_k\} \in C^{2,\alpha}(\overline{\Omega})$ converging uniformly on the C^2 norm on compacts subsets of Ω to a solution $u \in C^2(\Omega)$ of (1). Since $\{u^\pm_{k\,|\partial\Omega}\} = \{\varphi_k^\pm\}$ and $\{\varphi_k^\pm\}$ converge to φ, we conclude that u extends continuously to $\overline{\Omega}$ and $u_{|\partial\Omega} = \varphi$.

We finish this section given an 'geometric alternative' proof of Theorem 2 in a particular case. We notice that the core problem in the proof of Theorem 2 is obtaining the gradient estimates along $\partial\Omega$. These estimates were proved by means of the distance function d. We can say, in some sense, that the distance function allows to manage the convexity of $\partial\Omega$. However we want to use explicit examples of translating solitons, such as bowl solitons or grim reapers, as barriers to derive the estimates of $|Du|$ along $\partial\Omega$.

We prove the particular case of Theorem 2 when $\varphi = 0$ and we will use pieces of grim reapers as comparison surfaces. By Proposition 10, $C_1 < u \leq 0$. Consider the grim reaper $w(x, y) = -\log(\cos(y))$, $|y| < \pi/2$. We know that $w(x, 0) = 0$ and $\lim_{y\to\pi/2} w(x, y) = \infty$. Let $c > 0$ sufficiently close to $\pi/2$ with the property that $w(x, c) = -C_1$. Now we restrict w to the strip $|y| \leq c$ and consider $\Sigma_w^+ = \{w(x, y) : 0 \leq y \leq c\}$. Then Σ_w^+ is the half-grim reaper defined in the strip $0 \leq y \leq c$ with the property that the height of Σ_w^+, namely, $w(x, c) - w(x, 0)$, is $-C_1$. In particular, this height is strictly bigger that the one of Σ_u, that is,

$$\sup_\Omega u - \inf_\Omega u < -C_1.$$

Let us observe that the boundary of Σ_w^+ is formed by two horizontal parallel lines, namely,

$$\partial\Sigma_w^+ = L_1 \cup L_2 = \{(x, 0, 0) : x \in \mathbb{R}\} \cup \{(x, c, -C_1) : x \in \mathbb{R}\}.$$

Let p be an arbitrary fixed point of $\partial\Omega$. After a rotation about the z-axis and a horizontal translation, we suppose that Ω lies in the halfplane $y \leq 0$ and that the tangent line of $\partial\Omega$ at p is the x-axis of the xy-plane. We displace vertically down Σ_w^+ until that L_2 lies at the height $z = 0$, that is, L_2 coincides with the x-axis, and thus L_1 is situated to the height $z = C_1$. We keep the same notation for Σ_w^+, L_1 and L_2. We denote by $\tilde{w} = w + C_1$ the function whose graph is the displaced grim reaper Σ_w^+. We move horizontally Σ_w^+ in the direction of the vector $(0, 1, 0)$ until that Σ_w^+ does not intersect Σ_u. Then we come back Σ_w^+ in the direction of the vector $(0, -1, 0)$ until the first intersection point with Σ_u. Since $C_1 < u \leq 0$, it is not possible that L_1 touches touches $\partial\Sigma_u$. Thus, and by the touching principle, the first intersection point must occur between a point of L_2 and a boundary point Σ_u. Because $\partial\Omega$ is convex, necessarily this occurs along the x-axis. In particular, $p \in L_2 \cap \partial\Sigma_u$. Consequently, the function $\tilde{w} < u$ in a neighborhood of p. Since $\varphi = 0$, this means that the outward unit conormal derivative of $\tilde{w}$ is strictly bigger than the one of u at the point p, which can be expressed as

$$\langle \nu_u(p), \mathbf{a}\rangle < \langle \nu_{\tilde{w}}(p), \mathbf{a}\rangle,$$

where ν_u and $\nu_{\tilde{w}}$ are the outward unit conormal vectors along $\partial\Sigma_u$ and $\partial\Sigma_w^+$ respectively. In terms of the gradients of the functions, this inequality is

$$\frac{|Du|}{\sqrt{1 + |Du|^2}}(p) < \frac{|D\tilde{w}|}{\sqrt{1 + |D\tilde{w}|^2}}(p).$$

Hence, $|Du|(p) < |D\tilde{w}|(p)$. But $|D\tilde{w}|(p)|$ coincides with $|Dw(p)|$ for the points $p = (x, c)$ such that $w(p) = -C_1$. This gives the desired gradient a priori estimate of $|Du|$ along $\partial\Omega$.

This idea to get gradient a priori estimates by comparison the solution with known explicit solutions was used by the author to derive existence of the Dirichlet problem for the constant mean curvature equation in Euclidean space by using half-cylinders or spherical caps: see [25–27, 33].

5 The Dirichlet Problem for the Constant Weighted Mean Curvature

In this section we solve the Dirichlet problem for the case that H_ϕ is constant in (5):

$$\operatorname{div}\frac{Du}{\sqrt{1+|Du|^2}} = \frac{1}{\sqrt{1+|Du|^2}} + \mu \quad \text{in } \Omega \tag{27}$$

$$u = \varphi \quad \text{on } \partial\Omega, \tag{28}$$

where μ is a constant and $u \in C^2(\Omega) \cap C^0(\overline{\Omega})$. The motivation of this problem is twofold. First, Eq. (27) is the analogous to the constant mean curvature equation in Euclidean space in the context of manifolds with density, whereas (1) corresponds with the minimal surface equation. Second, the solvability of the constant mean curvature equation holds for any φ if $\kappa \geq 2|H| \geq 0$ [40] and the next result for (27)–(28) establishes a similar result.

Theorem 3 *Let $\Omega \subset \mathbb{R}^2$ be a bounded $C^{2,\alpha}$-domain with inward curvature κ. If $\kappa \geq \mu \geq 0$ and $\varphi \in C^0(\partial\Omega)$, then there is a unique solution of (27)–(28).*

Notice that the solvability of this Dirichlet problem was proved in [24] in a context more general where the domain may be not bounded. As a difference, the present proof uses a comparison argument with rotational surfaces in order to obtain the C^0 estimates. The uniqueness is again a consequence of the maximum principle for the Eq. (27). After Theorem 2, we assume that $\mu > 0$. The proof now differs only in minor details from than of the preceding section, which are left to the reader. We point out that the assumption $\mu > 0$ will be used strongly.

Lemma 1 *If $\Omega \subset \mathbb{R}^2$ is a bounded domain with diam$(\Omega) < 1/\mu$, then there is a constant $C_1 = C_1(\varphi, \Omega)$ such that if u is a solution of (27)–(28), then*

$$C_1 \leq u \leq \max_{\partial\Omega} \varphi \quad \text{in } \Omega. \tag{29}$$

Proof Using $\mu > 0$, the right-hand side of (27) is non-negative, the maximum principle implies $\sup_\Omega u = \max_{\partial\Omega} u = \max_{\partial\Omega} \varphi$, proving the inequality in the right-hand side of (29). The lower estimate for u in (29) is obtained by using radial solutions of (27). It was proved in [28] that if $\mu > 0$, any radial solution intersecting the rotational axis (and necessarily perpendicularly) converges to a right circular cylinder of radius $1/\mu$. More exactly, let $D_r \subset \mathbb{R}^2$ be a disc centered at the origin of radius

r. If $\mu > 1/2$, there is a radial solution of (27) on D_r for some $r_0 > 1/\mu$ and if $0 < \mu \leq 1/2$, there is a radial solution of (27) on D_r for any $r < 1/\mu$.

Since $\mathrm{diam}(\Omega) < 1/\mu$, let $r > 0$ such that $\mathrm{diam}(\Omega) < r < 1/\mu$ and denote by v the radial solution of (27) on D_r with $v = 0$ on ∂D_r. After a horizontal translation if necessary, we suppose $\Omega \subset D_r$. Now the argument works the same as in Proposition 10 with the graph Σ_v, where now $C_1 = v(0) + \min_{\partial\Omega} \varphi$.

Lemma 2 (Interior gradient estimates) *If u is a solution of (27)–(28), then*

$$\sup_{\Omega} |Du| = \max_{\partial\Omega} |Du|.$$

Proof Now the corresponding Eq. (17) for (27) is

$$\left(\left(1 + |\nabla v|^2\right) \delta_{ij} - v_i v_j\right) z^k_{i;j} + 2\left(v_i \Delta v - v_{i;j} u_i - v_i - \frac{3}{2}\mu(1 + |\nabla v|^2)\right) z^k_i = 0, \tag{30}$$

and the arguments are similar.

Lemma 3 (Boundary gradient estimates) *Let $\Omega \subset \mathbb{R}^2$ be a bounded domain with C^2-boundary, $\kappa \geq \mu > 0$ and let $\varphi \in C^2(\partial\Omega)$. If $u \in C^2(\Omega) \cap C^1(\overline{\Omega})$ is a solution of (27)–(28), then there is a constant $C_2 = C_2(\Omega, C_1, \|\varphi\|_2)$ such that*

$$\max_{\partial\Omega} |Du| \leq C_2.$$

Proof We consider the operator

$$Q[u] = a_{ij} u_{i;j} - (1 + |Du|^2) - \mu(1 + |Du|^2)^{3/2}. \tag{31}$$

The minimal solution v^0 is an upper barrier for u. For the lower barrier for u, we use again the function $w = -h \circ d + \varphi$. Now

$$Q[w] = -h'' a_{ij} d_i d_j - h' a_{ij} d_{i;j} + a_{ij} \varphi_{i;j} - (1 + |Dw|^2) - \mu(1 + |Dw|^2)^{3/2}. \tag{32}$$

Taking into account $(1 + |Dw|^2)^{1/2} \leq 1 + |Dw|$ and $|Dw|^2 \leq h'^2 + |D\varphi|^2 + 2h'|D\varphi| \leq (h' + |D\varphi|)^2$, we deduce from (32)

$$\begin{aligned} Q[w] \geq& -h'' - h'(1 + |Dw|^2 - |D\varphi|^2)\Delta d - (1 + \sqrt{2}|D^2\varphi|)(1 + |Dw|^2) \\ & -\mu(1 + |Dw|^2)^{3/2} \geq -h'' - h'(1 + |Dw|^2 - |D\varphi|^2)\Delta d \\ & -(1 + \sqrt{2}|D^2\varphi|)(1 + |Dw|^2) - \mu(1 + |Dw|^2)(1 + |Dw|) \\ \geq& -h'' - h'(1 + |Dw|^2 - |D\varphi|^2)(\Delta d + \mu) - \mu h'|D\varphi|^2 \\ & - (\mu(1 + |D\varphi|) + 1 + \sqrt{2}|D^2\varphi|)(1 + |Dw|^2). \end{aligned}$$

Let $\beta = \mu(1 + \|D\varphi\|_0) + 1 + \sqrt{2}\|D^2\varphi\|_0$. Since $\Delta d + \mu \leq -\kappa + \mu \leq 0$, it follows

$$Q[w] \geq \frac{h'^2}{a} - \beta(1 + |Dw|^2) - \mu h'|D\varphi|^2$$
$$\geq \left(\frac{1}{a} - \beta\right) h'^2 - h'(2\beta\|D\varphi\|_0 + \mu\|D\varphi\|_0^2) - \beta(1 + \|D\varphi\|_0^2).$$

The rest of the proof runs as in Proposition 11.

With the help of the preceding three lemmas we can now prove Theorem 3.

Proof (of Theorem 3)
For the method of continuity, let

$$Q_t[u] = (1 + |Du|^2)\Delta u - u_i u_j u_{i;j} - (1 + |Du|^2) - t\mu(1 + |Du|^2)^{3/2},$$

and

$$\mathcal{A} = \{t \in [0, 1] : \exists u_t \in C^{2,\alpha}(\overline{\Omega}),\ Q_t[u_t] = 0,\ u_{t|\partial\Omega} = \varphi\}.$$

The set $\mathcal{A}$ is not empty because the solution of Theorem 2 corresponds with the value $t = 0$. For the openness of $\mathcal{A}$, the computation of L leads to

$$Lv = (DQ_t)_u v = a_{ij}(Du)v_{i;j} + \mathcal{B}_i(Du, D^2u)v_i,$$

where $\mathcal{B}_i = 2(u_i\Delta u - u_j u_{i;j} - u_i - 3t(1 + |Du|^2)/2)$. Then the proof works again.

Finally, we show that the set $\mathcal{A}$ is closed in $[0, 1]$. For the height and gradient estimates for u_t, we use Lemmas 1–3. The arguments are similar once we prove that the solutions u_{t_i} are ordered in decreasing sense. If $t_1 < t_2$, then $Q_{t_1}[u_{t_1}] = 0$ and

$$Q_{t_1}[u_{t_2}] = (t_2 - t_1)\mu(1 + |Du_{t_2}|^2)^{3/2} > 0 = Q_{t_1}[u_{t_1}].$$

Since $u_{t_1} = u_{t_2}$ on $\partial\Omega$, the comparison principle yields $u_{t_2} < u_{t_1}$ in Ω.

6 The Dirichlet Problem in Unbounded Domains

We study in this section the Dirichlet problem (1)–(3) in unbounded convex domains contained in a strip. We have two cases depending if the domain is or is not a strip.

The first result assumes that Ω is a strip. In such a case, the motivation comes from the grim reapers that appeared in (8). For each θ, the surface Σ_{w_θ} is a graph defined in the (maximal) strip Ω^θ, with $w_\theta(x, y) \to +\infty$ as $|y| \to \pi/(2\cos\theta)$. If we narrow the strip to $|y| < b$, with $0 < b < \pi/(2\cos\theta)$, then the value of w_θ on $|y| = b$ is the linear function $x \mapsto \varphi(x, \pm b) = w_\theta(x, b)$ and $\partial\Sigma_{w_\theta}$ is formed by two parallel straight lines.

Our purpose is to consider the Dirichlet problem when Ω is a strip and φ is formed by two copies of a convex function. Let $\Omega_m = \{(x, y) \in \mathbb{R}^2 : -m < y < m\}$,

$m > 0$, be the strip of width $2m$. For each smooth convex function f defined in $\mathbb{R}$, we extend f to a function φ_f on $\partial\Omega_m$ by $\varphi_f(x, \pm m) = f(x)$. The result of existence is established by our next theorem [31].

Theorem 4 *If $m < \pi/2$, then for each convex function f, there is a solution of (1)–(3) for boundary values φ_f on $\partial\Omega_m$.*

The proof uses the classical Perron method of sub and supersolutions: see [12, pp. 306–312], [15, Sect. 6.3]). We consider the operator Q defined in (15), where we know that $Q[u] = 0$ if and only if u is a solution of the translating soliton equation. The existence result of Theorem 2 holds in disks, so we can proceed to apply the Perron process when the domain is a strip.

First we need a subsolution of (1)–(3). In the following result, f is not necessarily a convex function [9].

Proposition 12 *Let $\Omega_m \subset \mathbb{R}^2$ be a strip. If f is a continuous function defined in $\mathbb{R}$, then there is a solution v^0 of the Dirichlet problem*

$$\begin{aligned} \operatorname{div}\left(\frac{Du}{\sqrt{1+|Du|^2}}\right) &= 0 \quad \text{in } \Omega_m \\ u = \varphi_f \quad &\text{on } \partial\Omega_m \end{aligned} \tag{33}$$

with the property $f(x) < v^0(x, y)$ for all $(x, y) \in \Omega_m$.

Let $u \in C^0(\overline{\Omega_m})$ be a continuous function and let D be a closed round disk in Ω_m. We denote by $\bar{u} \in C^2(D)$ the unique solution of the Dirichlet problem

$$\begin{cases} Q[\bar{u}] = 0 & \text{in } D \\ \bar{u} = u & \text{on } \partial D, \end{cases}$$

whose existence and uniqueness is assured by Theorem 2. We extend $\bar{u}$ to Ω_m by continuity as

$$M_D[u] = \begin{cases} \bar{u} & \text{in } D \\ u & \text{in } \Omega_m \setminus D. \end{cases}$$

The function u is said to be a *supersolution* in Ω_m if $M_D[u] \leq u$ for every closed round disk D in Ω_m. For example, for any domain $\Omega \subset \mathbb{R}^2$, the function $u = 0$ in $\overline{\Omega}$ is a supersolution in Ω. Indeed, if $D \subset \Omega$ is a closed round disk, then $\bar{u} < 0$ since $Q[0] = -1 < 0 = Q[\bar{u}]$ and the comparison principle applies. Thus $M_D[u] \leq 0$.

On the other hand, for each $p \in \Omega$, there is a supersolution u with $u(p) < 0$. To this end, consider $D \subset \Omega$ a closed round disk centered at the origin of $\mathbb{R}^2$. Let $\mathbf{b} = \mathbf{b}(r)$ be the bowl soliton with $\mathbf{b}_{|\partial D} = 0$. Then the function u defined as $u = \mathbf{b}$ in D and $u = 0$ in $\overline{\Omega} \setminus D$ is a supersolution.

Definition 3 A function $u \in C^0(\overline{\Omega_m})$ is called a superfunction relative to f if u is a supersolution in Ω_m and $f \leq u$ on $\partial\Omega_m$. Denote by $\mathcal{S}_f$ the class of all superfunctions relative to f,

$$\mathcal{S}_f = \{u \in C^0(\overline{\Omega_m}) : M_D[u] \leq u \text{ for every closed disk } D \subset \Omega_m,\ f \leq u \text{ on } \partial\Omega_m\}.$$

Lemma 4 *The set $\mathcal{S}_f$ is not empty.*

Proof We claim that $v^0 \in \mathcal{S}_f$, where v^0 is the minimal solution given in Proposition 12. Let $D \subset \Omega_m$ be a closed round disk. Since v^0 is a minimal surface, $Q[v^0] = -(1 + |Dv^0|^2) < 0$ and because $\overline{v^0} = v^0$ in ∂D, the comparison principle implies $M_D[v^0] = \overline{v^0} \leq v^0$ in D. On the other hand, $v^0 = f$ on $\partial\Omega_m$, proving definitively that $v^0 \in \mathcal{S}_f$.

We now give some properties about superfunctions whose proofs are straightforward: in the case of the constant mean curvature equation, we refer [26]; in the context of translating solitons, see [22, Lemmas 4.2–4.4].

Lemma 5 *1. If $\{u_1, \ldots, u_n\} \subset \mathcal{S}_f$, then* $\min\{u_1, \ldots, u_n\} \in \mathcal{S}_f$.
2. The operator M_D is increasing in $\mathcal{S}_f$.
3. If $u \in \mathcal{S}_f$ and D is a closed round disk in Ω_m, then $M_D[u] \in \mathcal{S}_f$.

Consider the family of grim reaper w_θ of (8). Since w_θ is defined in the strip Ω^θ and, by assumption, $m < \pi/2$, then $\Omega_m \subset \Omega^0 \subset \Omega^\theta$ for any θ. Thus it makes sense to restrict w_θ to the strip Ω_m and we keep the same notation for its restriction in Ω_m. Consequently w_θ is a linear function on $\partial\Omega_m$ and $\partial\Sigma_{w_\theta}$ consists of two parallel lines.

Consider the subfamily of grim reapers

$$\mathcal{G} = \{w_\theta : w_\theta \leq f \text{ on } \partial\Omega_m, \theta \in (-\pi/2, \pi/2)\}.$$

Notice that the set $\mathcal{G}$ is not empty because f is convex. Furthermore, the minimal surface v^0 with $v^0 = f$ on $\partial\Omega_m$ satisfies $Q[v^0] < 0 = Q[w_\theta] = 0$ for all $w_\theta \in \mathcal{G}$. Hence, the comparison principle asserts that $w_\theta < v^0$ in Ω_m for all $\theta \in (-\pi/2, \pi/2)$.

We now construct a solution of Eq. (1) between the grim reapers of $\mathcal{G}$ and the minimal surface v^0. Let

$$\mathcal{S}_f^* = \{u \in \mathcal{S}_f : w_\theta \leq u \leq v^0, \text{ for every } w_\theta \in \mathcal{G}\}.$$

We point out that $\mathcal{S}_f^*$ is not empty because $v^0 \in \mathcal{S}_f^*$. By using the maximum principle, it is not difficult to see that set $\mathcal{S}_f^*$ is stable for the operator M_D, that is, if $u \in \mathcal{S}_f^*$, then $M_D[u] \in \mathcal{S}_f^*$. The key point is the next proposition.

Proposition 13 (Perron process) *The function $v : \Omega_m \to \mathbb{R}$ given by*

$$v(x, y) = \inf\{u(x, y) : u \in \mathcal{S}_f^*\}$$

is a solution of (1) with $v = \varphi_f$ on $\partial\Omega_m$.

Proof The proof consists of two parts.

Claim 1. *The function v is a solution of Eq.* (1).

The proof is standard and here we follow [15]. Let $p \in \Omega_m$ be an arbitrary fixed point of Ω_m. Consider a sequence $\{u_n\} \subset \mathcal{S}_f^*$ such that $u_n(p) \to v(p)$ when $n \to \infty$. Let D be a closed round disk centered at p and contained in Ω_m. For each n, define on $\overline{\Omega_m}$ the function

$$v_n(q) = \min\{u_1(q), \dots, u_n(q)\}, \quad q \in \overline{\Omega_m}.$$

Then $v_n \in \mathcal{S}_f^*$ by Lemma 5. Since $M_D[v_n] \in \mathcal{S}_f^*$, we deduce $M_D[v_n](p) \to v(p)$ as $n \to \infty$. Set $V_n = M_D[v_n]$. Then $\{V_n\}$ is a decreasing sequence bounded from below by w_θ for all $w_\theta \in \mathcal{G}$ and satisfying (1) in the disk D. It turns out that the functions V_n are uniformly bounded on compact sets K of D. In each compact set K, the norms of the gradients $|DV_n|$ are bounded by a constant depending only on K and using Hölder estimates of Ladyzhenskaya and Ural'tseva, there exist uniform $C^{1,\beta}$ estimates for the sequence $\{V_n\}$ on K [17]. By compactness, there is a subsequence of V_n, that we denote V_n again, such that $\{V_n\}$ converges on K to a C^2 function V in the C^2 topology and by continuity, V satisfies (1). Moreover, by construction, at the fixed point p we have $V(p) = v(p)$.

It remains to prove that $V = v$ in $\text{int}(D)$. For $q \in \text{int}(D)$, the same argument as before gives the existence of $\{\tilde{u}_n\} \subset \mathcal{S}_f^*$ with $\tilde{u}_n(q) \to v(q)$. Let $\tilde{v}_n = \min\{V_n, \tilde{u}_n\}$ and $\tilde{V}_n = M_D[\tilde{v}_n]$. Again $\tilde{V}_n$ converges on D in the C^2 topology to a C^2 function $\tilde{V}$ satisfying (1) and $\tilde{V}(q) = v(q)$. By construction, $\tilde{V}_n \leq \tilde{v}_n \leq V_n$, hence $\tilde{V} \leq V$. In view that $v \leq \tilde{V}$, we infer $\tilde{V}(p) = v(p) = V(p)$. Thus V and $\tilde{V}$ coincide at an interior point of D, namely, the point p, and both functions V and $\tilde{V}$ satisfy the translating soliton equation. Because $\tilde{V} \leq V$, the touching principle implies $V = \tilde{V}$ in $\text{int}(D)$. In particular, $V(q) = \tilde{V}(q) = v(q)$. This shows that $V = v$ in $\text{int}(D)$ and the claim is proved.

In order to finish the proof of Theorem 4, we prove that the function v takes the value φ_f on $\partial\Omega_m$ and consequently, v is continuous up to $\partial\Omega_m$ proving that $v \in C^2(\Omega_m) \cap C^0(\overline{\Omega_m})$. In contrast to with the proof of Theorem 2, here we will find local barriers for each boundary point $p \in \partial\Omega_m$.

Claim 2. *The function v is continuous up to $\partial\Omega_m$ with $v = \varphi_f$ on $\partial\Omega_m$.*

The graph of φ_f consists of two copies of f,

$$\Gamma_{\varphi_f} = \Gamma_1 \cup \Gamma_2 = \{(x, m, f(x)) : x \in \mathbb{R}\} \cup \{(x, -m, f(x)) : x \in \mathbb{R}\}.$$

Let $p = (x_0, m) \in \partial\Omega_m$ be a boundary point (similar argument if $p = (x_0, -m)$). Because of the convexity of f, in the plane of equation $y = m$ the tangent line L_p to the planar curve Γ_1 leaves Γ_1 above L_p. We choose the number θ such that the grim reaper w_θ takes the values L_p on $\partial\Omega_m$: exactly, θ is chosen so $\tan\theta$ is the slope of L_p. Recall that all rulings of this grim reaper are parallel to L_p. Let $w_\theta^p = w_\theta$ denote this grim reaper in order to indicate its dependence on the point p.

Taking into account the symmetry of φ_f and the convexity of f, we have $w_\theta^p(p) = f(x_0)$ and $w_\theta^p < f$ in $\Gamma_{\varphi_f} \setminus \{(x_0, m, f(x_0)), (x_0, -m, f(x_0))\}$, or in other words,

$\partial \Sigma_{w_\theta^p}$ lies strictly below $\partial \Sigma_v$, except at the points $(x_0, m, f(x_0)$ and $(x_0, -m, f(x_0))$, where both graphs coincide.

Therefore the function w_θ^p and the minimal surface v^0 form a modulus of continuity in a neighborhood of p, namely, $w_\theta^p \leq v \leq v^0$. Because $w_\theta^p(p) = v^0(p) = f(p)$, we infer that $v(p) = f(p)$ and this completes the proof of Theorem 4.

We finish this section with the second type of domains, that is, when Ω is an unbounded convex domain contained in a strip. Under this situation, we will suppose $\varphi = 0$ on $\partial\Omega$.

Theorem 5 *Let Ω be an unbounded convex domain contained in a strip of width strictly less than π. Then there is a solution of the translating soliton Eq. (1) in Ω with $\varphi = 0$ on $\partial\Omega$.*

Proof If Ω is a strip, then the result was established in Theorem 4. In fact, if $\Omega = \Omega_m$, $m < \pi/2$, the solution is $w(x, y) = -\log(\cos(y)) + \log(\cos(m))$.

Suppose now that Ω is not a strip. After a change of coordinates, we assume that the narrowest strip containing Ω is Ω_m. Since Ω is an unbounded domain contained in a strip, then $\partial\Omega$ has two branches asymptotic to the boundary set $\partial\Omega_m$ and the x-coordinate function is bounded in $\partial\Omega$ from above or from below.

We follow the same reasoning as in Theorem 4, and we only point out the differences. The subsolution is the function $v^0 = 0$, which is a solution of the minimal surface equation. We consider the family of operators M_D and

$$\mathcal{S} = \{u \in C^0(\overline{\Omega}) : M_D[u] \leq u \text{ for every closed round disk } D \subset \Omega, 0 \leq u \text{ on } \partial\Omega\}.$$

Let the grim reaper $w(x, y) = -\log(\cos(y))$ whose domain is the strip Ω^0 of width π and define $\omega(x, y) = -\log(\cos(y)) + \log(\cos(m))$. Note that $\omega = 0$ on $\partial\Omega_m$ and $\omega < 0$ on $\partial\Omega$ because $\Omega \subset \Omega_m$. We construct a solution of Eq. (1) between the grim reaper ω and the minimal surface v^0. Let

$$\mathcal{S}^* = \{u \in \mathcal{S} : \omega \leq u \leq 0\}.$$

Note that $\mathcal{S}^*$ is not empty because $0 \in \mathcal{S}^*$: indeed, $Q[0] = -1 < 0 = Q[\omega]$ and $\omega < v^0$ in $\partial\Omega$, hence $\omega < 0$ in Ω by the comparison principle. Again, the function

$$v(x, y) = \inf\{u(x, y) : u \in \mathcal{S}^*\} = \inf\{M_D[u](x, y) : u \in \mathcal{S}^*\}$$

is a solution of (1) and it remains to prove that the function v is continuous up to $\partial\Omega$ with $v = 0$ on $\partial\Omega$. Here the barrier construction in the proof of Theorem 4 can be adapted to provide boundary modulus of continuity estimates.

Let $p = (x_0, y_0) \in \partial\Omega$ be a boundary point of Ω. We rotate Ω with respect to the z-axis and translate along a horizontal direction if necessary, in such way that the tangent line L to Ω at p is one of the boundaries of Ω_m and a neighborhood U_p of p in Ω is contained in Ω_m: this is possible by the convexity of Ω. There is no loss of generality in assuming that $L = \{(x, m, 0) : x \in \mathbb{R}\}$. We take now the restriction of ω, $\omega^* = \omega_{|\Omega_m^*}$, in the half-strip $\Omega_m^* = \{(x, y) \in \mathbb{R}^2 : 0 < y \leq m\}$. Let Σ_{ω^*} be the graph of ω^*. Notice that $\partial\Sigma_{\omega^*}$ is formed by two parallel lines, one is L and the other one is $L' = \{(x, 0, \omega(0)) : x \in \mathbb{R}\}$.

Let $\mathbf{n}(p) = (0, 1, 0)$ be the unit outward normal vector to $\partial\Omega$ at p. Let us move horizontally Σ_{ω^*} in the direction $\mathbf{n}(p)$ until Σ_{ω^*} does not intersect Σ_v. Then we come back in the direction $-\mathbf{n}(p)$ until the first touching point q between Σ_{ω^*} and Σ_v. Since $\omega < v < 0$ in Ω, it is not possible that $q \in L'$. By the tangency principle, $q \in L$ and by the convexity of Ω, the point q coincides with p. Accordingly, we have proved that in the interior of the neighborhood U, we have $\omega < v < 0$. Since $\omega(p) = 0$, the functions ω and 0 are a modulus of continuity in U of p, hence $v(p) = 0$. This completes the proof of Theorem 5.

We point out that the domain Ω is not necessarily strictly convex. Thus in the last part of the above proof, the intersection between Σ_{ω^*} and Σ_v at the first touching point, may occur along a segment of L. In any case, we can take that a first contact point is the very point p.

References

1. Alexandrov, A.D.: Uniqueness theorems for surfaces in the large I. Vestnik Leningrand Univ. Math. **11**, 5–17 (1956)
2. Altschuler, S.J., Wu, L.F.: Translating surfaces of the non-parametric mean curvature flow with prescribed contact angle. Calc. Var. **2**, 101–111 (1994)
3. Bergner, M.: The Dirichlet problem for graphs of prescribed anisotropic mean curvature in R^{n+1}. Analysis (Munich) **28**, 149–166 (2008)
4. Bernstein, S.: Sur la généralisation du problème de Dirichlet. Math. Ann. **69**, 82–136 (1910)
5. Bernstein, S.: Sur les surfaces définies au moyen de leur courbure moyenne ou totale. Ann. Sci. École Norm. Sup. **27**, 233–256 (1910)
6. Bernstein, S.: Conditions nécessaires et suffisantes pour la possibilité du problème de Dirichlet. C. R. Math. Acad. Sci. Paris. **150**, 514–515 (1910)
7. Bereanu, C., Jebelean, P., Mawhin, J.: Radial solutions for some nonlinear problems involving mean curvature operators in Euclidean and Minkowski spaces. Proc. Am. Math. Soc. **137**, 161–169 (2009)
8. Bueno, A., Gálvez, J. A., Mira, P.: The global geometry of surfaces with prescribed mean curvature in ρ^3. Preprint arXiv:1802.08146 [math.DG]
9. Collin, P.: Deux exemples de graphes de courbure moyenne constante sur une band de R^2. C. R. Acad. Sci. Paris. **311**, 539–542 (1990)
10. Collin, P., Krust, R.: Le problème de Dirichlet pour l'équation des surfaces minimales sur des domaines non bornés. Bull. Soc. Math. France **119**, 443–462 (1991)
11. Corsato, C., De Coster, C., Omari, P.: Radially symmetric solutions of an anisotropic mean curvature equation modeling the corneal shape. Discrete Contin. Dyn. Syst. Dynamical systems, differential equations and applications. 10th AIMS Conference Supplement, 297–303 (2015)
12. Courant, R., Hilbert, D.: Methods of Mathematical Physics, vol. II. Interscience, New York (1962)
13. Clutterbuck, J., Schnürer, O., Schulze, F.: Stability of translating solutions to mean curvature flow. Calc. Var. **29**, 281–293 (2007)
14. Enache, C., López, R.: Minimum principles for equation of translating soliton type. Nonlinear Anal. **187**, 352–364 (2019)

15. Gilbarg, D., Trudinger, N. S.: Elliptic partial differential equations of second order. Classics in Mathematics. Reprint of the 1998 Edition. Springer, Berlin (2001)
16. Gromov, M.: Isoperimetry of waists and concentration of maps. Geom. Func. Anal **13**, 178–215 (2003)
17. Gui, C., Jian, H.-Y., Ju, H.-J.: Properties of translating solutions to mean curvature flow. Discret. Contin. Dyn. Syst. **28B**, 441–453 (2010)
18. Halldorsson, H.P.: Helicoidal surfaces rotating/translating under the mean curvature flow. Geom. Dedicata **162**, 45–65 (2013)
19. Huisken, G., Sinestrari, C.: Mean curvature flow singularities for mean convex surfaces. Calc. Var. **8**, 1–14 (1999)
20. Huisken, G., Sinestrari, C.: Convexity estimates for mean curvature flow and singularities of mean convex surfaces. Acta Math. **183**, 45–70 (1999)
21. Ilmanen, T.: Elliptic regularization and partial regularity for motion by mean curvature. Mem. Am. Math. Soc. 108, x+90 (1994)
22. Jian, H.-Y., Ju, H.-J.: Existence of translating solutions to the flow by powers of mean curvature on unbounded domains. J. Differ. Equ. **250**, 3967–3987 (2011)
23. Ju, H., Liu, Y.: Long-time existence of mean curvature flow with external force fields. Pac. J. Math. **234**, 311–325 (2008)
24. Ju, H., Liu, Y.: Dirichlet problem for anisotropic prescribed mean curvature equation on unbounded domains. J. Math. Anal. Appl. **439**, 709–724 (2016)
25. López, R.: Constant mean curvature surfaces with boundary in Euclidean three-space. Tsukuba J. Math. **23**, 27–36 (1999)
26. López, R.: Constant mean curvature graphs on unbounded convex domains. J. Differ. Equ. **171**, 54–62 (2001)
27. López, R.: An existence theorem of constant mean curvature graphs in Euclidean space. Glasgow Math. J. **44**, 455–461 (2002)
28. López, R.: Compact λ-translating solitons with boundary. Mediterranean J. Math. **15** (2018). https://doi.org/10.1007/s00009-018-1241-6
29. López, R.: Invariant surfaces in Euclidean space with a log-linear density. Adv. Math. **339**, 285–309 (2018)
30. López, R.: Some geometric properties of translating solitons in Euclidean space. J. Geom. **109** (2018). https://doi.org/10.1007/s00022-018-0444-0
31. López, R.: The Dirichlet problem on a strip for the α-translating soliton equation. C. R. Math. Acad. Sci. Paris. **356**, 1179–1187 (2018)
32. López. R.: Invariant singular minimal surfaces. Ann. Global Anal. Geom. **53**, 521–541 (2018)
33. López, R., Montiel, S.: Constant mean curvature surfaces with planar boundary. Duke Math. J. **85**, 583–604 (1996)
34. Marquardt, T.: Remark on the anisotropic prescribed mean curvature equation on arbitrary domain. Math. Z. **264**, 507–511 (2010)
35. Pérez-García, J.: Some results on translating solitons of the mean curvature flow (2016). arXiv:1601.07287 [math.DG]
36. Pyo, J.: Compact translating solitons with non-empty planar boundary. Differ. Geom. App. **47**, 79–85 (2016)
37. Schnurer, O., Smoczyk, K.: Evolution of hypersurfaces in central force fields. J. Reine Angew. Math. **550**, 77–95 (2002)
38. Schulze, F.: Evolution of convex hypersurfaces by powers of the mean curvature. Math. Z. **251**, 721–733 (2005)
39. Schulze, F.: Nonlinear evolution by mean curvature and isoperimetric inequalities. J. Differ. Geom. **79**, 197–241 (2008)
40. Serrin, J.: The problem of Dirichlet for quasilinear elliptic differential equations with many independent variables. Philos. Trans. R. Soc. Lond. **264**, 413–496 (1969)
41. Serrin, J.: The Dirichlet problem for surfaces of constant mean curvature. Proc. Lond. Math. Soc. **21**, 361–384 (1970)

42. Sheng, W., Wu, C.: On asymptotic behavior for singularities of the powers of mean curvature flow. Chin. Ann. Math. Ser. B **30**, 51–66 (2009)
43. Spruck, J., Xiao, L.: Complete translating solitons to the mean curvature flow in R^3 with nonnegative mean curvature. Am. J. Math. (To appear)
44. White, B.: The nature of singularities in mean curvature flow of mean convex surfaces. J. Am. Math. Soc. **16**, 123–138 (2003)

Constant Mean Curvature Surfaces for the Bessel Equation

Eduardo Mota

Abstract In this note we construct a family of immersions with constant mean curvature of the twice-punctured Riemann sphere into $\mathbb{R}^3$ from the Bessel equation.

Keywords Differential geometry · Constant mean curvature surfaces · Bessel equation

Introduction

The generalised Weierstrass representation due to Dorfmeister, Pedit and Wu [3] can be used to construct locally conformal constant mean curvature (CMC) immersions in Euclidean 3-space from a holomorphic 1-form ξ on a Riemann surface Σ. Several examples of CMC surfaces have been made using these techniques by Dorfmeister, Wu [4], Haak [2], Kilian, McIntosh, Schmitt [6], Kobayashi, Rossmann [5] and Traizet [11], among others. Also, more recently [12], Traizet uses an *opening nodes* method in the generalised Weierstrass representation set-up to obtain embedded CMC surfaces with Delaunay ends, with arbitrary genus and number of ends.

In the spirit of [7], our aim is to employ a second order differential equation, the Bessel equation, to find a new family of cylinders with constant mean curvature. This family possess one asymptotically Delaunay end and one irregular end, corresponding respectively to the regular and irregular singularities in the ODE. The first step in the construction of CMC surfaces is to write down a suitable potential ξ. That one of the ends of the surfaces is asymptotic to half Delaunay surface follows from the fact that at this end ξ is a perturbation of the potential of a Delaunay surface [8]. Since Σ is the twice-punctured Riemann sphere, it is enough to guarantee closing conditions at the Delaunay end in order to solve the period problem.

In Sect. 1 we outline the constructing method for CMC surfaces, due to [3]. The section number Sect. 2 has as purpose introducing the Bessel equation in our recipe for surfaces with constant mean curvature, along with a brief explanation of its features. Section 3 recalls the very well-known theory regarding CMC surfaces with Delaunay ends and perturbations of Delaunay potentials (see [5, 8], among others).

E. Mota (✉)
Department of Mathematics, University College Cork, Cork, Ireland
e-mail: eduardo.motasanchez@ucc.ie
URL: https://www.eduardo-mota.es

T. Hoffmann et al. (eds.), *Minimal Surfaces: Integrable Systems and Visualisation*, Springer Proceedings in Mathematics & Statistics 349,
https://doi.org/10.1007/978-3-030-68541-6_12

In the main section in this note, Sect. 4, we write down the potential that constructs CMC cylinders via the Bessel equation and put together in Theorem 1 the different ingredients for the construction. Finally, in Sect. 5 we prove the existence of a reflectional symmetry in the CMC cylinders that we have constructed. Graphics in Figs. 1 and 2 were produced with CMCLab [10].

1 The Generalised Weierstrass Representation

Let us briefly recall the generalized Weierstrass representation [3] to set the notation. We refer the reader to [5] for details on this method, which comprises the following steps:

1. On a connected Riemann surface Σ, let ξ be a holomorphic 1-form, called *potential*, with values in the loop algebra of maps $\mathbb{S}^1 \to \mathfrak{sl}_2(\mathbb{C})$. The potential ξ has to have a simple pole in its upper right entry in the loop parameter λ at $\lambda = 0$, and has no other poles for $\lambda < 1$. Moreover, the upper-right entry of ξ is non-zero on Σ. Let Φ be a solution of
$$\begin{cases} \mathrm{d}\Phi = \Phi\,\xi \\ \Phi(z_0) = \Phi_0. \end{cases} \tag{1}$$
2. Let $\Phi = F\,B$ be the point-wise Iwasawa factorization on the universal cover $\widetilde{\Sigma}$.
3. Then, the Sym–Bobenko formula $f := \mathrm{Sym}[F_\lambda] = (\partial_\lambda F)\,F^{-1}$ gives an associated family of conformal CMC immersions $\widetilde{\Sigma} \to \mathfrak{su}_2 \cong \mathbb{R}^3$.

If g is a map on Σ with values in a positive loop group of $\mathrm{SL}_2(\mathbb{C})$, then $\xi.g$ is again a potential. The gauge action is defined by $\xi.g := g^{-1}\xi g + g^{-1}\mathrm{d}g$. If $\mathrm{d}\Phi = \Phi\,\xi$, then Φg solves $\mathrm{d}\Psi = \Psi(\xi.g)$.

Now let $\Sigma = \mathbb{C}^*$ be the twice-punctured Riemann sphere. Let Δ denote the group of deck transformations of the universal cover $\widetilde{\Sigma}$, that is, $\Delta = \mathbb{Z}$ and $\widetilde{\Sigma} = \mathbb{C}$. The group of deck transformations Δ is generated by

$$\tau : \log z \mapsto \log z + 2\pi i. \tag{2}$$

Let ξ be a holomorphic potential on Σ. Let Φ be a solution of $\mathrm{d}\Phi = \Phi\xi$. Since Σ is not simply connected, in general Φ is only defined on $\widetilde{\Sigma}$. Let τ be a loop around the puncture $z = 0$, we define the *monodromy matrix* M of Φ along τ by

$$M = (\tau^*\Phi)\,\Phi^{-1}. \tag{3}$$

In order to construct CMC cylinders, the sufficient closing conditions for the monodromy that ensure that the immersion is well-defined are (see [5]):

$$M \in \mathrm{SU}_2, \text{ for all } \lambda \in \mathbb{S}^1, \tag{4a}$$

$$M|_{\lambda=1} = \pm\mathbb{1}, \tag{4b}$$

$$\partial_\lambda M|_{\lambda=1} = 0. \tag{4c}$$

Since $\Sigma = \mathbb{C}^*$, the monodromy group is infinite cyclic. Thus the fundamental group of the twice-punctured Riemann sphere has 1 generator and we only need to solve the monodromy problem (4) for the loop τ around the puncture $z = 0$.

2 The Bessel Equation

In this section we show how the Bessel equation constructs CMC surfaces with two ends. We start by prescribing this ODE in the first step (1) of the Weierstrass recipe [3] and then we point out the relevant features of this differential equation for the purpose of our work, namely, we describe its singularities' behaviour.

Let us consider without loss of generality a potential ξ of the form

$$\xi = \begin{pmatrix} 0 & \nu(z,\lambda) \\ \rho(z,\lambda) & 0 \end{pmatrix} dz. \tag{5}$$

The strategy to associate a scalar second order ODE to our constructing algorithm is illustrated in the following

Lemma 1 *Solutions of* $\mathrm{d}\Phi = \Phi\xi$ *are of the form*

$$\begin{pmatrix} y_1'/\nu & y_1 \\ y_2'/\nu & y_2 \end{pmatrix} \tag{6}$$

where y_1 *and* y_2 *are a fundamental system of the scalar* ODE

$$y'' - \frac{\nu'}{\nu} y' - \rho\nu y = 0. \tag{7}$$

By means of Lemma 1, we can choose functions ν and ρ in the potential (5) so that the Bessel equation becomes the associated ODE (7) in the initial value problem (1).

The *Bessel equation* [1] is the linear second-order ODE given by

$$z^2 y'' + zy' + (z^2 - \alpha^2)y = 0. \tag{8}$$

Equivalently, dividing through by z^2,

$$y'' + \frac{1}{z}y' + \left(1 - \frac{\alpha^2}{z^2}\right) y = 0. \tag{9}$$

The parameter α is an arbitrary complex number. This equation has one regular singularity at $z = 0$ and one irregular singularity of rank 2 at ∞ (see [9]), and for this reason the Bessel equation does not belong to the class of *Fuchsian equations*. Each of these punctures will correspond to one end on the CMC surface; the regular singularity generates a *regular* end (asymptotic to a Delaunay surface) while the irregular singularity generates an *irregular* end.

Consider again the off-diagonal potential in (5) and let us choose functions

$$\nu := \frac{1}{z} \quad \text{and} \quad \rho := -z + \frac{\alpha^2}{z}. \tag{10}$$

Plugging ν and ρ into Eq. 7, one obtains the Bessel equation and, in particular, prescribes this ODE in the algorithm for CMC surfaces seen in Sect. 1.

3 The $z^A P$ Lemma and Perturbations at a Simple Pole

Consider a differential equation $d\Phi = \Phi\xi$ for which the potential can be written as $\xi = A\frac{dz}{z-z_k} + \mathcal{O}(z^0)\,dz$, that is, it has a simple pole at $z = z_k$. A basic result in ODE theory can be extended to the context of loops. This is known as the $z^A P$ *lemma* (see [8, Lemma 2.3]) and states that under certain conditions on the eigenvalues of A, there exists a solution of the form $\Phi = Cz^A P = C\exp(A\log z)\,P$, where P extends holomorphically to $z = z_k$ and C is in the loop group of $\mathrm{SL}_2(\mathbb{C})$.

Lemma 2 ([5, Lemma 14]) *Let* $A : \mathbb{C}^* \to \mathfrak{sl}_2(\mathbb{C})$ *be an analytic map with non-constant eigenvalues* $\pm\mu(\lambda)$. *Let* $\xi = A(\lambda)dz/z + \mathcal{O}(z^0)dz$ *in a neighbourhood of* $z = 0$.

Let $M : \mathbb{C}^* \to \mathrm{SL}_2(\mathbb{C})$ *be a monodromy of* ξ *associated to a once-wrapped closed curve around* $z = 0$. *Then* $\operatorname{tr} M = 2\cos(2\pi\mu)$ *on* $\mathbb{C}^*$.

A *Delaunay residue* is a meromorphic $\mathfrak{sl}_2(\mathbb{C})$-valued map of the form

$$A = \begin{pmatrix} c & a\lambda^{-1} + \bar{b} \\ \bar{a}\lambda + b & -c \end{pmatrix}, \tag{11}$$

with $a, b \in \mathbb{C}^*$ and $c \in \mathbb{R}$. Any Delaunay surface in $\mathbb{R}^3$, up to rigid motion, can be derived [6] from an off-diagonal Delaunay residue (11) with $a, b \in \mathbb{R}^*$ satisfying the closing condition $a + b = 1/2$, so that the resulting surface's *necksize* depends on ab: when $ab > 0$, the surface is an *unduloid*, when $ab < 0$, it is a *nodoid*, and when $a = b$, it is a round cylinder (see for instance [8]). On the other hand, a *perturbation* of a Delaunay potential is a potential of the form

$$\xi = A\frac{dz}{z} + \mathcal{O}(z^0)dz \tag{12}$$

where A is as in (11). That a CMC end obtained from a perturbation of a Delaunay potential is asymptotic to a half-Delaunay surface is proven in [8, Theorem 5.9]. We will use this result in the next section.

4 Constructing CMC Cylinders

Constructing CMC surfaces with two ends via the Bessel equation is in the following two steps:

- Write down a potential on $\mathbb{C}^*$ which prescribes the Bessel equation as associated ODE and which is locally gauge-equivalent to a perturbation of a Delaunay potential at $z = 0$ (Definition 1).
- Show that the monodromy representation M is unitary in $\mathbb{S}^1$ (Sect. 4.3).

4.1 *Cylinder Potential*

In this part the potential which will be used to produce cylinders with one irregular end and one asymptotic Delaunay end is defined. Near the puncture $z = 0$ the potential is a local perturbation of a Delaunay potential via gauge equivalence.

Definition 1 Let $\Sigma = \mathbb{C}^*$ and let $r \in (-\infty, 1) \setminus \{0\}$. Define the $\mathfrak{sl}_2(\mathbb{C})$-valued cylinder potential by

$$\xi_c = \begin{pmatrix} 0 & \lambda^{-1} \\ \lambda Q_t & 0 \end{pmatrix} dz, \tag{13}$$

where $Q_t = \left(-\frac{r}{4z^2}t - 1\right)$ and $t := -\frac{1}{4}\lambda^{-1}(\lambda - 1)^2$, for all $\lambda \in \mathbb{S}^1$.

Remark 1 Note that if $\lambda = 1$, then $t = 0$ and the potential (13) becomes holomorphic so the monodromy $M(\lambda = 1) = \mathbb{1}$. Hence, parts (4b) and (4c) of the monodromy problem are automatically solved.

4.2 *Local Gauge*

In this part we use the theory of gauging to see that the potential in Sect. 2 with associated ODE the Bessel equation, is equivalent to the potential defined in Definition 1. Then, it is shown that the double pole of the potential in (13) can be gauged to a simple pole with Delaunay residue.

Consider the potential ξ introduced in Sect. 2, namely

$$\xi = \begin{pmatrix} 0 & 1/z \\ -z + \alpha^2/z & 0 \end{pmatrix} dz. \tag{14}$$

Note that for

$$g_1 = \begin{pmatrix} (1/z)^{1/2} & 0 \\ 0 & (1/z)^{-1/2} \end{pmatrix}, \quad g_2 = \begin{pmatrix} 1 & 0 \\ \frac{1}{2z} & 1 \end{pmatrix}. \tag{15}$$

the gauge $\xi.(g_1 g_2)$ gives

$$\xi = \begin{pmatrix} 0 & 1 \\ -1 + \frac{-1+4\alpha^2}{4z^2} & 0 \end{pmatrix} dz. \tag{16}$$

Putting $\alpha := \frac{1}{2}\sqrt{1-rt}$ (note that the parameter α depends on λ) and gauging by $\Lambda = \mathrm{diag}(\lambda^{1/2}, \lambda^{-1/2})$ we obtain the constructing potential ξ_c from Definition 1.

Lemma 3 *Let ξ_c a potential as in (13). Then there exists a neighbourhood U of $z = 0$ and a positive gauge g such that the expansion of $\xi_c.g$ is*

$$A\frac{dz}{z} + \mathcal{O}(z^0)dz, \quad \text{where } A = \begin{pmatrix} 0 & a\lambda^{-1} + b \\ a\lambda + b & 0 \end{pmatrix} \tag{17}$$

for some $a, b \in \mathbb{R}$ with $a + b = 1/2$.

Proof Let

$$g_{1c} = \begin{pmatrix} z^{1/2} & 0 \\ 0 & z^{-1/2} \end{pmatrix}, \quad g_{2c} = \begin{pmatrix} 1 & 0 \\ -\frac{1}{2}\lambda & a + b\lambda \end{pmatrix}. \tag{18}$$

Taking the real values $a = \frac{1}{4}\left(1 + \sqrt{1-r}\right)$ and $b = \frac{1}{4}\left(1 - \sqrt{1-r}\right)$, then $\xi_c.(g_{1c}g_{2c})$ has a simple pole at $z = 0$ and is of the form (17). Therefore, $g = g_{1c}g_{2c}$ is the required gauge. □

4.3 Unitary Monodromy on the Twice-Puncture Riemann Sphere

Since the fundamental group of the twice-punctured Riemann sphere has only 1 generator, we just need to solve the monodromy problem (4) at $z = 0$. Recall that, with our choice of potential (13), (4b) and (4c) of the monodromy problem are already solved.

For $\Phi_0 = \mathbb{1}$ and $z_0 = 1$, consider a solution Φ of (1) for $\xi = A$ with A as in Lemma 3, that is, satisfying that $a, b \in \mathbb{R}$ and $a + b = 1/2$. Then $\Phi = z^A$ and the monodromy is given by

$$M = \exp(2\pi i A) = \cos(2\pi\mu)\mathbb{1} + \frac{1}{\mu}\sin(2\pi\mu)A, \tag{19}$$

where it holds for the eigenvalues that

$$\mu^2 = a^2 + b^2 + ab(\lambda^{-1} + \lambda). \tag{20}$$

Thus, since $a + b = 1/2$, M is unitary for all $\lambda \in \mathbb{S}^1$ and then also (4a) is solved. The immersion will be well defined.

4.4 Main Theorem

Theorem 1 *Let $\Sigma = \mathbb{C}^*$ and let $r \in (-\infty, 1) \setminus \{0\}$. Then, there exists a conformal* CMC *immersion $f : \Sigma \to \mathbb{R}^3$ with one end which is asymptotic to half Delaunay surface and one irregular end.*

Proof Let ξ_c be a cylinder potential as in Definition 1. A solution Φ of (1) for ξ_c, $\Phi_0 = \mathbb{1}$ and $z_0 = 1$ can be found with the $z^A P$ lemma [5]. Let M be the monodromy of Φ at the puncture $z = 0$. By the remarks in Sect. 4.3, M is unitary and the monodromy problem (4) at $z = 0$ is solved. Then, the general Weierstrass representation [3] constructs a CMC immersion f in $\mathbb{R}^3$ which has two ends corresponding to the singularities from the Bessel equation—the ODE associated to $\mathrm{d}\Phi = \Phi\xi_c$. By Lemma 3, the potential is locally gauge-equivalent to a Delaunay potential and thus, by the asymptotics theorem of [8], the end at $z = 0$ is asymptotic to half Delaunay surface. □

5 Symmetry

In this last part, we explore some symmetries appearing in the surfaces constructed in Sect. 4. We prove that these symmetries in the resultant surfaces can be tracked to the level of the potentials, which are transformed under automorphisms on the domain.

A transformation ϕ is an involution if $\phi^2 = \mathrm{id}$ but $\phi \neq \mathrm{id}$. The following proposition holds for elements of $\mathrm{Iso}(\mathbb{R}^3)$ that are involutions.

Proposition 1 *The involutory isometries are the reflections and the half-turns (rotations by π).*

In what follows, to lighten notation, we may denote the dependence on λ with a subscript, that is,

$$\begin{aligned} \xi_\lambda &= \xi(z, \lambda), \\ \Phi_\lambda &= \Phi(z, \lambda), \\ F_\lambda &= F(z, \lambda). \end{aligned} \tag{21}$$

5.1 Cylinders with One Reflection

Consider the orientation reversing automorphism of Σ given by $\sigma(z) = \bar{z}$. It defines a symmetry that reflects the domain across the real axis. We prove the following

Theorem 2 *Consider a potential ξ that generates via the Weierstrass representation a* CMC *family of immersions f. Suppose that ξ satisfies the symmetries*

$$\begin{aligned} \xi_{1/\lambda} &= \overline{\sigma^* \xi_{1/\bar{\lambda}}}, \\ G^{-1} \xi_\lambda G &= \overline{\sigma^* \xi_{1/\bar{\lambda}}}, \end{aligned} \tag{22}$$

where $G = \mathrm{diag}(1/\lambda, \lambda)$. *Then, the induced immersion* $\breve{f} = \mathrm{Sym}\left[\overline{\sigma^* F_{1/\bar{\lambda}}}\right]$ *possesses reflective symmetry by a plane.*

Proof Let Φ_λ be the solution of $\mathrm{d}\Phi_\lambda = \Phi_\lambda \xi_\lambda$, $\Phi_\lambda(z_0) = \Phi_0$, with $z_0 \in \Sigma$ and Φ_0 diagonal. Naturally, the transformation $\overline{\sigma^* \Phi_{1/\bar{\lambda}}}$ defines a solution to the differential equation $\mathrm{d}\left(\overline{\sigma^* \Phi_{1/\bar{\lambda}}}\right) = \left(\overline{\sigma^* \Phi_{1/\bar{\lambda}}}\right)\left(\overline{\sigma^* \xi_{1/\bar{\lambda}}}\right)$, which in view of (22) reads as

$$\mathrm{d}\left(\overline{\sigma^* \Phi_{1/\bar{\lambda}}}\right) = \left(\overline{\sigma^* \Phi_{1/\bar{\lambda}}}\right)(\xi_\lambda \,.\, G)\,. \tag{23}$$

Since this ODE is also solved by $\Phi_\lambda G$, i.e.

$$\mathrm{d}\left(\Phi_\lambda G\right) = \left(\Phi_\lambda G\right)(\xi_\lambda \,.\, G)\,, \tag{24}$$

both solutions only differ by a matrix R in the loop group of $\mathrm{SL}_2(\mathbb{C})$. Hence, Φ_λ has the symmetry

$$R\Phi_\lambda = \overline{\sigma^* \Phi_{1/\bar{\lambda}}} G^{-1}, \tag{25}$$

for some z-independent R.

Evaluation at the fixed point 1 of σ using (25) yields $R = \overline{\Phi_0(1/\bar{\lambda})} G^{-1} \Phi_0(\lambda)^{-1}$. Since Φ_0 is diagonal, then by a simple calculation one gets that R is unitary for all $\lambda \in \mathbb{S}^1$.

Let us write the Iwasawa splitting $\Phi_\lambda = FB$. When Iwasawa decomposing the solution $\overline{\sigma^* \Phi_{1/\bar{\lambda}}}$ a unitary term U must be introduced, obtaining

$$\overline{\sigma^* F_{1/\bar{\lambda}}}\,\overline{\sigma^* B_{1/\bar{\lambda}}} = \overline{\sigma^* \Phi_{1/\bar{\lambda}}} = R\Phi_\lambda G = RFUU^{-1}BG. \tag{26}$$

The uniqueness of this splitting allows us to identify unitary and positive parts respectively, obtaining that

$$\overline{\sigma^* F_{1/\bar{\lambda}}} = RFU. \tag{27}$$

This implies that, using the generalised Weierstrass representation, $\overline{\sigma^* \xi_{1/\bar{\lambda}}}$ produces on the one hand the family of immersions given by plugging $\overline{\sigma^* F_{1/\bar{\lambda}}}$ in the Sym–

Bobenko formula and on the other hand the one obtained using RFU. Consequently, these two surfaces coincide.

At the level of the immersion, the symmetry (27) of the unitary frame appears in the Sym–Bobenko formula $\check{f} = \text{Sym}\left[\overline{\sigma^* F_{1/\bar{\lambda}}}\right]$ as follows:

$$\begin{aligned}\check{f} &= \left(\partial_\lambda \overline{\sigma^* F_{1/\bar{\lambda}}}\right) \left(\overline{\sigma^* F_{1/\bar{\lambda}}}\right)^{-1} \\ &= \partial_\lambda (RFU)\,(RFU)^{-1} \\ &= \left((\partial_\lambda R) R^{-1} + R(\partial_\lambda F) F^{-1} R^{-1}\right).\end{aligned} \tag{28}$$

It is left to prove that this symmetry is a reflection. To do so, we show that the transformation is an involution. It is easy to check that each of the symmetries seen so far remain the same if we 'reapply' the transformations done in (22) and used throughout this proof. Let us denote by $\check{\check{f}} = \text{Sym}\left[\overline{\sigma^* \left(\overline{\sigma^* F_{1/\bar{\lambda}}}\right)_{1/\bar{\lambda}}}\right]$ the resultant

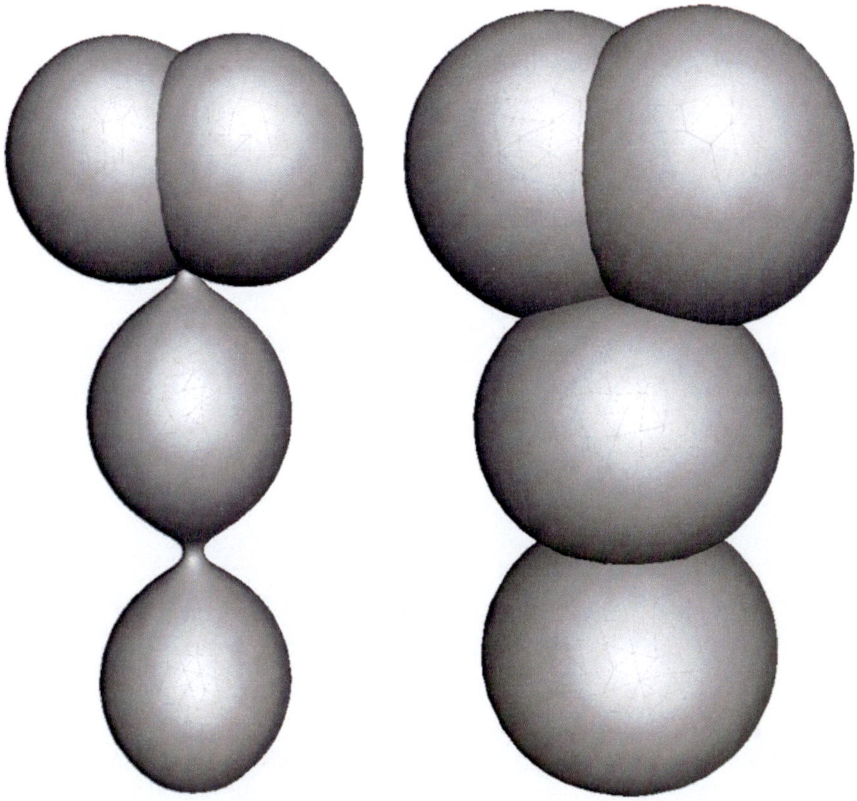

Fig. 1 Cylinders with one irregular end and one Delaunay end. The regular end weights are $1/3$ and $-1/4$, so the surfaces have one unduloid end and one nodoid end respectively

immersion of applying again those transformations, and by $'$ the derivative with respect to ∂_λ. Then,

$$\begin{aligned}\breve{\breve{f}} &= \overline{\sigma^* R_{1/\bar{\lambda}}}\, \partial_\lambda \left(\overline{\sigma^* F_{1/\bar{\lambda}}}\right) \overline{\sigma^* F^{-1}_{1/\bar{\lambda}} \sigma^* R^{-1}_{1/\bar{\lambda}}} \\ &= \bar{R}_{1/\bar{\lambda}} \left(R' R^{-1} + R F' F^{-1} R^{-1}\right) \bar{R}^{-1}_{1/\bar{\lambda}} + \bar{R}'_{1/\bar{\lambda}} \bar{R}^{-1}_{1/\bar{\lambda}} \\ &= \bar{R}_{1/\bar{\lambda}} R F' F^{-1} R^{-1} \bar{R}^{-1}_{1/\bar{\lambda}} + \bar{R}_{1/\bar{\lambda}} R' R^{-1} \bar{R}^{-1}_{1/\bar{\lambda}} + \bar{R}'_{1/\bar{\lambda}} \bar{R}^{-1}_{1/\bar{\lambda}}\end{aligned} \tag{29}$$

An easy computation yields that $\bar{R}_{1/\bar{\lambda}} R = \mathbb{1}$. Hence, differentiating in this equation, one also gets that

$$\bar{R}_{1/\bar{\lambda}} R' R^{-1} \bar{R}^{-1}_{1/\bar{\lambda}} = -\bar{R}'_{1/\bar{\lambda}} \bar{R}^{-1}_{1/\bar{\lambda}}. \tag{30}$$

Putting together (29) and (30), we manage to derive that $\breve{\breve{f}} = (\partial_\lambda F)\, F^{-1} = \mathrm{Sym}[F_\lambda] = f$. That is, this symmetry is an involution. Since, it is also orientation reversing, by Proposition 1, this symmetry must be a reflection. □

Using Theorem 2, since the constructing potential in (13) satisfies the relations in (22), the resulting surfaces in Figs. 1 and 2 have a reflectional symmetry fixing their ends.

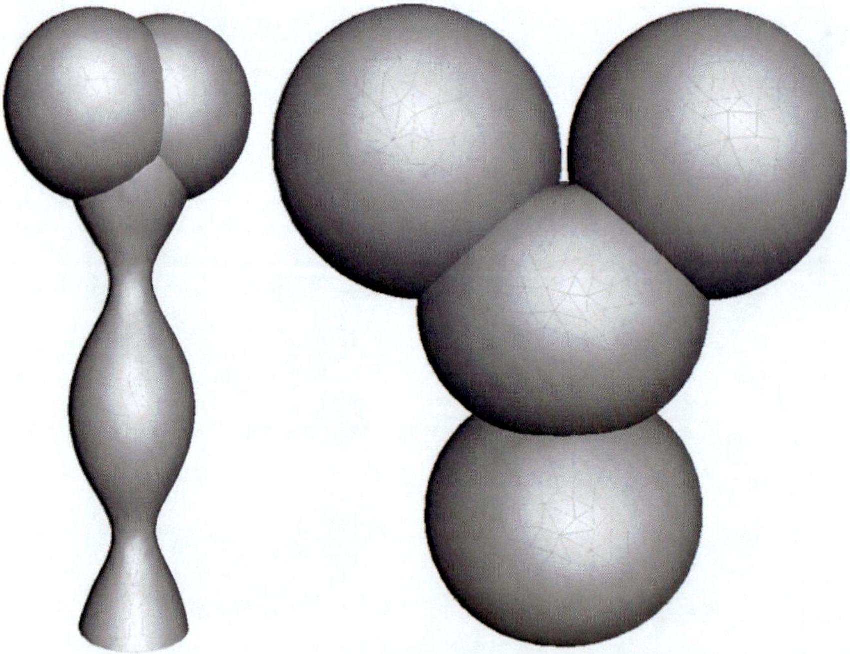

Fig. 2 Cylinders with one irregular end and one Delaunay end. The regular end weights are $1/\sqrt{2}$ and $-1/\pi$, so the surfaces have one unduloid end and one nodoid end respectively

References

1. Bowman, F.: Introduction to Bessel Functions. Dover Publications Inc., Mineola (1958)
2. Dorfmeister, J., Haak, G.: Construction of non-simply connected CMC surfaces via dressing. J. Math. Soc. Jpn. **55**(2), 335–364 (2003)
3. Dorfmeister, J., Pedit, F., Wu, H.: Weierstrass type representation of harmonic maps into symmetric spaces. Comm. Anal. Geom. **6**(4), 633–668 (1998)
4. Dorfmeister, J., Wu, H.: Construction of constant mean curvature n-noids from holomorphic potentials. Math. Z. **258**(4), 773–803 (2008)
5. Kilian, M., Kobayashi, S.-P., Rossman, W., Schmitt, N.: Constant mean curvature surfaces of any positive genus. J. Lond. Math. Soc. (2) **72**(1), 258–272 (2005)
6. Kilian, M., McIntosh, I., Schmitt, N.: New constant mean curvature surfaces. Experiment. Math. **9**(4), 595–611 (2000)
7. Kilian, M., Mota, E., Schmitt, N.: Constant mean curvature trinoids with one irregular end. Tohoku Math. J. (2018). arXiv:1808.05491
8. Kilian, M., Rossman, W., Schmitt, N.: Delaunay ends of constant mean curvature surfaces. Compos. Math. **144**(1), 186–220 (2008)
9. Kristensson, G.: Second Order Differential Equations-Special Functions and Their Classification. Springer, New York (2010)
10. Schmitt, N.: 'CMCLab', software package. University of Massachusetts, GANG (2000)
11. Traizet, M.: Construction of constant mean curvature n-noids using the DPW method. J. Reine Angew. Math. (2017). arXiv:1709.00924
12. Traizet, M.: Opening nodes and the DPW method (2018). arXiv:1808.01366

Minimal Surfaces Under Constrained Willmore Transformation

Áurea Casinhas Quintino

Abstract The class of constrained Willmore (CW) surfaces in space-forms constitutes a Möbius invariant class of surfaces with strong links to the theory of integrable systems, with a *spectral deformation* [8], defined by the action of a loop of flat metric connections, and *Bäcklund transformations* [9], defined by a dressing action by simple factors. Constant mean curvature (CMC) surfaces in 3-dimensional space-forms are [25] examples of CW surfaces, characterized by the existence of some *polynomial conserved quantity* [21, 22, 24]. Both CW spectral deformation and CW Bäcklund transformation preserve [21, 22, 24] the existence of such a conserved quantity, defining, in particular, transformations within the class of CMC surfaces in 3-dimensional space-forms, with, furthermore [21, 22, 24], preservation of both the space-form and the mean curvature, in the latter case. A classical result by Thomsen [28] characterizes, on the other hand, isothermic Willmore surfaces in 3-space as minimal surfaces in some 3-dimensional space-form. CW transformation preserves [8, 9] the class of Willmore surfaces, as well as the isothermic condition, in the particular case of spectral deformation [8]. We define, in this way, a CW spectral deformation and CW Bäcklund transformations of minimal surfaces in 3-dimensional space-forms into new ones, with preservation of the space-form in the latter case. This paper is dedicated to a reader-friendly overview of the topic.

Keywords Willmore energy · Constrained Willmore surfaces · Constant mean curvature surfaces · Minimal surfaces · Isothermic surfaces · Bäcklund transformations · Spectral deformation · Polynomial conserved quantities

Á. C. Quintino (✉)
Faculdade de Ciências da Universidade de Lisboa, Centro de Matemática, Aplicações Fundamentais e Investigação Operacional, 1749-016 Lisboa, Portugal
e-mail: amquintino@ciencias.ulisboa.pt; aurea.quintino@novasbe.pt; aurea.quintino@universidadeeuropeia.pt

NOVA School of Business and Economics, NOVA University Lisbon, 2775-405 Carcavelos, Portugal

Universidade Europeia, 1500-210 Lisboa, Portugal

T. Hoffmann et al. (eds.), *Minimal Surfaces: Integrable Systems and Visualisation*, Springer Proceedings in Mathematics & Statistics 349,
https://doi.org/10.1007/978-3-030-68541-6_13

1 Introduction

A central theme in Mathematics is that of the search for the optimal representative within a certain class of objects, often driven by the minimization of some energy, reflecting what occurs in many physical processes. From the early 1960s, Thomas Willmore devoted particular attention to the quest for the optimal immersion of a given closed surface in Euclidean 3-space, regarding the minimization of some natural energy, motivated by questions on the elasticity of certain biological membranes and the energetic cost associated with membrane bending deformations.

We can characterize how much a membrane is bent at a particular point on the membrane by means of the curvature of the osculating circles of the planar curves obtained as perpendicular cross sections through the point. The curvature of these circles consists of the inverse of their radii, with a positive or negative sign depending on whether the membrane curves upwards or downwards, respectively. The minimal and maximal values of the radii of the osculating circles associated with a particular point on the membrane define the principal curvatures, k_1 and k_2, and, from these, the mean curvature, $H = (k_1 + k_2)/2$ and the Gaussian curvature, $K = k_1 k_2$, at the point.

In modern literature on the elasticity of membranes, a weighed sum

$$a \int H + b \int H^2 + c \int K,$$

of the total mean curvature, the total squared mean curvature and the total Gaussian curvature, is considered to be the elastic bending energy of a membrane. By physical considerations, the total mean curvature is neglected. On the other hand, from the perspective of critical points of energy, in deformations conserving the topological type, the total Gaussian curvature can be ignored, according to Gauss–Bonnet theorem. What's left is what Willmore considered to be the *elastic bending energy* of a compact, oriented Riemannian surface, without boundary, isometrically immersed in $\mathbb{R}^3$, nowadays known as the *Willmore energy*.

The Willmore energy had already made its appearance early in the nineteenth century, through the works of Marie-Sophie Germain [15, 16] and Siméon Poisson [20] and their pioneering studies on elasticity and vibrating properties of thin plates, with the claim that the elastic force of a thin plate is proportional to its mean curvature. Since then, the mean curvature has remained a key concept in the theory of elasticity. The Willmore energy appeared again in the 1920s, in the works of Wilhelm Blaschke [1] and Gerhard Thomsen [28], but their findings were forgotten and only brought to light after the increased interest on the subject motivated by the work of Thomas Willmore.

Willmore surfaces are the critical points of the Willmore energy functional. Minimal surfaces, in their turn, are defined variationally as the stationary configurations for the area functional, amongst all those spanning a given boundary.

Minimal surfaces were first considered by Joseph-Louis Lagrange [17], in 1762, who raised the question of existence of surfaces of least area among all those spanning a given closed curve in Euclidean 3-space as boundary. Earlier, Leonhard Euler [14] had already discussed minimizing properties of the surface now known as the catenoid, although he only considered variations within a certain class of surfaces. The problem raised by Lagrange became known as the Plateau's Problem, referring to Joseph Plateau [19], who first experimented with soap films.

A physical model of a minimal surface can be obtained by dipping a wire loop into a soap solution. The resulting soap film is minimal in the sense that it always tries to organize itself so that its surface area is as small as possible whilst spanning the wire contour. This minimal surface area is, naturally, reached for the flat position, which happens to be a position of vanishing mean curvature. This does not come as a particular feature of this rather simple example of minimal surface. In fact, the Euler–Lagrange equation of the variational problem underlying minimal surfaces turns out to be precisely the zero mean curvature equation, as discovered by Jean Baptiste Meusnier [18]. The flat position of the soap film is also the position in which the membrane is the most relaxed. These surfaces are elastic energy minimals and, in this way, examples of Willmore surfaces.

Unlike flat soap films, soap bubbles exist under a certain surface tension, in an equilibrium where slightly greater pressure inside the bubble is balanced by the area-minimizing forces of the bubble itself. With their spherical shape, soap bubbles are examples of area-minimizing surfaces under the constraint of volume enclosed. These are surfaces of (non-zero) constant mean curvature and examples of *constrained Willmore surfaces*, the generalization of Willmore surfaces that arises when we consider critical points of the Willmore functional only with respect to infinitesimally conformal variations.

A very interesting fact about the Willmore energy is that it is scale-invariant: if one dilates the surface by any factor, the Willmore energy remains the same. Think of a round sphere in $\mathbb{R}^3$ as an example: if one increases the radius, the surface becomes flatter and its squared mean curvature decreases, but, at the same time, the surface area gets larger, which increases the value of the total squared mean curvature over the surface. One can show that these two phenomena counterbalance each other on any surface. In fact, the Willmore energy has the remarkable property of being invariant under any conformal transformation of $\mathbb{R}^3$, as established in a paper by James White [30] and, actually, already known to Blaschke [1] and Thomsen [28].

The class of constrained Willmore surfaces in space-forms constitutes a Möbius invariant class of surfaces with strong links to the theory of integrable systems, with a *spectral deformation*, defined by Fran Burstall, Franz Pedit and Ulrich Pinkall [8], by the action of a loop of flat metric connections, and *Bäcklund transformations* [9], defined by a dressing action by simple factors.

Constant mean curvature surfaces in 3-dimensional space-forms are examples of constrained Willmore surfaces, as established by Jörg Richter [25], characterized by the existence of some *polynomial conserved quantity* [21, 22, 24]. Both constrained Willmore spectral deformation and constrained Willmore Bäcklund transformation preserve [21, 22, 24] the existence of such a conserved quantity, for special choices of parameters, defining, in particular, transformations within the class of constant mean curvature surfaces in 3-spaces, with, furthermore [21, 22, 24], preservation of both the space-form and the mean curvature, in the latter case.

A classical result by Thomsen [28] characterizes, on the other hand, isothermic Willmore surfaces in 3-space as minimal surfaces in some 3-dimensional space-form. Constrained Willmore transformation preserves [8, 9] the class of Willmore surfaces, as well as, in the particular case of spectral deformation [8], the isothermic condition.

We define, in this way, a constrained Willmore spectral deformation and constrained Willmore Bäcklund transformations of minimal surfaces in 3-dimensional space-forms into new ones, with preservation of the space-form, in the latter case. This paper is dedicated to a reader-friendly overview of the topic. A detailed account of elementary computations can be found in [22, 23].

Along this text, we shall make no explicit distinction between a bundle and its complexification, and move from real tensors to complex tensors by complex multilinear extension, preserving notation. Our theory is local and, throughout this text, restriction to a suitable non-empty open set shall be underlying. Underlying throughout will be, as well, the identification

$$\wedge^2\mathbb{R}^{n+1,1} \cong o(\mathbb{R}^{n+1,1})$$

of the exterior power $\wedge^2\mathbb{R}^{n+1,1}$ with the orthogonal algebra $o(\mathbb{R}^{n+1,1})$ via

$$u \wedge v(w) := (u, w)v - (v, w)u,$$

for $u, v, w \in \mathbb{R}^{n+1,1}$.

2 The Willmore Energy

Among the classes of Riemannian submanifolds, there is that of *Willmore surfaces*, named after Willmore [31], in the 1960s, although the topic was mentioned by Blaschke [1] and Thomsen [28], in the 1920s, as a variational problem of optimal realization of a given closed surface in Euclidean 3-surface, regarding the minimization of some natural energy, motivated by questions on the elasticity of certain biological membranes and vesicles.

In modern literature on the elasticity of membranes, a weighed sum of the total squared mean curvature and the total Gaussian curvature, is considered to be the elastic energy of a membrane. From the perspective of critical points of energy, in deformations conserving the topological type, the total Gaussian curvature can be ignored, according to Gauss–Bonnet theorem. What's left is what is defined as the *Willmore energy*,

$$\mathcal{W} = \int_M H^2 dA,$$

of a compact, oriented (Riemannian) surface M, without boundary, (isometrically) immersed in $\mathbb{R}^3$.

From the perspective of energy extremals, the Willmore functional can be extended to compact, oriented (Riemannian) surfaces (isometrically) immersed in a general Riemannian manifold $\hat{M}$ with constant sectional curvature, or *space-form*, by means of

$$\mathcal{W} = \int_M |\Pi_0|^2 dA,$$

the total squared norm of the trace-free part Π_0 of the second fundamental form: by the Gauss equation, relating the curvature tensors of M and $\hat{M}$, we have

$$|\Pi_0|^2 = 2(|\mathcal{H}|^2 - K + \hat{K}),$$

for $\mathcal{H}$ the mean curvature vector and K and $\hat{K}$ the sectional curvatures of M and $\hat{M}$, respectively, so that, in the particular case of surfaces in $\mathbb{R}^3$,

$$|\Pi_0|^2 = 2(H^2 - K),$$

and, therefore, the two functionals share critical points. *Willmore surfaces* are the critical points of the Willmore functional.

3 Conformal Invariance and the Central Sphere Congruence

It is well-known that the Levi-Civita connection is not a conformal invariant (see, for example, [32, Sect. 3.12]). Although the second fundamental form is not conformally invariant, under a conformal change of the metric, its trace-free part remains invariant (see [23, Sect. 2.1]), so the respective squared norm and the area element change in inverse ways, leaving the Willmore energy unchanged and establishing the class of Willmore surfaces as a conformally invariant class. There is then no reason for carrying a distinguished metric—instead, we consider a conformal class of metrics.

Our study is one of surfaces in n-dimensional space-forms, with $n \geq 3$, from a conformally invariant point of view. So let S^n be the conformal n-sphere, in which, by stereographic projection, we find, in particular, the Euclidean n-space, as well as two copies of hyperbolic n-space. Our surfaces are immersions

$$\Lambda : M \to S^n$$

of a compact, oriented surface M, which we provide with the conformal structure $\mathcal{C}_\Lambda$ induced by Λ and with the canonical complex structure (that is, 90° rotation in the positive direction in tangent spaces, a notion that is, obviously, invariant under conformal changes of the metric). We find a convenient setting in Darboux's light-cone model of the conformal n-sphere [11]. We follow the modern account presented in [3]. So consider the Lorentzian space $\mathbb{R}^{n+1,1}$ and its light-cone $\mathcal{L}$, and fix a unit time-like vector t_0. We identify $v \in S^n \subseteq \mathbb{R}^{n+1}$ with the light-line through $v + t_0$, identifying, in this way, S^n with the projectivized light-cone,

$$S^n \cong \mathbb{P}(\mathcal{L}).$$

For us, a surface is, in this way, a null line subbundle $\Lambda = \langle\sigma\rangle$ of the trivial bundle $\underline{\mathbb{R}}^{n+1,1} = M \times \mathbb{R}^{n+1,1}$, with $\sigma : M \to \mathcal{L}$ a never-zero section of Λ. For further reference, set

$$\Lambda^{1,0} := \Lambda \oplus d\sigma(T^{1,0}M), \quad \Lambda^{0,1} := \Lambda \oplus d\sigma(T^{0,1}M),$$

independently of the choice of a never-zero $\sigma \in \Gamma(\Lambda)$, and then

$$\Lambda^{(1)} := \Lambda^{1,0} + \Lambda^{0,1}.$$

A fundamental construction in conformal geometry of surfaces is the mean curvature sphere congruence, or central sphere congruence, the bundle of 2-spheres tangent to the surface and sharing with it mean curvature vector at each point (although the mean curvature vector is not conformally invariant, under a conformal change of the metric, it changes in the same way for the surface and the osculating 2-sphere). In the light-cone picture, 2-spheres correspond to (3, 1)-planes in $\mathbb{R}^{n+1,1}$ and, in this way, the central sphere congruence defines a map

$$S : M \to \mathrm{Gr}_{(3,1)}(\mathbb{R}^{n+1,1}),$$

into the Grassmannian $\mathcal{G} := \mathrm{Gr}_{(3,1)}(\mathbb{R}^{n+1,1})$ of (3, 1)-planes in $\mathbb{R}^{n+1,1}$. We have, therefore, a decomposition

$$\underline{\mathbb{R}}^{n+1,1} = S \oplus S^\perp$$

and then a decomposition of the trivial flat connection d as

$$d = \mathcal{D} + \mathcal{N},$$

for $\mathcal{D}$ the connection given by the sum of the connections induced by d on S and $S^\perp$, respectively, through orthogonal projection.

Given $\mu, \eta \in \Omega^1(S^*T\mathcal{G})$, let $(\mu \wedge \eta)$ be the 2-form defined from the metric on $S^*T\mathcal{G}$:

$$(\mu \wedge \eta)_{(X,Y)} = (\mu_X, \eta_Y) - (\mu_Y, \eta_X),$$

for all $X, Y \in \Gamma(TM)$. Next we present a manifestly conformally invariant formulation of the Willmore energy. It follows the definition presented in [7], in the quaternionic setting, for the particular case of $n = 4$. The intervention of the conformal structure restricts to the Hodge $*$-operator, which is conformally invariant on 1-forms over a surface.

Theorem 1 ([7])

$$\mathcal{W}(\Lambda) = \frac{1}{2}\int_M (dS \wedge *dS).$$

Note that

$$(dS \wedge *dS) = -(*dS \wedge dS) = (dS, dS)dA,$$

$(dS \wedge *dS)$ is a conformally invariant way of writing $(dS, dS)_g dA_g$, for $g \in \mathcal{C}_\Lambda$, with dA_g denoting the area element of (M, g) and $(\, , \,)_g$ denoting the Hilbert–Schmidt metric on $L((TM, g), S^*T\mathcal{G})$. It follows that the Willmore energy of Λ coincides with the Dirichlet energy of S with respect to any of the metrics in the conformal class $\mathcal{C}_\Lambda$,

$$\mathcal{W}(\Lambda) = E(S).$$

4 Constrained Willmore Surfaces and Harmonicity

Harmonic maps are the critical points of the Dirichlet energy functional. Willmore surfaces are closely related to harmonic maps via the central sphere congruence, in a key result established by Blaschke [1], for $n = 3$, and, independently, Ejiri [13] and Rigoli [26], for general n:

Theorem 2 ([1, 13, 26]) *Λ is a Willmore surface if and only if its central sphere congruence S is a harmonic map.*

The well-developed theory of harmonic maps into Grassmannians now applies. First of all, it provides a zero-curvature characterization of Willmore surfaces: for a map into a Grassmannian, the harmonicity amounts to the flatness of a certain[1] family of connections, as established by Uhlenbeck [29], and so does then the Willmore surface condition:

Theorem 3 *Λ is a Willmore surface if and only if $d^\lambda := \mathcal{D} + \lambda\mathcal{N}^{1,0} + \lambda^{-1}\mathcal{N}^{0,1}$ is a flat connection, for all $\lambda \in S^1$.*

A larger class of surfaces arises when one imposes the weaker requirement that a surface extremize the Willmore functional only with respect to infinitesimally conformal variations: these are the *constrained Willmore surfaces*. The introduction of a constraint in the variational problem equips surfaces Λ with *Lagrange multipliers*,

[1] In the literature, the associated family of flat connections corresponding to a different choice of orientation in M can also be found.

as first proven by Burstall–Pedit–Pinkall [8] and then given the following manifestly conformally invariant formulation by Burstall–Calderbank [4]:

Theorem 4 ([4, 8]) *Λ is a constrained Willmore surface if and only if there exists a real form $q \in \Omega^1(\Lambda \wedge \Lambda^{(1)})$ such that*

$$d_q^\lambda := \mathcal{D} + \lambda \mathcal{N}^{1,0} + \lambda^{-1}\mathcal{N}^{0,1} + (\lambda^2 - 1)q^{1,0} + (\lambda^{-2} - 1)q^{0,1}$$

is a flat connection, for all $\lambda \in S^1$. Such a form q is said to be a (Lagrange) multiplier *for Λ and Λ is said to be a q-constrained Willmore surface. At times, it will be convenient to make an explicit reference to the central sphere congruence of Λ, writing $d_S^{\lambda,q}$ for d_q^λ.*

Willmore surfaces are the constrained Willmore surfaces admitting the zero multiplier. This is not necessarily the only multiplier, as we shall see.

5 Isothermic Constrained Willmore Surfaces

Isothermic surfaces are classically defined by the existence of conformal curvature line coordinates. Although the second fundamental form is not conformally invariant, conformal curvature line coordinates are preserved under conformal changes of the metric and, therefore, so is the isothermic surface condition. The next result presents a manifestly conformally invariant formulation of the isothermic surface condition, established by Burstall–Donaldson–Pedit–Pinkall [6].

Proposition 1 ([6]) *Λ is an isothermic surface if and only if there exists a non-zero closed real 1-form $\eta \in \Omega^1(\Lambda \wedge \Lambda^{(1)})$. In this case, we say that Λ is a η-isothermic surface.*

If $q_1 \neq q_2$ are multipliers for Λ, then Λ is a $*(q_1 - q_2)$-isothermic surface, and, reciprocally, if Λ is a η-isothermic q-constrained Willmore surface, then the set of multipliers for Λ is the affine space $q + \langle *\eta \rangle_{\mathbb{R}}$. Hence:

Proposition 2 ([9]) *A constrained Willmore surface Λ admits a unique multiplier if and only if Λ is not an isothermic surface.*

A classical result by Thomsen [28] characterizes isothermic Willmore surfaces in 3-space as minimal surfaces in some 3-dimensional space-form. (In contrast to constrained Willmore surfaces, constant mean curvature surfaces are not conformally invariant objects, requiring a distinguished space-form to be considered.)

Theorem 5 ([28]) *Λ is a minimal surface in some 3-dimensional space-form if and only if Λ is an isothermic Willmore surface in 3-space.*

Constant mean curvature surfaces in 3-dimensional space-forms are examples of isothermic constrained Willmore surfaces, as proven by Richter [25]. However, isothermic constrained Willmore surfaces in 3-space are not necessarily constant mean curvature surfaces in some space-form, as established by an example due to Burstall, presented in [2], of a constrained Willmore cylinder that does not have constant mean curvature in any space-form.

6 Transformations of Constrained Willmore Surfaces

The zero-curvature characterization of constrained Willmore surfaces presented above allows one to deduce two types of symmetry.

Suppose that Λ is a q-constrained Willmore surface. The two types of transformations that we describe next apply to any choice of the multiplier q (when there is a choice to be made) and depend on it. In the particular case that Λ is a Willmore surface, we consider q to be the zero multiplier, without further reference.

6.1 Spectral Deformation

The simplest transformation of Λ into new constrained Willmore surfaces arises from exploiting a scaling freedom in the spectral parameter, as follows.

For each $\lambda \in S^1$, the flatness of the metric connection d_q^λ establishes, at least locally, the existence of an isometry of bundles

$$\phi_\lambda : (\underline{\mathbb{R}}^{n+1,1}, d_q^\lambda) \to (\underline{\mathbb{R}}^{n+1,1}, d),$$

preserving connections, defined on a simply connected component of M and unique up to a Möbius transformation. We define a *spectral deformation* of $\Lambda = \langle \sigma \rangle$ by setting, for each $\lambda \in S^1$,

$$\Lambda_\lambda := \phi_\lambda \Lambda = \langle \phi_\lambda \sigma \rangle.$$

For each $\lambda \in S^1$, set

$$q_\lambda := \phi_\lambda \circ (\lambda^2 q^{1,0} + \lambda^{-2} q^{0,1}) \circ (\phi_\lambda)^{-1}.$$

The central sphere congruence of Λ_λ is $\phi_\lambda S$ and, given $\mu \in S^1$, we have

$$d_{\phi_\lambda S}^{\mu, q_\lambda} = \phi_\lambda \circ d_S^{\mu\lambda, q} \circ (\phi_\lambda)^{-1},$$

establishing the flatness of $d_{\phi_\lambda S}^{\mu, q_\lambda}$ from the one of $d_S^{\mu\lambda, q}$ (note that $\mu\lambda \in S^1$). It follows that:

Theorem 6 *Λ_λ is a q_λ-constrained Willmore surface, for all $\lambda \in S^1$.*

In particular, this spectral deformation preserves the zero multiplier.

Corollary 1 *If Λ is a Willmore surface, then so is Λ_λ, for all $\lambda \in S^1$.*

This spectral deformation coincides, up to reparameterization, with the one presented in [8], in terms of the *Hopf differential* and the *Schwarzian derivative* (see [22, Sect. 6.4.1]).

The isothermic surface condition is known [8] to be preserved under constrained Willmore spectral deformation. In our setting, one can verify (see [23, Sect. 2.3.5]) that, if Λ is also a η-isothermic surface, then Λ_λ is a η_λ-isothermic surface, for

$$\eta_\lambda := \phi_\lambda \circ (\lambda\eta^{1,0} + \lambda^{-1}\eta^{0,1}) \circ (\phi_\lambda)^{-1}.$$

Proposition 3 ([8]) *If Λ is an isothermic surface, then so is Λ_λ, for all $\lambda \in S^1$.*

From Theorem 5, it follows that:

Corollary 2 *If Λ is a minimal surface in some 3-dimensional space-form, then so is Λ_λ, for each $\lambda \in S^1$ (although not necessarily with preservation of the space-form).*

As we shall see later in this text, this spectral deformation preserves, as well, the class of constant mean curvature surfaces in 3-dimensional space-forms, for special choices of the spectral parameter.

6.2 *Bäcklund Transformation*

Having exploited the equivalence of $d_S^{\lambda,q}$ to the trivial flat connection, as flat metric connections, by means of

$$d_S^{\lambda,q} = (\phi_\lambda)^{-1} \circ d \circ \phi_\lambda,$$

we now explore equivalences starting from $d_S^{\lambda,q}$, i.e., equivalences given by

$$d_{S^*}^{\lambda,q^*} = r(\lambda) \circ d_S^{\lambda,q} \circ r(\lambda)^{-1},$$

for some q^* and some S^*, with $r(\lambda) \in \Gamma(O(\underline{\mathbb{R}}^{n+1,1}))$, so that the flatness of $d_S^{\lambda,q}$ establishes that of $d_{S^*}^{\lambda,q^*}$. The difficulties involved are of two different orders, namely, the preservation of the algebraic shape of $d_S^{\lambda,q}$, together with ensuring that S^* is the central sphere congruence of some surface, so that the family of flat connections $d_{S^*}^{\lambda,q^*}$ is the associated family to some constrained Willmore surface. A version of the Terng–Uhlenbeck [27] dressing action by simple factors proves to offer a simple construction, out of two parameters, a complex number α and a null line bundle L, parallel with respect to $d_S^{\alpha,q}$, from which we define, respectively, the eigenvalues and the eigenspaces of two different types of linear fractional transformations, out of which we define $r(\lambda)$, as follows.

Let ρ denote reflection across S,

$$\rho = \pi_S - \pi_{S^\perp},$$

for π_S and $\pi_{S^\perp}$ the orthogonal projections of $\underline{\mathbb{R}}^{n+1,1}$ onto S and $S^\perp$, respectively. Given $\alpha \in \mathbb{C}$ and L a null line subbundle of $\underline{\mathbb{R}}^{n+1,1}$ such that $\rho L \cap L^\perp = 0$, set

$$p_{\alpha,L}(\lambda) := I \begin{cases} \frac{\alpha-\lambda}{\alpha+\lambda} & \text{on } L \\ 1 & \text{on } (L \oplus \rho L)^\perp \\ \frac{\alpha+\lambda}{\alpha-\lambda} & \text{on } \rho L \end{cases},$$

for $\lambda \in \mathbb{C}\backslash\{\pm\alpha\}$ and $I \in \Gamma(O(\underline{\mathbb{R}}^{n+1,1}))$ the identity map of $\underline{\mathbb{R}}^{n+1,1}$. Let $q_{\alpha,L}$ denote the map obtained from $p_{\alpha,L}$ by considering the additive inverses of the eigenvalues associated to the eigenspaces L and ρL, respectively. Define $p_{\alpha,L}(\infty)$ and $q_{\alpha,L}(\infty)$ by holomorphic extension of

$$p_{\alpha,L}, q_{\alpha,L} : \mathbb{C}\backslash\{\pm\alpha\} \to \Gamma(O(\underline{\mathbb{R}}^{n+1,1}))$$

respectively.

Now consider $\alpha \in \mathbb{C}\backslash(S^1 \cup \{0\})$ and L a $d_S^{\alpha,q}$-parallel null line subbundle of $\underline{\mathbb{R}}^{n+1,1}$ such that $\rho L \cap L^\perp = 0$ (whose existence is established in [9]). Set $\alpha^* := \overline{\alpha}^{-1}$, $L' := p_{\alpha,L}(\alpha^*)\overline{L}$ and, for each $\lambda \in \mathbb{P}^1\backslash\{\pm\alpha\}$,

$$r(\lambda) := q_{\alpha^*,L'}(\lambda)\, p_{\alpha,L}(\lambda).$$

Set, furthermore,

$$\Lambda^* := (r(1)^{-1} r(0)\Lambda^{1,0}) \cap (r(1)^{-1} r(\infty)\Lambda^{0,1}).$$

Theorem 7 ([9]) *Λ^* is a q^*-contrained Willmore surface, for*

$$q^* := r(1)^{-1} \circ (r(\infty) \circ q^{1,0} \circ r(\infty)^{-1} + r(0) \circ q^{0,1} \circ r(0)^{-1}) \circ r(1),$$

with central sphere congruence

$$S^* := r(1)^{-1} S;$$

said to be the Bäcklund transform of Λ of parameters α, L.

In particular, Bäcklund transformation preserves the zero multiplier.

Corollary 3 *If Λ is a Willmore surface, then so is Λ^*.*

It is not clear that if Λ is an isothermic surface, then so is Λ^*. So far, it is not clear either that Bäcklund transformation preserves the class of minimal surfaces in 3-dimensional space-forms. However, as we shall see later, that proves to be the case. We shall see, furthermore, that Bäcklund transformation preserves the class of constant mean curvature surfaces in 3-dimensional space-forms, for special choices of parameters, with preservation of both the mean curvature and the curvature of space.

7 Polynomial Conserved Quantities for Constrained Willmore Surfaces

The isothermic surface condition amounts [6], just as well, to the flatness of a certain family ∇^t of connections, indexed in $\mathbb{R}$. In [10], the classical notion of *special isothermic surface*, introduced by Darboux [12], is given a simple explanation in terms of the integrable systems approach to isothermic surfaces. They are realized as a particular case of a hierarchy of classes of isothermic surfaces filtered by an integer d. Here is the basic idea: The theory of ordinary differential equations ensures that one can find ∇^t-parallel sections depending smoothly on the spectral parameter t. The existence of such sections with polynomial dependence of degree d on t is of particular geometric significance, as first observed by Burstall–Calderbank [5], and gave rise to the notion of *polynomial conserved quantity of type d*, developed in [10], in the isothermic context, where the notion of *special isothermic surface of type d* is introduced, having the classical notion as a particular case ($d = 2$).

We are in this way led to the notion of *special constrained Willmore surface of type d*, presented in [24]:

Definition 1 Let Λ be a q-constrained Willmore surface and $d \in \mathbb{N}_0$. A Laurent polynomial

$$p(\lambda) = \overline{p_d}\,\lambda^{-d} + \cdots + \overline{p_1}\,\lambda^{-1} + p_0 + p_1\,\lambda + \cdots + p_d\,\lambda^d$$

with

$$p_d \in \Gamma(S^\perp)$$

and $p_k \in \Gamma(S^\perp)$ if and only if k and d have the same parity, or, otherwise, $p_k \in \Gamma(S)$; is said to be a *polynomial conserved quantity of type d of Λ* if

$$p(1) \neq 0$$

and

$$d_q^\lambda\, p(\lambda) = 0,$$

for all $\lambda \in S^1$. We say that Λ is a *special constrained Willmore surface of type d* if it admits a polynomial conserved quantity of type d.

The case $d = 1$ recovers the notion of conserved quantity presented in [21, 22], an idea by Burstall–Calderbank [5].

The fact that $p(\lambda)$ is a polynomial conserved quantity of type d of Λ establishes, in particular, that $p(1)$ is real and constant, that is, $p(1) \in \mathbb{R}^{n+1,1}$. As we shall see, $p(1)$ carries very important information regarding both the curvature of space in which, under some conditions, Λ proves to have constant mean curvature, and the mean curvature of the surface Λ in such a space.

In the isothermic context, type 1 characterizes [5] *H-generalised surfaces*, surfaces admitting a parallel unit normal vector field which has constant inner product with the mean curvature vector (see also [10]). In the constrained Willmore context, type 1 with parallel top term characterizes surfaces with parallel mean curvature vector:

Theorem 8 ([24]) *Λ is a special constrained Willmore surface of type* 1*, admitting a polynomial conserved quantity with parallel top term, if and only if Λ has parallel mean curvature vector in some space-form.*

In codimension 1, the condition of parallelism of the top term of a polynomial conserved quantity of type 1 proves [24] to be vacuous. It follows that, in codimension 1, type 1 characterizes constant mean curvature surfaces, in both contexts, recovering, in particular, a result established in [21, 22]:

Theorem 9 ([21, 22, 24]) *Suppose that $\Lambda \subset S^3$. Then Λ is a special constrained Willmore surface of type* 1 *if and only if Λ has constant mean curvature in some space-form.*

Furthermore:

Theorem 10 ([21, 22, 24]) *Suppose that $\Lambda \subset S^3$. If $p(\lambda)$ is a polynomial conserved quantity of type* 1 *of Λ, then Λ has constant mean curvature H, with*

$$H^2 = |\pi_{S^\perp}(p(1))|^2, \tag{1}$$

in a space-form with sectional curvature

$$K = -(p(1), p(1)). \tag{2}$$

Reciprocally, if Λ has constant mean curvature H in some space-form with sectional curvature K, then Λ admits a polynomial conserved quantity $p(\lambda)$, of type 1*, satisfying (1) and (2).*

Corollary 4 *Suppose that $\Lambda \subset S^3$. Then Λ is a minimal surface in some space-form if and only if Λ is a constrained Willmore surface admitting a polynomial conserved quantity $p(\lambda)$ of type* 1 *with*

$$p(1) \in \Gamma(S).$$

8 Transformations of Special Constrained Willmore Surfaces

The class of special constrained Willmore surfaces of any given type d is preserved under both spectral deformation and Bäcklund transformation, defining, in particular, for special choices of parameters, as established in [21, 22] ($d = 1$) and [24] (general d), as follows.

Let Λ be a q-constrained Willmore surface.

Theorem 11 ([21, 22, 24]) *Let λ be in S^1 and $\phi_\lambda : (\underline{\mathbb{R}}^{n+1,1}, d_q^\lambda) \to (\underline{\mathbb{R}}^{n+1,1}, d)$ be an isometry of bundles, preserving connections. Suppose that $p(\mu)$ is a polynomial conserved quantity of type d of Λ, with $p(\lambda)$ non-zero. Then*

$$p_\lambda(\mu) := \phi_\lambda\, p(\lambda\mu)$$

is a polynomial conserved quantity of type d of the spectral deformation $\phi_\lambda\Lambda$, of parameter λ, of Λ.

As for Bäcklund transformation of special constrained Willmore surfaces:

Theorem 12 ([21, 22, 24]) *Suppose that $p(\lambda)$ is a polynomial conserved quantity of type d of Λ. Suppose that α, L are Bäcklund transformation parameters for Λ with*

$$p(\alpha) \perp \overline{L}.$$

Then

$$p^*(\lambda) := r(1)^{-1} r(\overline{\lambda}^{-1}) p(\lambda)$$

is a polynomial conserved quantity of type d of the Bäcklund transform Λ^* of Λ, of parameters α, L.

Note that

$$p^*(1) = p(1),$$

establishing the preservation of the curvature of space, when carrying a distinguished one.

9 Constant Mean Curvature Surfaces Under Constrained Willmore Transformation

From Theorems 9 and 11, we conclude that the class of constant mean curvature surfaces in 3-dimensional space-forms is preserved under spectral deformation, for special choices of the spectral parameter. Recall, furthermore, that, for each $\lambda \in S^1$, the central sphere congruence of Λ_λ is $\phi_\lambda S$. According to Theorem 10, it follows that:

Corollary 5 *If Λ is a constant mean curvature surface in some 3-dimensional space-form, then so is Λ_λ (although not necessarily with preservation of the space-form), for special choices of the parameter $\lambda \in S^1$. Furthermore: if $p(\mu)$ is a polynomial conserved quantity of type 1 of Λ, with $p(\lambda)$ non-zero, then Λ_λ has constant mean curvature H_λ with*

$$H_\lambda^2 = |\pi_{S^\perp}(p(\lambda))|^2$$

in a space-form with sectional curvature

$$K_\lambda = -(p(\lambda), p(\lambda)).$$

Both constrained Willmore spectral deformation and Bäcklund transformation prove [24] to preserve, furthermore, the parallelism of the top term of a polynomial conserved quantity, for special choices of parameters. Hence:

Theorem 13 ([24]) *The class of parallel mean curvature surfaces in space-forms is preserved under both spectral deformation and Bäcklund transformation, for special choices of parameters, with preservation of the space-form, in the latter case.*

In particular, the class of constant mean curvature surfaces in 3-dimensional space-forms is preserved under Bäcklund transformation, for special choices of parameters, with preservation of the space-form. Furthermore, recovering a result established in [21, 22]:

Theorem 14 ([21, 22, 24]) *If Λ is a constant mean curvature surface in some 3-dimensional space-form, then so is Λ^*, for special choices of parameters, with preservation of both the space-form and the mean curvature.*

Corollary 6 *If Λ is a minimal surface in some 3-dimensional space-form, then so is Λ^*, for special choices of parameters, with preservation of the space-form.*

Acknowledgements Most of the research work involved in [9, 21, 22] was carried out as part of the Ph.D. studies of the author at the University of Bath, UK, under the supervision of Professor Fran Burstall, to whom the author is deeply grateful.

The author was supported by FCT—Fundação para a Ciência e a Tecnologia, with a Ph.D. scholarship, and by FFCUL—Fundação da Faculdade de Ciências da Universidade de Lisboa, with a postdoctoral scholarship. The research work in [24] was additionally supported by National Funding from FCT—Fundação para a Ciência e a Tecnologia, under the project UID/MAT/04561/2013.

The author would like to take this opportunity to thank the organisers of the Winter 2018 M:iv Workshop for the invitation to deliver a talk and for the warm welcome to Granada.

References

1. Blaschke, W.J.E.: Vorlesungen über Differentialgeometrie III: Differentialgeometrie der Kreise und Kugeln. Die Grundlehren der Mathematischen Wissenschaften in Einzeldarstellungen, band XXIX. Verlag von Julius Springer in Berlin (1929)
2. Bohle, C., Peters, G.P., Pinkall, U.: Constrained Willmore surfaces. Calc. Var. Part. Differ. Equ. **32**(2), 263–277 (2008)
3. Burstall, F.E.: Isothermic surfaces: conformal geometry, Clifford algebras and integrable systems. In: Terng, C.-L. (ed.) Integrable Systems, Geometry and Topology. AMS/IP Studies in Advanced Mathematics, vol. 36, pp. 1–82. American Mathematical Society, Providence (2006)
4. Burstall, F.E., Calderbank, D.M.J.: Conformal submanifold geometry I-III (2010). arXiv:1006.5700 [math.DG]
5. Burstall, F.E., Calderbank, D.M.J.: Conformal submanifold geometry IV-V. (In preparation)
6. Burstall, F.E., Donaldson, N.M., Pedit, F., Pinkall, U.: Isothermic submanifolds of symmetric R-spaces. Journal für die reine und angewandte Mathematik **660**, 191–243 (2011)
7. Burstall, F.E., Ferus, D., Leschke, K., Pedit, F., Pinkall, U.: Conformal Geometry of Surfaces in S^4 and Quaternions. Lecture Notes in Mathematics, vol. 1772. Springer, Berlin (2002)
8. Burstall, F.E., Pedit, F., Pinkall, U.: Schwarzian derivatives and flows of surfaces. In: Guest, M., Miyaoka, R., Ohnita, Y. (eds.) Differential Geometry and Integrable Systems. Contemporary Mathematics Series, vol. 308, pp. 39–61. American Mathematical Society, Providence (2002)
9. Burstall, F.E., Quintino, A.C.: Dressing transformations of constrained Willmore surfaces. Commun. Anal. Geom. **22**, 469–518 (2014)
10. Burstall, F.E., Santos, S.D.: Special isothermic surfaces of type d. J. Lond. Math. Soc. **85**, 571–591 (2012)
11. Darboux, J.-G.: Leçons sur la Théorie Générale des Surfaces et les Applications Géométriques du Calcul Infinitésimal. Gauthier-Villars, Paris (1887)
12. Darboux, J.-G.: Sur les surfaces isothermiques. Comptes Rendus Hebdomadaires des Séances de l'Académie des Sciences **128**, 1299–1305 (1899)
13. Ejiri, N.: Willmore surfaces with a duality in $S^n(1)$. Proc. Lond. Math. Soc. (Series 3) **57**(2), 383–416 (1988)
14. Euler, L.: Methodus Inveniendi Lineas Curvas Maximi Minimive Proprietate Gaudentes, sive, Solutio Problematis Isoperimetrici Latissimo Sensu Accepti. Marcum-Michaelem Bousquet & Socios, Lausannæ & Genevæ (1744)
15. Germain, M.-S.: Recherches sur la Théorie des Surfaces Élastiques. Mme Ve Courcier. Libraire pour les Sciences, Paris (1821)
16. Germain, M.-S.: Remarques sur la Nature, les Bornes et l'Étendue de la Question des Surfaces Élastiques, et Équation Générale de ces Surfaces. Imprimerie de Huzard-Courcier, Paris (1826)
17. Lagrange, J.-L.: Essai d'une nouvelle méthode pour déterminer les maxima et les minima des formules intégrales indéfinies. Mélanges de Philosophie et de Mathématique de la Société Royale de Turin (Miscellanea Taurinensia), vol. 2, pp. 335–362 (1762)
18. Meusnier, J.B.M.C.: Mémoire sur la courbure des surfaces. Mémoires de Mathématique et de Physique, présentés à l'Académie Royale des Sciences de Paris, par divers Savans, & lûs dans ses Assemblées, vol. 10, pp. 477–510 (1785)
19. Plateau, J.A.F.: Statique Expérimentale et Théorique des Liquides Soumis aux Seules Forces Moléculaires. Gauthier-Villars, Paris (1873)
20. Poisson, S.D.: Mémoire sur les surfaces élastiques. Mémoires de la Classe des Sciences Mathématiques et Physiques de l'Institut National de France, Année 1812, Seconde Partie, pp. 167-225 (1812)
21. Quintino, A.C.: Constrained Willmore Surfaces: Symmetries of a Möbius Invariant Integrable System. Ph.D. Thesis, University of Bath, UK (2008)
22. Quintino, A.C.: Constrained Willmore Surfaces: Symmetries of a Möbius Invariant Integrable System - Based on the author's Ph.D. thesis. London Mathematical Society Lecture Note Series, vol. 465. Cambridge University Press, Cambridge (2021)

23. Quintino, A.C.: Transformations of generalized harmonic bundles and constrained Willmore surfaces. In: Toda, M.D. (ed.) Willmore Energy and Willmore Conjecture, pp. 9–47. Chapman & Hall/CRC Monographs and Research Notes in Mathematics. CRC Press, Taylor and Francis Group, Boca Raton, Florida (2017)
24. Quintino, A.C., Santos, S.D.: Polynomial conserved quantities for constrained Willmore surfaces (2015). arXiv:1507.01253v1 [mathDG]
25. Richter, J.: Conformal maps of a Riemann surface into the space of quaternions. Ph.D. Thesis, Technische Universität Berlin (1997)
26. Rigoli, M.: The conformal Gauss map of submanifolds of the Möbius space. Ann. Global Anal. Geom. **5**(2), 97–116 (1987)
27. Terng, C.-L., Uhlenbeck, K.: Bäcklund transformations and loop group actions. Commun. Pure Appl. Math. **53**(1), 1–75 (2000)
28. Thomsen, G.: Über konforme Geometrie I: Grundlagen der konformen Flächentheorie. Abhandlungen aus dem Mathematischen Seminar der Universität Hamburg **3**, 31–56 (1923)
29. Uhlenbeck, K.: Harmonic maps into Lie groups: classical solutions of the chiral model. J. Diff. Geom. **30**, 1–50 (1989)
30. White, J.H.: A global invariant of conformal mappings in space. Proc. Amer. Math. Soc. **38**, 162–164 (1973)
31. Willmore, T.J.: Note on embedded surfaces. Analele Ştiinţifice ale Universitaţii "Alexandru Ioan Cuza" din Iaşi (Serie Noua), Sect. I a Mat. **11B**, 493–496 (1965)
32. Willmore, T.J.: Riemannian Geometry. Oxford Science Publications. Clarendon Press/Oxford University Press, New York (1993)

On an Enneper–Weierstrass-Type Representation of Constant Gaussian Curvature Surfaces in 3-Dimensional Hyperbolic Space

Graham Smith

Abstract For all $k \in]0, 1[$, we construct a canonical bijection between the space of ramified coverings of the sphere of hyperbolic type and the space of complete immersed surfaces in 3-dimensional hyperbolic space of finite area and of constant extrinsic curvature equal to k. We show, furthermore, that this bijection restricts to a homeomorphism over each stratum of the space of ramified coverings of the sphere.

Keywords Teichmüller theory · Ramified coverings · Immersed surfaces · Gaussian curvature · Extrinsic curvature · Hyperbolic space

Classifications 30F60 · 53C42

1 Introduction

1.1 Background

The Enneper–Weierstrass (EW) representation describes minimal surfaces in $\mathbb{R}^3$ in terms of a meromorphic function and a holomorphic 1-form. This remarkable tool of minimal surface theory has played a key role in the construction of minimal surfaces possessing unexpected and often surprising properties (see, for example, [4] and Chap. 8 of [9]). For this reason, various authors have also interested themselves in the study of EW-type representations of other types of surfaces (see, for example, [1, 3, 5]). In a similar vein, in [10], following the work [7, 8] of Labourie, we constructed an EW-type representation for immersed surfaces of constant positive extrinsic curvature in 3-dimensional hyperbolic space $\mathbb{H}^3$ in terms of one meromor-

G. Smith (✉)
Instituto de Matemática, Universidad Federal do Rio de Janeiro, Av. Athos da Silveira Ramos 149, Centro de Tecnologia, Bloco C, Cidade Universitária, Ilha do Fundão, Caixa Postal 68530, 21941-909,
Rio de Janeiro, RJ, Brazil
e-mail: grahamandrewsmith@gmail.com

T. Hoffmann et al. (eds.), *Minimal Surfaces: Integrable Systems and Visualisation*, Springer Proceedings in Mathematics & Statistics 349,
https://doi.org/10.1007/978-3-030-68541-6_14

phic function defined over a Riemann surface. More precisely, we showed that, for all $k \in]0, 1[$, every holomorphic local homeomorphism from the Poincaré disk into the Riemann sphere is the Weierstrass map (defined below) of a unique simply-connected immersed surface in $\mathbb{H}^3$ of constant extrinsic curvature equal to k which is complete in a certain sense. In the current paper, building upon our subsequent results of [11] we show that, for all $k \in]0, 1[$, the Weierstrass map defines a homeomorphism between the space of complete immersed surfaces in $\mathbb{H}^3$ of finite area and of constant extrinsic curvature equal to k, on the one hand, and the space of pointed ramified coverings of hyperbolic type of the Riemann sphere on the other. We believe this result should lead to new research directions along the frontier between the theory of immersed surfaces on the one hand and Teichmüller theory on the other. These are themes which we propose to investigate in depth in forthcoming work.

1.2 Main Result

In order to define the EW-type representation, we first recall the definition of the horizon map. Let $\mathrm{T}\mathbb{H}^3$ be the tangent bundle over $\mathbb{H}^3$ and let $\mathrm{U}\mathbb{H}^3 \subseteq \mathrm{T}\mathbb{H}^3$ be the subbundle of unit vectors over $\mathbb{H}^3$. Let $\partial_\infty\mathbb{H}^3$ be the ideal boundary of $\mathbb{H}^3$ (see [2]). The **horizon map** $\vec{n} : \mathrm{U}\mathbb{H}^3 \to \partial_\infty\mathbb{H}^3$ is defined such that, for all $X \in \mathrm{U}\mathbb{H}^3$,

$$\vec{n}(X) := \gamma(+\infty) = \operatorname*{Lim}_{t\to+\infty} \gamma(t), \tag{A}$$

where $\gamma : \mathbb{R} \to \mathbb{H}^3$ is the unique geodesic such that $\partial_t\gamma(0) = X$. Informally, $\vec{n}(X)$ is the point in $\partial_\infty\mathbb{H}^3$ towards which X points.

Let $\Sigma := (i, S)$ be an **immersed surface** in $\mathbb{H}^3$. We recall that this means that S is a surface and $i : S \to \mathbb{H}^3$ is an immersion. In the sequel, all functions will be taken to be smooth and all manifolds will be taken to be oriented. Let $N : S \to \mathbb{H}^3$ be the unit normal vector field over i compatible with the orientation.

Let $A \in \Gamma(TS)$ be the shape operator of i. We define $K_i : S \to \mathbb{R}$, the **extrinsic curvature** function of i, by

$$K_i := \mathrm{Det}(A), \tag{B}$$

and we define $\varphi_i : S \to \partial_\infty\mathbb{H}^3$, the **Weierstrass map** of i, by

$$\varphi_i := \vec{n} \circ N. \tag{C}$$

We say that i is **locally strictly convex** (LSC) whenever its shape operator is at all points positive definite. This means in particular that K is everywhere positive. Furthermore, when i is LSC, we chose the orientation such that at every point N points outwards from the convex set bounded locally by the image of i at that point. In this case, it follows from classical hyperbolic geometry that the Weierstrass map is a local homeomorphism (see [2]). In particular, since $\partial_\infty\mathbb{H}^3$ carries the conformal structure of the Riemann sphere, upon pulling back this conformal structure through φ_i, we may suppose that φ_i is a locally conformal mapping. We prove

Theorem 1 *Let $\tilde{S}$ be a compact Riemann surface. Let $P \subseteq \tilde{S}$ be a finite set of points such that $S := \tilde{S} \setminus P$ is of hyperbolic type. Let $\varphi : \tilde{S} \to \partial_\infty \mathbb{H}^3$ be a ramified covering with ramification points contained in P. Then, for all $k \in]0, 1[$, there exists a unique complete LSC immersion $i : S \to \mathbb{H}^3$ of finite area and of constant extrinsic curvature equal to k such that φ is the Weierstrass map of i.*

Conversely, if $\Sigma := (i, S)$ is a complete immersed LSC surface in $\mathbb{H}^3$ of finite area and constant extrinsic curvature equal to k, for some $k \in]0, 1[$, and if $\varphi : S \to \partial_\infty \mathbb{H}^3$ is the Weierstrass map of i, then the Riemann surface $(S, \varphi^* \partial_\infty \mathbb{H}^3)$ is conformally equivalent to a compact Riemann surface $\tilde{S}$ with a finite set P of points removed. Furthermore, φ extends to a ramified covering of the sphere by $\tilde{S}$ with ramification points contained in P.

Finally, for all $k \in]0, 1[$, this mapping defines a homeomorphism between the set of complete immersed LSC surfaces in $\mathbb{H}^3$ of finite area and of constant extrinsic curvature equal to k, on the one hand, and the set of pointed ramified coverings of hyperbolic type of the Riemann sphere, on the other.

The bijection constructed in Theorem 1 can now be visualised schematically as follows:

$$\left\{\begin{array}{c}(i, S) \text{ s.t.}\\ i : S \to \mathbb{H}^3 \text{ an immersion,}\\ i \text{ complete,}\\ i \text{ finite area \&}\\ K_i = k.\end{array}\right\} \longleftrightarrow \left\{\begin{array}{c}(\varphi, \tilde{S}, P) \text{ s.t.}\\ \tilde{S} \setminus P \text{ hyperbolic,}\\ \varphi : S \to \hat{\mathbb{C}} \text{ holomorphic,}\\ \varphi \text{ non-constant,}\\ P \subseteq S \text{ finite \&}\\ \mathrm{Crit}(\varphi) \subseteq P.\end{array}\right\}$$

We say that a pointed ramified covering $(\varphi, \tilde{S}, P)$ is **exceptional** whenever $\tilde{S} \setminus P$ is not hyperbolic. Up to equivalence, there are only 3 exceptional cases, namely $(z, \hat{\mathbb{C}}, \emptyset)$, $(z, \hat{\mathbb{C}}, \{0\})$ and $(z^m, \hat{\mathbb{C}}, \{0, \infty\})$, for some non-negative integer m. The latter two cases may actually be considered as the images of degenerate limits of sequences of complete immersed LSC surfaces of finite area and of constant extrinsic curvature equal to k. Indeed, $(z, \hat{\mathbb{C}}, \{0\})$ corresponds to the degenerate limit consisting of the single point $\{0\}$ on the ideal boundary whilst $(z^m, \hat{\mathbb{C}}, \{0, \infty\})$ corresponds to the degenerate limit consisting of the complete geodesic in $\mathbb{H}^3$ joining 0 to ∞ weighted by a factor of m.

The author is grateful to Asun Jiménez Grande, David Dumas, Lucio Rodriguez and Harold Rosenberg for helpful conversations.

1.3 Labourie's Compactness Theorem

Labourie's compactness theorem (c.f. [7]) presents a powerful tool for the study of LSC immersed surfaces of constant extrinsic curvature inside 3-dimensional manifolds. We first require a few definitions. Let $M := (M, g)$ be a complete 3-dimensional Riemannian manifold. Let TM be the tangent bundle of M, let $UM \subseteq TM$ be the unit sphere bundle and let $\pi : UM \to M$ be the canonical projection. Observe that π is distance non-increasing. Let (i, S) be an immersed surface in M. Let $N : S \to UM$ be the unit normal vector field over Σ compatible with the orientation. We define the **Gauss lift** $\hat{\Sigma}$ of Σ by $(\hat{\imath}, S)$, where $\hat{\imath} = N$. This terminology merely distinguishes between the mapping N, considered as a section of i^*UM over i, and the mapping $\hat{\imath}$, considered as an immersion in its own right into the total space of UM. For $k > 0$, following [7], we say that Σ is a k-**surface** whenever Σ has constant extrinsic curvature equal to k and $\hat{\Sigma}$ is complete.

Let $\Gamma \subseteq M$ be a complete geodesic. Let $\mathrm{N}\Gamma \subseteq UM$ be the bundle of unit normal vectors over Γ. If $(\hat{\imath}, S)$ is an immersed surface in UM, then we say that $\hat{\Sigma}$ is a **tube** whenever $\hat{\imath}$ defines a covering map from S onto $\mathrm{N}\Gamma$ for some complete geodesic Γ. Furthermore, we say that $\hat{\Sigma}$ is a tube of **finite order** whenever this covering is of finite order.

Let $(\hat{\imath}_n, S_n, x_n)$ be a sequence of complete pointed immersed surfaces in UM. We say that $(\hat{\imath}_n, S_n, x_n)$ converges towards the complete pointed immersed surface $(\hat{\imath}_\infty, S_\infty, x_\infty)$ in the **pointed Cheeger-Gromov** sense whenever there exists a sequence (α_n) of mappings such that:

(1) for all n, $\alpha_n : S_\infty \to S_n$;
(2) for all n, $\alpha_n(x_\infty) = x_n$;
(3) for every relatively compact open subset $\Omega \subseteq S_\infty$, there exists $N \in \mathbb{N}$ such that for all $n \geqslant N$, the restriction of α_n to Ω is a diffeomorphism onto its image; and
(4) the sequence $(\hat{\imath}_n \circ \alpha_n)$ converges in the C^∞_{loc} sense to $\hat{\imath}_\infty$.

We refer to (α_n) as a **sequence of convergence maps** for the sequence $(\hat{\imath}_n, S_n, x_n)$ with respect to the limit $(\hat{\imath}_\infty, S_\infty, x_\infty)$.

Labourie's compactness theorem is now stated as follows:

Theorem 2 (Labourie 1997) *Let M be a complete 3-dimensional Riemannian manifold. Let (i_n, S_n, x_n) be a sequence of pointed k-surfaces in M for some fixed $k > 0$. For all n, let $\hat{\imath}_n$ be the Gauss Lift of i_n. If there exists a compact set $K \subseteq M$ such that $i_n(x_n) \in K$ for all n, then there exists a complete pointed immersed surface $(\hat{\imath}_\infty, S_\infty, x_\infty)$ towards which $(\hat{\imath}_n, S_n, x_n)$ subconverges in the pointed Cheeger-Gromov sense. Furthermore, either:*

(1) $(\hat{\imath}_\infty, S_\infty)$ is a tube; or
(2) $i_\infty := \pi \circ \hat{\imath}_\infty$ is an immersion.

Remark 1 Care should be taken in interpreting the limit obtained in the second case of Theorem 2. Indeed, although $\hat{\imath}_\infty$ is complete, there is no reason to suppose that i_∞ is too. This often leads to counterintuitive phenomena. □

1.4 General Properties of k-Surfaces

Lemma 1 *Let (M, g) be a complete 3-dimensional Riemannian manifold whose isometry group acts co-compactly. Let (i, S) be a k-surface in M for some $k > 0$. If (S, i^*g) has finite area then for all $B > 0$ there exists a compact subset $K \subseteq S$ such that for all $x \in S \setminus K$, $\|A(x)\| > B$.*

Proof Suppose the contrary. There exists $B > 0$ and a diverging sequence $(x_n) \in S$ such that $\|A(x_n)\| \leqslant B$ for all n. Let $\hat{g}$ be the Sasaki metric over UM. For all n, let B_n be the unit ball about x_n in S with respect to the metric $\hat{\imath}^*\hat{g}$. Upon extracting a subsequence, we may suppose that all the (B_n) are disjoint. For all n, let Area_n be the area of B_n with respect to the metric i^*g. We claim that the sequence (Area_n) is uniformly bounded below. Indeed, let $K \subseteq M$ be a compact fundamental domain for the isometry group. For all n, let $\alpha_n : M \to M$ be an isometry such that $M_n(i(x_n)) \in K$. For all n, denote $i_n = \alpha_n \circ i$ and let $\hat{\imath}_n$ be the Gauss lift of i_n. By Labourie's Compactness Theorem, upon extracting a subsequence, there exists a complete pointed immersed surface $(\hat{\imath}_\infty, S_\infty, x_\infty)$ towards which $(\hat{\imath}_n, S, x_n)$ converges. Let (β_n) be a sequence of convergence mappings for $(\hat{\imath}_n, S, x_n)$ with respect to the limit $(\hat{\imath}_\infty, S_\infty, x_\infty)$. Observe that $(i_n \circ \beta_n)$ converges in the C^∞_{loc} sense to $i_\infty := \pi \circ \hat{\imath}_\infty$. However, since $(\|A(x_n)\|)$ is uniformly bounded, $(\hat{\imath}_\infty, S_\infty)$ is not a tube, and so i_∞ is an immersion. In particular, if B_∞ is the unit ball about x_∞ in S_∞ with respect to the metric $\hat{\imath}^*_\infty\hat{g}$, and if Area_∞ is the area of B_∞ with respect to the metric $i^*_\infty g$, then $\text{Area}_\infty > 0$. Thus:

$$\underset{n\to\infty}{\text{LimInf}}\, \text{Area}_n \geqslant \text{Area}_\infty > 0.$$

In particular, the sequence (Area_n) is uniformly bounded below, as asserted. However, since the sequence (B_n) consists of disjoint balls, it follows that S has infinite area. This is absurd, and the result follows. □

Now consider the case where $M := \mathbb{H}^3$ is 3-dimensional hyperbolic space. For any closed subset $X \subseteq \mathbb{H}^3 \cup \partial_\infty \mathbb{H}^3$, let $\text{Conv}(X)$ be its convex hull and let $\overline{B}_r(X)$ be the closure in $\mathbb{H}^3 \cup \partial_\infty \mathbb{H}^3$ of the set of all points in $\mathbb{H}^3$ lying at a distance no greater than r from X.

Lemma 2 *Let (i, S) be a compact LSC immersed surface in $\mathbb{H}^3$ of constant extrinsic curvature equal to $k \in]0, 1[$. Then:*

$$i(S) \subseteq \overline{B}_r(\mathit{Conv}(i(\partial S))),$$

where $r = \mathit{Tanh}^{-1}(k)$.

Remark 2 In particular, any such surface must have non-trivial boundary.

Proof Suppose the contrary. Let $X := \text{Conv}(i(\partial S))$. Let $x \in S$ be such that $i(x)$ lies at a distance greater than r from X. In particular, x is an interior point of S. Let $p \in X$ be the closest point to $i(x)$. Let P be the supporting plane to X at p normal to the geodesic joining p to $i(x)$. Let (P_s) be the foliation of $\mathbb{H}^3$ by equidistant planes to P

parametrised by (signed) distance from P. Let $r' := \mathrm{Sup}\{s \mid i(S) \cap P_s \neq \emptyset\}$. Since S is compact, $r' < \infty$ and we may suppose that $i(x) \in P_{r'}$. In particular, $i(S)$ is an interior tangent to $P_{r'}$ at this point. It follows by the geometric maximum principle that the extrinsic curvature of i at z is at least $\mathrm{Tanh}(r') > \mathrm{Tanh}(r) = k$. This is absurd, and the result follows. □

1.5 Limit Points of the Immersion

Henceforth, we will assume that (i, S) is a complete finite-area immersed surface in $\mathbb{H}^3$ of constant extrinsic curvature equal to k for some fixed $k \in]0, 1[$. We continue to denote by g the metric over $\mathbb{H}^3$, by $\hat{g}$ the Sasaki metric over $\mathrm{U}\mathbb{H}^3$ and by $\pi : \mathrm{U}\mathbb{H}^3 \to \mathbb{H}^3$ the canonical projection.

By Gauss' equation, (S, i^*g) has constant *intrinsic* curvature equal to $k - 1 < 0$. Since it has finite area, it follows by Hüber's Theorem (c.f. [6]) that (S, i^*g) is conformally equivalent to a compact surface $\tilde{S}$ with a finite set P of points removed. Furthermore, if we denote by $\mathbb{D}$ the Poincaré disk, then for every point $p \in P$ there exists $\epsilon \in]0, 1[$ and a conformal mapping $\alpha : \mathbb{D} \to \tilde{S}$ such that $\alpha(0) = p$ and:

$$(i \circ \alpha)^* g_{ij} = \frac{1}{|z|^2 \mathrm{Log}(\epsilon\, |z|)^2} \delta_{ij}, \tag{D}$$

where δ denotes the Euclidean metric over $\mathbb{D}$. The metric on the right-hand side of (D) is the standard metric of a finite area hyperbolic cusp.

We henceforth identify $\mathbb{D}^*$ with its image in S and suppress α in what follows. For all $r \in]0, 1[$, let C_r be the Euclidean circle of radius r about 0 in $\mathbb{D}^*$, and let $\mathrm{Length}(C_r)$ be its length with respect to the cusp metric (D). Observe that $(\mathrm{Length}(C_r))$ tends to 0 as r tends to 0.

Proposition 1 *There exists a sequence (r_n) converging to 0 and a point $p_\infty \in \mathbb{H}^3 \cup \partial_\infty \mathbb{H}^3$ such that the sequence (C_{r_n}) converges to $\{p_\infty\}$ in the Hausdorff sense.*

Proof By compactness of the family of compact sets, there exists a sequence (r_n) converging to 0 and a subset C_∞ of $\partial_\infty \mathbb{H}^3 \cup \mathbb{H}^3$ such that the sequence (C_{r_n}) converges to C_∞ in the Hausdorff sense. Since the sequence $(\mathrm{Length}(C_{r_n}))$ converges to 0, C_∞ consists of a single point, and the result now follows. □

For all n, we henceforth denote $C_n := C_{r_n}$. For all $n < m$, let $A_{n,m} \subseteq \mathbb{D}^*$ be the anulus bounded by C_n and C_m and for all n, let D_n^* be the pointed disk bounded by C_n. Let $\hat{\imath}$ be the Gauss lift of i. Observe that since i^*g is complete at 0, so too is $\hat{\imath}^*\hat{g}$. For all n, denote $X_n := \overline{B}_r(\mathrm{Conv}(i(C_n) \cup \{p_\infty\}))$.

Proposition 2 *For all n, $i(D_n) \subseteq X_n$.*

Proof Indeed, for all $n < m$, let $X_{n,m} := \overline{B}_r(\mathrm{Conv}(i(C_n) \cup i(C_m)))$. By Lemma 2, for all $n < m$, $i(A_{n,m}) \subseteq X_{n,m}$. For all n, $(X_{n,m})$ converges to $X_n := \overline{B}_r(\mathrm{Conv}(i(C_n)$

$\cup \{p_\infty\}))$ in the Hausdorff sense as m tends to infinity. The result follows upon taking limits. □

Proposition 3 $p_\infty \in \partial_\infty \mathbb{H}^3$.

Proof Suppose the contrary. Let $\hat{\imath}$ be the Gauss lift of i. Let $z_n \in \mathbb{D}^*$ be a sequence converging to 0. We may suppose that for all n, $z_n \in D_n$. Furthermore, since $\hat{\imath}^*\hat{g}$ is complete at 0, we may suppose that z_n lies at a distance of at least n from C_n with respect to this metric. In particular, for all $R > 0$, there exists $N \in \mathbb{N}$ such that for $n \geqslant N$, $\Omega_{R,n} \subseteq D_n$, where $\Omega_{R,n}$ is the open ball of radius R about z_n in $\mathbb{D}^*$ with respect to the metric $\hat{\imath}^*\hat{g}$.

For all n, by Proposition 2, $i(D_n) \subseteq X_n$. Thus, since (X_n) converges to $\overline{B}_r(p_\infty)$ in the Hausdorff sense as n tends to infinity, for sufficiently large n, $i(D_n) \subseteq \overline{B}_{2r}(p_\infty)$. However, by Labourie's Compactness Theorem, there exists a complete pointed immersed surface $(\hat{\imath}_\infty, S_\infty, z_\infty)$ towards which $(\hat{\imath}, \mathbb{D}^*, z_n)$ subconverges. By Lemma 1, $(\hat{\imath}_\infty, S_\infty, z_\infty)$ is a tube. By the preceeding discussion, and taking limits, for all $R > 0$, $(\pi \circ \hat{\imath}_\infty)(\Omega_{R,\infty}) \subseteq \overline{B}_{2r}(y_\infty)$, where $\Omega_{R,\infty}$ is the open ball of radius R about z_∞ in S_∞ with respect to the metric $\hat{\imath}_\infty^*\hat{g}$. In other words, $(\pi \circ \hat{\imath}_\infty)(S_\infty) \subseteq \overline{B}_{2r}(y_\infty)$. This is absurd, since $(\pi \circ \hat{\imath}_\infty)(S_\infty)$ is a complete geodesic, and the result follows. □

Proposition 4 *$(i(z))$ tends to p_∞ as z tends to 0.*

Proof By Proposition 3, $p_\infty \in \partial_\infty \mathbb{H}^3$. The result now follows by Proposition 2 since (X_n) converges to $\{p_\infty\}$ in the Hausdorff sense as n tends to ∞. □

1.6 Limit Points of the Weierstrass Map

Proposition 5 *Let $(x_n) \in \mathbb{H}^3$ be a sequence converging towards $x_\infty \in \partial_\infty \mathbb{H}^3$. Let $\Gamma \subseteq \mathbb{H}^3$ be a geodesic with end-point x_∞. Let y_∞ be the other end-point of Γ and let z be any point of Γ. If (α_n) is a sequence of isometries of $\mathbb{H}^3$ such that for all n, $\alpha_n(x_\infty) = x_\infty$ and $\alpha_n(x_n) = z$, then for every compact subset $K \subseteq \partial_\infty \mathbb{H}^3 \setminus \{x_\infty\}$, $(\alpha_n(K))$ converges to $\{y_\infty\}$ in the Hausdorff sense as n tends to ∞.*

Proof We identify $\mathbb{H}^3$ with the upper half-space in $\mathbb{R}^3$. Upon applying an isometry, we may suppose that $x_\infty = 0$, $y_\infty = \infty$ and $z = (0, 0, 1)$. For all n, let $x_n := (\xi_n, \eta_n, \tau_n)$. For all n, the mapping α_n is given in these coordinates by:

$$\alpha_n(x, y, t) = \frac{1}{\tau_n}(x - \xi_n, y - \eta_n, t).$$

The result follows. □

Proposition 6 *$(\overrightarrow{n} \circ \hat{\imath})(z)$ tends to p_∞ as z tends to 0.*

Proof Suppose the contrary. There exists a sequence (z_n) converging to 0 such that the sequence $(q_n) := ((\overrightarrow{n} \circ \hat{\imath})(z_n))$ converges to $q_\infty \neq p_\infty$. We may suppose that for all n, $z_n \in D_n$. Furthermore, since $\hat{\imath}^*\hat{g}$ is complete at 0, we may suppose that, for all n, z_n lies at a distance of at least n from C_n with respect to this metric. In particular, for all $R > 0$, there exists $N \in \mathbb{N}$ such that for all $n \geqslant N$, $\Omega_{R,n} \subseteq D_n$, where $\Omega_{R,n}$ is the open ball of radius R about z_n in $\mathbb{D}^*$ with respect to the metric $\hat{\imath}^*\hat{g}$. Let r_∞ be any other point of $\hat{\mathbb{C}}$ distinct from both p_∞ and q_∞. Let Γ be the geodesic joining p_∞ and r_∞. Let P be a totally geodesic plane in $\mathbb{H}^3$ normal to Γ. For all n, let α_n be an isometry of $\mathbb{H}^3$ such that $\alpha_n(p_\infty) = p_\infty$, $\alpha_n(C_n) \cap \Gamma \neq \emptyset$ and $\alpha_n(i(z_n)) \in P$. For all n, let $i_n := \alpha_n \circ i$ and let $\hat{\imath}_n$ be the Gauss lift of i_n.

Let $r := \mathrm{Tanh}^{-1}(k)$. Since $(\mathrm{Length}(C_n))$ tends to 0, we may suppose that, for all n, $\mathrm{Length}(C_n) < r$. Since $\overline{B}_r(\Gamma)$ is convex, it follows that for all n, $\mathrm{Conv}(i_n(C_n) \cup \{y_\infty\}) \subseteq \overline{B}_r(\Gamma)$. Thus, by Lemma 2, for all n, $i_n(D_n) \subseteq \overline{B}_r(\mathrm{Conv}(i_n(C_n)) \cup \{p_\infty\}) \subseteq \overline{B}_{2r}(\Gamma)$. In particular, for all n, $i_n(z_n) \in \overline{B}_{2r}(\Gamma) \cap P$, which is a compact set. Thus, by Labourie's Compactness Theorem, there exists a complete pointed immersed surface $(\hat{\imath}_\infty, S_\infty, z_\infty)$ towards which $(\hat{\imath}_n, \mathbb{D}^*, z_n)$ converges. Let (β_n) be a sequence of convergence maps for $(\hat{\imath}_n, \mathbb{D}^*, z_n)$ with respect to $(\hat{\imath}_\infty, S_\infty, z_\infty)$. Observe that $(i_n \circ \beta_n)$ converges towards $i_\infty := \pi \circ \hat{\imath}_\infty$ in the C^∞_{loc} sense. By Lemma 1, $(\hat{\imath}_\infty, S_\infty)$ is a tube and so $i_\infty(S_\infty)$ is a complete geodesic. However, for all $R > 0$, there exists $N \in \mathbb{N}$ such that for $n \geqslant N$, $i_n(\Omega_{R,n}) \subseteq i_n(D_n) \subseteq \overline{B}_{2r}(\Gamma)$. Taking limits, it follows that for all $R > 0$, $i_\infty(\Omega_{R,\infty}) \subseteq \overline{B}_{2r}(\Gamma)$, where $\Omega_{R,\infty}$ is the open ball of radius R about z_∞ in S_∞ with respect to the metric $\hat{\imath}^*_\infty\hat{g}$. In other words, $i_\infty(S_\infty) \subseteq \overline{B}_{2r}(\Gamma)$. That is, $i_\infty(S_\infty)$ is a complete geodesic lying at constant distance from Γ. This geodesic therefore coincides with Γ. In particular, $\hat{\imath}_\infty(z_\infty)$ is a unit normal vector to Γ at the point $\Gamma \cap P$. That is, $\hat{\imath}_\infty(z_\infty)$ is tangent to P at this point. Since P is totally geodesic, it follows that $q_\infty := (\overrightarrow{n} \circ \hat{\imath}_\infty)(z_\infty) \in \partial_\infty P$. However, $(\alpha_n(q_n)) = ((\alpha_n \circ \overrightarrow{n} \circ \hat{\imath})(z_n)) = ((\overrightarrow{n} \circ \hat{\imath}_n)(z_n))$ converges to q_∞. Furthermore, by Proposition 4, $(i(z_n))$ converges to p_∞ as n tends to ∞. Since, by hypothesis, (q_n) remains uniformly bounded away from p_∞, it follows by Proposition 5 that $(\alpha_n(q_n))$ converges to r_∞ as n tends to ∞. In particular, $r_\infty = q_\infty$. This is absurd, and it follows that $(\overrightarrow{n} \circ \hat{\imath})(z)$ converges to p_∞ as z tends to 0, as desired. □

We now obtain the second part of Theorem 1:

Proposition 7 *$(S, \varphi^*\hat{\mathbb{C}})$ is conformally equivalent to a compact Riemann surface $\hat{S}$ with a finite set P of points removed. Furthermore, φ extends to a meromorphic map from $\hat{S}$ to $\partial_\infty\mathbb{H}^3 = \hat{\mathbb{C}}$. That is, φ is a ramified covering.*

Proof Let $\tilde{S}$ and P be as at the beginning of this section. Choose $p \in P$. Let U be a neighbourhood of p homeomorphic to a disk such that p is the only point of P in U. By definition $\varphi := \overrightarrow{n} \circ \hat{\imath}$. Upon reducing U if necessary, $(U, \varphi^*\hat{\mathbb{C}})$ is conformally equivalent to the annulus $A_c := \{z \mid c < |z| < 1\}$ where $c \in [0, 1[$. Observe that φ defines a locally conformal mapping $\tilde{\varphi} : A_c \to \hat{\mathbb{C}}$. Furthermore, by Proposition 6, there exists $q \in \hat{\mathbb{C}}$ towards which $\tilde{\varphi}(z)$ converges as $|z|$ converges to c. If $c > 0$, then, by classical complex analysis, $\tilde{\varphi}$ is constant. This is absurd, and it follows that $c = 0$. Furthermore, by Cauchy's removable singularity theorem, $\tilde{\varphi}$ extends to a holomorphic mapping from A_0 into $\hat{\mathbb{C}}$. Since $p \in P$ is arbitrary, the result follows. □

1.7 Systoles

The proof of the first part of Theorem 1 is a fairly straightforward consequence of the results of [10, 11]. Indeed, let $\tilde{S}$ be a compact Riemann surface and let $\varphi : \tilde{S} \to \hat{\mathbb{C}}$ be a ramified covering of the sphere. Let $P \subseteq \tilde{S}$ be a finite subset containing the ramification points of φ such that $\tilde{S} \setminus P$ is of hyperbolic type. Choose $k \in]0, 1[$. We identify $\hat{\mathbb{C}}$ with $\partial_\infty \mathbb{H}^3$ in the canonical manner, and, denoting $S := \tilde{S} \setminus P$, we define (i_k, S) to be the unique k-surface in $\mathbb{H}^3$ whose Weierstrass map is φ. The existence and uniqueness of $\hat{\imath}_k$ are proven in Theorem 1.4 of [10].

Proposition 8 *$(S, i_k^* g)$ is complete.*

Proof By Theorem 1.2 of [11], for all $p \in P$, $i_k(x)$ converges to $\varphi(p) \in \partial_\infty \mathbb{H}^3$ as x converges to p. In particular, i_k is proper, and completeness follows. □

It remains to show that $(S, i_k^* g)$ has finite area. However, recall that $(S, i_k^* g)$ has constant intrinsic curvature equal to $k - 1 < 0$. It is therefore sufficient to show that for all $p \in P$, $(S, i_k^* g)$ has a hyperbolic cusp at p. This is readily proven using systoles. We recall that for all $x \in S$, and for any metric h over S, the **systole** of h at x, which we denote by $\mathrm{Sys}(h, x)$ is defined to be the length of the shortest topologically non-trivial curve in S passing through x.

Proposition 9 *There exists $B > 0$ such that for all $x \in S$, the systole of $i_k^* g$ at x is no greater than B.*

Proof It suffices to prove the result in a neighborhood of every point of P. Choose $p \in P$. By Theorem 1.2 of [11], Σ is "asymptotically tubular" at p. That is, if $\hat{\imath}_k$ denotes the Gauss Lift of i_k, then, near p, $\hat{\imath}_k$ is a graph over a finite order tube (as defined in Sect. 2.1) of a section which converges to zero to every order near infinity. Formally, there exists a tube $(\hat{\jmath}, \mathbb{R} \times S^1)$ in $U\mathbb{H}^3$ of finite order; a smooth section σ of $\hat{\jmath}^* TU\mathbb{H}^3$ over $]0, \infty[\times S^1$; a neighbourhood Ω of $p \in S$ homeomorphic to the disk; and a homeomorphism $\alpha :]0, \infty[\times S^1 \to \Omega \setminus \{p\}$ such that:

(1) $\alpha(s, \theta)$ tends to p as s tends to $+\infty$;
(2) for all $(s, \theta) \in]0, \infty[\times S^1$, $(\hat{\imath}_k \alpha)(s, \theta) = \mathrm{Exp}(\sigma(s, \theta))$, where Exp here denotes the exponential map of the total space of $U\mathbb{H}^3$; and
(3) for all $k \geqslant 0$, $\nabla^k \sigma(s, \theta)$ tends to 0 as $s \to +\infty$.

Since $S := \tilde{S} \setminus P$ is of hyperbolic type, in particular, it is not contractible. Thus, for all $(s, \theta) \in]0, \infty[\times S^1$, $\alpha(\{s\} \times S^1)$ is a closed curve passing through $\alpha(s, \theta)$ which is topologically non-trivial in S. Furthermore, $\mathrm{Length}(\alpha(\{s\} \times S^1); \hat{\imath}_k^* \hat{g})$ converges to $2\pi N$ as s tends to $+\infty$, where N is the order of the tube $(\hat{\jmath}, \mathbb{R} \times S^1)$, and $\hat{g}$

denotes the Sasaki metric over $\mathrm{U}\mathbb{H}^3$. In particular, upon reducing Ω if necessary, for all $x \in \Omega \setminus \{p\}$, we have:

$$\mathrm{Sys}_x(\hat{i}_k^*\hat{g}, x) \leqslant 2\pi N + 1.$$

Since $\pi : \mathrm{U}\mathbb{H}^3 \to \mathbb{H}^3$ is distance non-increasing, in particular, $i_k^*g = (\pi \circ \hat{i}_k)^*g \leqslant \hat{i}_k^*\hat{g}$. Thus:

$$\mathrm{Sys}(i_k^*g, x) \leqslant 2\pi N + 1.$$

Since $p \in P$ is arbitrary, the result follows. □

Proposition 10 *(S, i_k^*g) has finite area.*

Proof Choose $p \in P$. Let Ω be a neighbourhood of p in $\tilde{S}$ which is homeomorphic to the disk. Since i_k^*g is complete and has constant instrinsic curvature equal to $k - 1 < 0$, (S, i_k^*g) either has a cusp or a funnel at p. Since $\mathrm{Sys}(i_k^*g, x)$ is bounded, the singularity at p cannot be a funnel. It is therefore a cusp. In particular, it has finite area. Since $p \in P$ is arbitrary, we conclude that (S, i_k^*g) has finite area as desired. □

We now prove Theorem 1:

Proof The first assertion follows from Theorem 1.4 of [10] and Propositions 8 and 10. The second assertion follows from Proposition 7. The final assertion follows from Theorem 1.5 of [10], and this completes the proof. □

References

1. Aledo, J.A., Espinar, J.M.: A conformal representation for linear Weingarten surfaces in the de Sitter space. J. Geom. Phys. **57**(8), 1669–1677 (2007)
2. Ballmann, W., Gromov, M., Schroeder, V.: Manifolds of nonpositive curvature, Progress in Mathematics, vol. 61. Birkhäuser Boston Inc., Boston, MA (1985)
3. Corro, A.V., Martínez, A., Milán, F.: Complete flat surfaces with two isolated singularities in hyperbolic 3-space. J. Math. Anal. Appl. **366**(2), 582–592 (2010)
4. Costa C.J.: Example of a complete minimal immersion in $\mathbb{R}^3$ of genus one and three embedded ends. Bol. Soc. Bras. Mat. **15**(1–2), 47–54 (1984)
5. Gálvez, J.A., Martínez, A., Milán, F.: Complete constant Gaussian curvature surfaces in the Minkowski space and harmonic diffeomorphisms onto the hyperbolic plane. Tohoku Math. J. **55**(4), 467–614 (2003)
6. Huber, A.: On subharmonic functions and differential geometry in the large. Comment. Math. Helv. **52**, 13–72 (1957)
7. Labourie F.: Problèmes de Monge-Ampère, courbes pseudo-holomorphes et laminations (French). G.A.F.A. **7**, 496–534 (1997)
8. Labourie, F.: Un lemma de Morse pour les surfaces convexes (French). Invent. Math. **141**(2), 239–297 (2000)
9. Osserman R.: A survey of minimal surfaces. Courier Dover Publications, Mineola (2002)
10. Smith G.: Hyperbolic plateau problems, to appear in Geom. Dedicata
11. Smith, G.: Pointed *k*-surfaces. Bull. Soc. Math. France **134**(4), 509–557 (2006)

Loss of Initial Data Under Limits of Ricci Flows

Peter M. Topping

Abstract We construct a sequence of smooth Ricci flows on T^2, with standard uniform C/t curvature decay, and with initial metrics converging to the standard flat unit-area square torus g_0 in the Gromov-Hausdorff sense, with the property that the flows themselves converge not to the static Ricci flow $g(t) \equiv g_0$, but to the static Ricci flow $g(t) \equiv 2g_0$ of twice the area.

Keywords Ricci flow

1 Introduction

When tasked with starting a Ricci flow with singular initial data, the standard approach is to approximate the singular initial data by smooth initial Riemannian metrics, then to flow each of the smooth metrics and take a smooth limit of the smooth resulting flows. As an example, in [8, 11, 13, 14, 16] one flows so-called Ricci limit spaces, obtained as non-collapsed Gromov-Hausdorff limits (X, d_X) of sequences of smooth 3-manifolds (M_i, g_i) satisfying uniform lower Ricci bounds. Each (M_i, g_i) gives rise to a Ricci flow $(M_i, g_i(t))$ of one form or another, with uniform curvature bounds $|\mathrm{Rm}|_{g_i(t)} \leq C/t$ for $t \in (0, T)$, and Hamilton-Cheeger-Gromov compactness allows one to extract a subsequence that converges to a smooth limit Ricci flow $(M, g(t))$ for $t \in (0, T)$.

The challenge then is to show that the desired initial data is not lost in the limit $i \to \infty$. In other words, we require that the smooth limit Ricci flow $(M, g(t))$ converges weakly to the desired initial data (X, d_X) as $t \downarrow 0$. This amounts to showing that the Riemannian distance $d_{g(t)}$ has a uniform limit d_0 as $t \downarrow 0$, and that (M, d_0) is a metric space that is isometric to (X, d_X). In [14, 16], this was achieved by proving uniform lower Ricci bounds on the flows $g_i(t)$. In particular, the so-called

P. M. Topping (✉)
Mathematics Institute, University of Warwick, Coventry, UK
e-mail: P.M.Topping@warwick.ac.uk
URL: https://homepages.warwick.ac.uk/~maseq

T. Hoffmann et al. (eds.), *Minimal Surfaces: Integrable Systems and Visualisation*, Springer Proceedings in Mathematics & Statistics 349,
https://doi.org/10.1007/978-3-030-68541-6_15

double bootstrap technique of [15] gives the local lower Ricci control that implies the necessary control on the evolution of distances, and higher-dimensional versions in the presence of stronger curvature hypotheses can be found in [1, 10], with the closest analogue (also being purely local) in [9].

In this note we clarify that without uniform lower Ricci bounds this programme will fail completely in general. Indeed, this loss of initial data can occur even when (X, d_X) is a smooth manifold of any dimension $n \geq 2$ and the convergence is much stronger than Gromov-Hausdorff.

Theorem 1 *Let (T^2, g_0) be the standard flat square torus arising as the quotient $\mathbb{R}^2/\mathbb{Z}^2$, and let $d_0 : T^2 \times T^2 \to [0, \infty)$ be the Riemannian distance corresponding to g_0. Then there exists a sequence of smooth Ricci flows $g_i(t)$, $t \in [0, \infty)$ on T^2, with the property that $|K_{g_i(t)}| \leq c_0/t$ for some uniform c_0, all $i \in \mathbb{N}$, and all $t \in (0, \infty)$, and so that*

1. *$d_{g_i(0)} \to d_0$ uniformly as $i \to \infty$, but*
2. *the Ricci flows $g_i(t)$ converge smoothly locally on $T^2 \times (0, \infty)$ to the flat metric $2g_0$ of twice the volume.*

Here K_g denotes the Gauss curvature of a metric g.

Variations on Theorem 1 show that it is not even necessary for a limit Ricci flow and a limit initial metric to share the same conformal structure. What is most important is that the limit Ricci flow should just be *larger* than the limit initial metric.

Estimates of Hamilton-Perelman [7, 12] (see [15] for the local version) tell us that the uniform c_0/t decay on the curvature implies lower semicontinuity at $t = 0$ in the sense that

$$d_{g(t)}(x, y) \geq d_0(x, y) - \beta\sqrt{c_0 t},$$

for some universal $\beta < \infty$ and for $t \geq 0$. Clearly there is no analogous upper semicontinuity in this example, contrary to the situation in which there is a uniform lower Ricci bound.

In situations where we do have uniform estimates relating d_0 or $d_{g_i(0)}$ with $d_{g_i(t)}$ for $t > 0$, one might expect better behaviour. Indeed, one might be able to analyse the limit alone if one had a positive answer to the following, cf. [2, 13]:

Problem 1 Suppose (M, g_0) is a smooth compact Riemannian manifold and $g(t)$ is a smooth Ricci flow on M for $t \in (0, T)$ with the property that $d_{g(t)} \to d_{g_0}$ uniformly as $t \downarrow 0$. Is it then true that $g(t)$ extends to a smooth Ricci flow for $t \in [0, T)$ with $g(0) = g_0$?

Thus the question is whether attainment of initial data in a metric sense implies attainment of the initial data smoothly.

This problem is open even in extremely simple situations such as when (M, g_0) is the flat unit square torus as before. The problem is then to show that the only Ricci flow $g(t)$ on M for $t \in (0, \varepsilon)$ with $d_{g(t)} \to d_{g_0}$ uniformly as $t \downarrow 0$ is the stationary flow $g(t) \equiv g_0$. It is not even immediate that such a Ricci flow is conformally equivalent to g_0. What *is* currently understood in this situation, thanks to the work of Richard [13], is that if we additionally impose a hypothesis of a lower Ricci bound for $g(t)$ (equivalent here to a lower Gauss curvature bound) then we must indeed have $g(t) \equiv g_0$. The higher dimensional case is addressed in concurrent work of Deruelle, Schulze and Simon [3] in the presence of both an upper c_0/t curvature bound and a uniform lower Ricci bound.

2 The Construction

With respect to coordinates on T^2 coming from the Euclidean coordinates x, y, we can write $g_0 = dx^2 + dy^2$. Given an initial metric $u_0 g_0$, where $u_0 : T^2 \to (0, \infty)$ is smooth, there exists a unique subsequent Ricci flow $g(t)$ of the form $u g_0$, where the smooth function $u : T^2 \times [0, \infty) \to (0, \infty)$ solves

$$\begin{cases} \dfrac{\partial u}{\partial t} = \Delta \log u & \text{on } T^2 \times (0, \infty) \\ u(\cdot, 0) = u_0 & \text{on } T^2, \end{cases}$$

as described by Hamilton [6] (see [4, 18] for the general theory on surfaces). By Gauss-Bonnet, the area is constant, because

$$\frac{d}{dt} \text{Area}(T^2, g(t)) = -2 \int K \, dV = 0,$$

see e.g. [17, (2.5.8)]. It is often convenient to work with the function $v := \frac{1}{2} \log u$, which then satisfies the equation $\frac{\partial v}{\partial t} = e^{-2v} \Delta v$, and we view this solution lifted to $\mathbb{R}^2$ whenever convenient. A solution v whose norm is initially bounded by Λ retains this bound by the maximum principle, and parabolic regularity theory gives C^k bounds at, say, time $t = 1$, depending only on k and Λ. (See, for example, the discussion in Appendix B of [5].) In particular, by applying this estimate to rescaled solutions $(x, t) \mapsto v(\lambda x, \lambda^2 t)$, for $\lambda > 0$, we find that

$$|D^k v|(\cdot, t) \leq \frac{C(k, \Lambda)}{t^{k/2}},$$

and in particular we can control the Gauss curvature $K = -e^{-2v} \Delta v$ by

$$|K_{g(t)}| \leq \frac{c_0(\Lambda)}{t}, \tag{1}$$

(In fact, we always have $K_{g(t)} \geq -\frac{1}{2t}$, see e.g. [17, Corollary 3.2.5].)

The specific Ricci flows $g_i(t)$ will be determined by their initial data $g_i(0)$, which in turn will be chosen to be an appropriate i-dependent function times g_0. Thus as above we can write $g_i(t) = u_i(t)g_0$, for a one parameter family of functions $u_i(t) : T^2 \to (0, \infty)$. Our task is to choose the functions $u_i(0)$ appropriately.

We will choose the functions $u_i(0)$ to lie always within $[1, 2]$. As above, this property is then preserved by the flow, i.e. $u_i(t) \in [1, 2]$ throughout T^2 and for each $t \geq 0$.

For each i, consider the lattice L of points in T^2 represented by points $(a/i, b/i)$ in $\mathbb{R}^2$, where $a, b \in \{0, 1, \ldots, i-1\}$. For each pair of points in this lattice, choose a minimising geodesic connecting them within (T^2, g_0). Denote the union of the images of this finite number of geodesics by $\Sigma \subset T^2$. We ask that $u_i(0)$ takes the value 1 on the whole of Σ. We can then extend to a smooth function $u_i(0) : T^2 \to [1, 2]$ with almost-maximal area in the sense that $\text{Area}(T^2, g_i(0)) \geq 2 - 1/i$. (Note that the area would be exactly 2 if we could choose $u_i(0) \equiv 2$.) In particular,

$$\|u_i(0) - 2\|_{L^1(T^2, g_0)} \leq 1/i.$$

Since $g_i(0) \geq g_0$, the distance $d_{g_i(0)}(p, q)$ between any two points $p, q \in T^2$ with respect to $g_i(0)$ must be at least $d_0(p, q)$. On the other hand, we can always find a point P in the lattice L such that the distance from p to P is less than $1/i$ when measured with respect to g_0 or even with respect to $2g_0$ or $g_i(0)$. Similarly we can find a lattice point Q close to q. Since $u_i(0) = 1$ on Σ, the distance between lattice points with respect to $g_i(0)$ is equal to the distance with respect to g_0. Thus

$$\begin{aligned} d_{g_i(0)}(p, q) &\leq d_{g_i(0)}(p, P) + d_{g_i(0)}(P, Q) + d_{g_i(0)}(Q, q) \\ &\leq d_{g_0}(P, Q) + \frac{2}{i} \\ &\leq d_{g_0}(P, p) + d_{g_0}(p, q) + d_{g_0}(q, Q) + \frac{2}{i} \\ &\leq d_{g_0}(p, q) + \frac{4}{i} \end{aligned} \tag{2}$$

and we see that $d_{g_i(0)}(p, q)$ converges uniformly to $d_0(p, q)$ as $i \to \infty$, as required.

We now turn to the subsequent flows $g_i(t) = u_i(t)g_0$. By the discussion above, we have $\text{Area}(T^2, g_i(t)) = \text{Area}(T^2, g_i(0)) \geq 2 - 1/i$, or equivalently

$$\|u_i(t) - 2\|_{L^1(T^2, g_0)} \leq 1/i. \tag{3}$$

Moreover, the flows $g_i(t)$ satisfy a uniform Gauss curvature estimate $|K_{g_i(t)}| \leq \frac{c_0}{t}$, for some universal c_0, as required. Their conformal factors also enjoy uniform C^k bounds for any $k \in \mathbb{N}$ over $T^2 \times [\delta, \infty)$, any $\delta > 0$, where the bounds depend on k and δ. Thus a subsequence will converge smoothly locally on $T^2 \times (0, \infty)$ (as tensors) to a limit Ricci flow $g(t) = u(t)g_0$. By passing (3) to the limit, we find that $g(t) \equiv 2g_0$.

References

1. Bamler, R., Cabezas-Rivas, E., Wilking, B.: The Ricci flow under almost non-negative curvature conditions. Inventiones **217**, 95–126 (2019). http://arxiv.org/abs/1707.03002
2. Deruelle, A.: Private communication (2014)
3. Deruelle, A., Schulze, F., Simon, M.: On the regularity of Ricci flows coming out of metric spaces. http://arxiv.org/abs/1904.11870
4. Giesen, G., Topping, P.M.: Existence of Ricci flows of incomplete surfaces. Commun. Partial Differ. Equ. **36**, 1860–1880 (2011)
5. Giesen, G.: Instantaneously complete Ricci flows on surfaces. Ph.D. thesis, University of Warwick (2012)
6. Hamilton, R.S.: The Ricci flow on surfaces. Mathematics and general relativity (Santa Cruz, CA, : 71 Contemporary Mathematics, 237–262, p. 1988. American Mathematical Society, Providence, RI (1986)
7. Hamilton, R.S.: The formation of singularities in the Ricci flow. Surveys in differential geometry, Vol. II Cambridge, MA, : 7–136, p. 1995. Press, Cambridge, MA, Internat (1993)
8. Hochard, R.: Short-time existence of the Ricci flow on complete, non-collapsed 3-manifolds with Ricci curvature bounded from below. http://arxiv.org/abs/1603.08726v1
9. Hochard, R.: Théorèmes d'existence en temps court du flot de Ricci pour des variétés non-complètes, non-éffondrées, à courbure minorée. Ph.D. thesis, Université de Bordeaux (2019)
10. Lai, Y.: Ricci flow under local almost non-negative curvature conditions., Adv. Math. **343**, 353–392 (2019). https://doi.org/10.1016/j.aim.2018.11.006
11. McLeod, A.D., Topping: Global regularity of three-dimensional Ricci limit spaces. T.A.M.S. (Series B) (To appear). http://arxiv.org/abs/1803.00414
12. Perelman, G.: The entropy formula for the Ricci flow and its geometric applications. http://arXiv.org/abs/math/0211159v1 (2002)
13. Richard, T.: Flot de Ricci sans borne supérieure sur la courbure et géométrie de certains espaces métriques. Université de Grenoble, Thesis (2012)
14. Simon, M.: Ricci flow of non-collapsed three manifolds whose Ricci curvature is bounded from below. Journal für die reine und angewandte Mathematik. **662**, 59–94 (2012)
15. Simon, M., Topping, P.M.: Local control on the geometry in 3D Ricci flow. J. Differ. Geom. (To appear). http://arxiv.org/abs/1611.06137
16. Simon, M., Topping, P.M.: Local mollification of Riemannian metrics using Ricci flow, and Ricci limit spaces. Geom. Topol. (To appear). http://arxiv.org/abs/1706.09490
17. Topping, P.M. : Lectures on the Ricci flow. L.M.S. Lect. Notes Ser. **325** C.U.P. (2006). http://www.warwick.ac.uk/~maseq/RFnotes.html
18. Topping, P.M.: Uniqueness of instantaneously complete Ricci flows. Geom. Topol. **19**, 1477–1492 (2015)

Semi-discrete Maximal Surfaces with Singularities in Minkowski Space

Masashi Yasumoto

Abstract We investigate semi-discrete maximal surfaces with singularities in Minkowski 3-space. In the smooth case, maximal surfaces (spacelike surfaces with mean curvature identically 0) in Minkowski 3-space admit a Weierstrass-type representation and they generally have singularities. In this paper, we first describe semi-discrete isothermic maximal surfaces in Minkowski 3-space and give a Weierstrass-type representation for them determined from integrable system principles. Furthermore, we show that semi-discrete isothermic maximal surfaces admit associated one-parameter families of deformations whose mean curvature remains identically 0. Finally we give a criterion that naturally describes the unified scheme of the "singular set" for these semi-discrete maximal surfaces, including the associated family.

Keywords Discrete differential geometry · Surface theory · Weierstrass-type representation · Singularity

1 Introduction

1.1 Overview

The study of surfaces with certain curvature conditions is a classical and important research subject in differential geometry. In particular, since minimal surfaces and constant mean curvature (CMC, for short) surfaces have a variety of important properties such as curvature conditions, variational characterizations, integrabilities and so on, they are central objects in this subject.

In this paper we consider spacelike surfaces with vanishing mean curvature in 3-dimensional Minkowski space $\mathbb{R}^{2,1}$, which are called (spacelike) *maximal surfaces*. Like in the case of minimal surfaces in the Euclidean 3-space $\mathbb{R}^3$, maximal surfaces

M. Yasumoto (✉)
Osaka City University Advanced Mathematical Institute, 3-3-138, Sugimoto, Sumiyoshi-ku, Osaka 558-8585, Japan
e-mail: yasumoto@sci.osaka-cu.ac.jp

T. Hoffmann et al. (eds.), *Minimal Surfaces: Integrable Systems and Visualisation*, Springer Proceedings in Mathematics & Statistics 349,
https://doi.org/10.1007/978-3-030-68541-6_16

in $\mathbb{R}^{2,1}$ are critical points of total area, more precisely, they are locally maximizers of their area. Also, a Weierstrass-type representation for conformal maximal surfaces in $\mathbb{R}^{2,1}$ was derived by Kobayashi [10], which is similar to the one for conformal minimal surfaces. With this formula, our next task is to construct concrete examples and to analyze their global behaviors. A major difference between minimal and maximal surfaces is that maximal surfaces in $\mathbb{R}^{2,1}$ generally have singularities. This property naturally leads us to considering maximal surfaces with *singularities*, and there are a large number of interesting examples of maximal surfaces with singularities. Furthermore, recent progress on the treatment of singularities using differential geometric approaches enables us to analyze what types of singularities appear on maximal surfaces. Details can be found in [5–7, 10, 11, 19] for example.

Apart from the smooth case, the related research field of *discrete differential geometry* has been developed by various approaches. Here we briefly review recent progress on this subject. In particular, Bobenko and Pinkall [1] described a discrete version of isothermic surfaces (surfaces with conformal curvature line coordinates) in $\mathbb{R}^3$ and they derived a Weierstrass representation for discrete isothermic minimal surfaces. More recently, by Burstall, Hertrich-Jeromin, Rossman [3], a discrete surface theory in other 3-dimensional spaceforms based on integrable transformations has been described. Due to these works, we are able to treat more general discrete surfaces that are not necessarily discrete isothermic. Continuing from the work in [1], during the last decade, *semi-discrete surface theory* has been established. It is expected that this theory plays a pivotal role in understanding similarities and differences between discrete and smooth objects. Müller and Wallner [15] (see also [14]) introduced semi-discrete isothermic surfaces and semi-discrete minimal surfaces in $\mathbb{R}^3$, and Rossman and the author [17] gave a Weierstrass representation for semi-discrete isothermic minimal surfaces in $\mathbb{R}^3$. Furthermore, initiated by [4, 8, 12] (see also [13]), a discrete or semi-discrete surface theory independent of the choice of "coordinate system" has been launched.

1.2 Previous Works

Here we briefly review smooth maximal surfaces. For more theory, see [10, 19, 20], for example. Let

$$\mathbb{R}^{2,1} := (\{(x_1, x_2, x_0)^t | x_j \in \mathbb{R}\}, \langle \cdot, \cdot \rangle)$$

be 3-dimensional Minkowski space with the Lorentz metric

$$\langle (x_1, x_2, x_0)^t, (y_1, y_2, y_0)^t \rangle = x_1 y_1 + x_2 y_2 - x_0 y_0,$$

and $\|X\|^2 = \langle X, X \rangle$ for all $X \in \mathbb{R}^{2,1}$. Note that $\|X\|^2$ can be zero or negative. For fixed $d \in \mathbb{R}$ and vector $n \in \mathbb{R}^{2,1} \setminus \{0\}$, a plane $\mathcal{P} = \{X \in \mathbb{R}^{2,1} \mid \langle X, n \rangle = d\}$ is *spacelike* or *timelike* or *lightlike* when n is timelike or spacelike or lightlike, respectively. Furthermore, a smooth surface in $\mathbb{R}^{2,1}$ is spacelike if its tangent planes are spacelike.

Let $f : \Sigma \to \mathbb{R}^{2,1}$ be a conformal immersion, where Σ is a simply-connected domain in $\mathbb{C}$ with complex coordinate $z = u + \sqrt{-1}v$ $(u, v \in \mathbb{R})$. f is a maximal surface if it is spacelike (which follows automatically from the conformality condition) with mean curvature identically 0. Defining

$$\mathbb{H}^2_+ := \{X = (x_1, x_2, x_0)^t \in \mathbb{R}^{2,1} | \langle X, X\rangle = -1,\ x_0 > 0\},$$

$$\mathbb{H}^2_- := \{X = (x_1, x_2, x_0)^t \in \mathbb{R}^{2,1} | \langle X, X\rangle = -1,\ x_0 < 0\},$$

we have the following proposition, analogous to the case of smooth minimal surfaces in $\mathbb{R}^3$ (and having a similar proof):

Proposition 1 *Away from umbilic points, smooth maximal surfaces lie in the class of isothermic surfaces. In addition, a spacelike immersion $f = f(u, v)$ is a maximal surface if and only if it has a dual surface f^*, solving*

$$f^*_u = \frac{f_u}{\|f_u\|^2},\quad f^*_v = -\frac{f_v}{\|f_v\|^2},$$

contained in $\mathbb{H}^2_+ \cup \mathbb{H}^2_-$. This dual surface is the Gauss map of the maximal surface.

We can locally construct any smooth maximal surface f with isothermic coordinates u, v from a smooth holomorphic function $g : \Sigma \to \mathbb{C}$ by solving

$$f_u = \mathrm{Re}\left(\frac{1+g^2}{g_u}, \frac{\sqrt{-1}(1-g^2)}{g_u}, -\frac{2g}{g_u}\right)^t,$$

$$f_v = -\mathrm{Re}\left(\frac{1+g^2}{g_v}, \frac{\sqrt{-1}(1-g^2)}{g_v}, -\frac{2g}{g_v}\right)^t.$$

The function g is stereographic projection of the Gauss map. On the other hand, unlike the case of minimal surfaces in $\mathbb{R}^3$, smooth maximal surfaces in $\mathbb{R}^{2,1}$ have singularities when $|g| = 1$. Furthermore, we can construct a one-parameter family f^θ of an isothermic maximal surface f by solving

$$f^\theta_u = \mathrm{Re}\left(e^{\sqrt{-1}\theta}\left(\frac{1+g^2}{g_u}, \frac{\sqrt{-1}(1-g^2)}{g_u}, -\frac{2g}{g_u}\right)\right)^t,$$

$$f^\theta_v = -\mathrm{Re}\left(e^{\sqrt{-1}\theta}\left(\frac{1+g^2}{g_v}, \frac{\sqrt{-1}(1-g^2)}{g_v}, -\frac{2g}{g_v}\right)\right)^t \quad (\theta \in [0, 2\pi)).$$

Note that f^θ is not necessarily isothermic but still a maximal surface. This family is called the associated family of f.

1.3 Main Results

In this paper, we briefly review a theory of semi-discrete isothermic surfaces in $\mathbb{R}^{2,1}$ and describe a particular class of semi-discrete isothermic surfaces called semi-discrete isothermic maximal surfaces. In order to introduce our theorems, here we set up several terminologies and symbols. Let $x : \mathbb{Z} \times \mathbb{R}$ (or some subdomain) $\to \mathbb{R}^{2,1}$ be a semi-discrete surface parametrized by $k \in \mathbb{Z}$, $t \in \mathbb{R}$. In this paper we abbreviate

$$x = x(k,t),\ x_1 := x(k+1,t),\ \partial x = x' := \frac{dx}{dt},\ \Delta x := x_1 - x,$$

and $[x, x_1]$ denotes the edge with the two endpoints x and x_1.

Using the notion of Christoffel transforms for semi-discrete isothermic surfaces in $\mathbb{R}^{2,1}$, we derive a Weierstrass-type representation for semi-discrete isothermic maximal surfaces, which is our first result:

Theorem 1 *Any semi-discrete isothermic maximal surface x can be locally constructed using semi-discrete holomorphic functions g (defined in Sect. 2 here, along with the associated functions τ, σ, with the same domain as for x) to the complex plane $\mathbb{C}$ by solving*

$$\begin{aligned}\partial x &= -\frac{\tau}{2}\mathrm{Re}\left(\frac{1+g^2}{g'}, \frac{\sqrt{-1}(1-g^2)}{g'}, -\frac{2g}{g'}\right)^t =: -\frac{\tau}{2}\mathrm{Re}(\phi),\\ \Delta x &= \frac{\sigma}{2}\mathrm{Re}\left(\frac{1+gg_1}{\Delta g}, \frac{\sqrt{-1}(1-gg_1)}{\Delta g}, -\frac{g+g_1}{\Delta g}\right)^t =: \frac{\sigma}{2}\mathrm{Re}(\psi).\end{aligned} \tag{1}$$

Furthermore, semi-discrete isothermic maximal surfaces in $\mathbb{R}^{2,1}$ admit one-parameter families x^θ of semi-discrete surfaces having vanishing mean curvature by multiplying ϕ and ψ by $\lambda = e^{\sqrt{-1}\theta}$ ($\theta \in [0, 2\pi)$), and solving

$$\partial x^\theta = -\frac{\tau}{2}\mathrm{Re}(\lambda\phi),\quad \Delta x^\theta = \frac{\sigma}{2}\mathrm{Re}(\lambda\psi)\,. \tag{2}$$

In Sect. 5. it is shown that the mean curvature of x^θ is identically 0. Such a family is called an associated family of a semi-discrete isothermic maximal surface.

As mentioned before, smooth maximal surfaces generally have singularities. So it is natural to expect that semi-discrete maximal surfaces generally have certain "singularities". In the discrete case, it was shown in [20] that discrete isothermic maximal surfaces in $\mathbb{R}^{2,1}$ generally had singularities (singularities of discrete isothermic maximal surfaces are called *singular faces*), and we analyzed such singularities. More recently, Lam and the author [13] investigated another type of discrete maximal surface in $\mathbb{R}^{2,1}$ and analyzed their singularities.

In the smooth case, the condition for singular points of maximal surfaces to appear is invariant under the deformation parameter λ of the associated family. So the natural

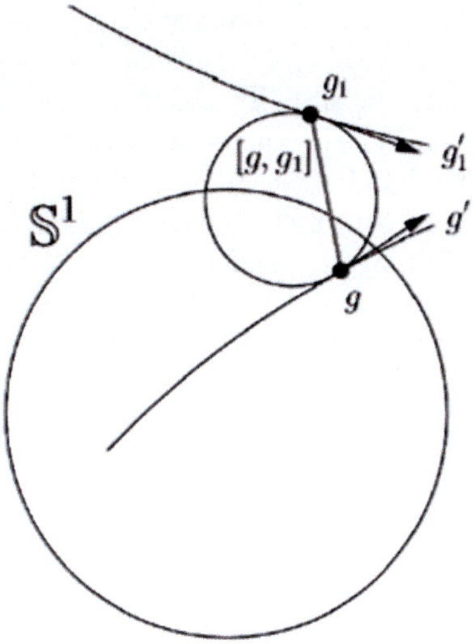

Fig. 1 An example for which the tangent circle $\mathscr{C}$ through g, g_1 intersects $\mathbb{S}^1$, and thus for which the singular edge $[x, x_1]$ appears

question arising from this property is to find a characterization of singularities of the associated family x^θ given by Eq. (2). Singularities of x^θ are called *singular edges* (Sect. 6), and we show the following result:

Theorem 2 *Let* $g : \mathbb{Z} \times \mathbb{R} \to \mathbb{C}$ *be a semi-discrete holomorphic function, and let* x^θ *be a semi-discrete maximal surface determined from* g*. Then the edge* $[x^\theta, x_1^\theta]$ *is a singular edge if and only if the tangent circle* $\mathscr{C}$ *at* g, g_1 *intersects* $\mathbb{S}^1 = \{z \in \mathbb{C} \mid |z| = 1\}$.

As will be seen in Sect. 6, any semi-discrete maximal admits a unique "unit normal vector field" from $\mathbb{Z} \times \mathbb{R}$ to the unit 2-sphere $\mathbb{S}^2 \subset \mathbb{R}^3$ globally. Thus semi-discrete maximal surfaces can be regarded as a semi-discrete version of "frontals" that often appear in the context of singularity theory (Fig. 1).

2 Semi-discrete Isothermic Surfaces in $\mathbb{R}^{2,1}$

Here we define semi-discrete conjugate and circular nets in $\mathbb{R}^{2,1}$. The notion of semi-discrete surfaces here is almost the same as in $\mathbb{R}^3$ (see [15]). In Definition 2, "circle" denotes the intersection of a translated light cone $\mathbb{L}^2_p$ and a plane, where $\mathbb{L}^2_p := \{x \in \mathbb{R}^{2,1} | \langle x - p, x - p \rangle = 0\}$ for fixed $p \in \mathbb{R}^{2,1}$.

Definition 1 Let $x : \mathbb{Z} \times \mathbb{R} \to \mathbb{R}^{2,1}$ be a semi-discrete surface.

- x is a semi-discrete conjugate net if ∂x, Δx and $\partial \Delta x$ are linearly dependent.
- x is a semi-discrete circular net if there exists a circle $\mathscr{C}$ passing through x and x_1 that is tangent to ∂x, ∂x_1 there (for all k, t).

Semi-discrete circular nets can be regarded as a semi-discrete version of space-like surfaces in $\mathbb{R}^{2,1}$ parametrized by curvature line coordinates. So semi-discrete isothermic surfaces must belong to the class of semi-discrete circular nets. In [15], semi-discrete isothermic surfaces are defined as semi-discrete circular nets having Christoffel duals. It is a natural description of semi-discrete isothermic surfaces, since Christoffel dualizability characterizes isothermic surfaces in the smooth case. Following [2], we can derive another characterization of semi-discrete isothermic surfaces in $\mathbb{R}^{2,1}$. Identifying elements in $\mathbb{R}^{2,1}$ with 2×2 matrices by the following identification

$$\mathbb{R}^{2,1} \ni (x_1, x_2, x_0)^t \cong \begin{pmatrix} \sqrt{-1}x_0 & x_1 - \sqrt{-1}x_2 \\ x_1 + \sqrt{-1}x_2 & -\sqrt{-1}x_0 \end{pmatrix} \in \mathrm{su}_{1,1},$$

where $\mathrm{su}_{1,1}$ is the Lie algebra of the Lie group $\mathrm{SU}_{1,1}$, we have the following equivalent condition for a semi-discrete conjugate net to be semi-discrete circular. For simplicity, for a given semi-discrete surface x, $X : \mathbb{Z} \times \mathbb{R} \to \mathrm{su}_{1,1}$ is expressed as a 2×2 matrix corresponding to x, that is, we use the following expression

$$\begin{aligned}\mathbb{R}^{2,1} &\ni x(k,t) = (x_1(k,t), x_2(k,t), x_0(k,t))^t \\ &\cong \begin{pmatrix} \sqrt{-1}x_0(k,t) & x_1(k,t) - \sqrt{-1}x_2(k,t) \\ x_1(k,t) + \sqrt{-1}x_2(k,t) & -\sqrt{-1}x_0(k,t) \end{pmatrix} = X(k,t) \in \mathrm{su}_{1,1}.\end{aligned}$$

Using the notion of *tangent cross ratio*, we have the following proposition.

Proposition 2 *Let x be a semi-discrete conjugate net in $\mathbb{R}^{2,1}$. Then x is semi-discrete circular if and only if*

$$Q(x, x_1) := \partial X \cdot (\Delta X)^{-1} \cdot \partial X_1 \cdot (\Delta X)^{-1} = cI,$$

where I is the 2×2 identity matrix, $c \in \mathbb{R}$ and $c < 0$. If x satisfies the above condition, c is called a tangent cross ratio *of x.*

As a particular class of semi-discrete circular nets, semi-discrete isothermic surfaces in $\mathbb{R}^{2,1}$ are as follows:

Definition 2 Let x be a semi-discrete circular net. Then x is called *semi-discrete isothermic* if the tangent cross ratio c of x is $c = -\dfrac{\tau(t)}{\sigma(k)} < 0$, where $\tau(t)$ (resp. $\sigma(k)$) is a positive scalar function depending only on the smooth parameter t (resp. discrete parameter k). In particular, a semi-discrete isothermic surface g is a *semi-discrete holomorphic function* if the image of g lies in $\mathbb{R}^2 \cong \mathbb{C}$.

Remark 1 We have two remarks here.

1. Definition 2 is slightly different from the original one in [15]. In [15], semi-discrete isothermic surfaces in $\mathbb{R}^3$ are defined as semi-discrete circular nets satisfying $\langle \Delta x, \Delta x \rangle = \sigma \nu \nu_1$, $\langle \partial x, \partial x \rangle = \tau \nu^2$, where ν is a positive-valued function

depending on (k, t). On the other hand, it was shown in [2] that this property was equivalent to Definition 2. Here we define semi-discrete isothermic surfaces in $\mathbb{R}^{2,1}$ as in Definition 2.

2. For any semi-discrete holomorphic function g, Definition 2 is equivalent to

$$\frac{g' \cdot g_1'}{(\Delta g)^2} = -\frac{\tau}{\sigma}. \tag{3}$$

In the smooth case, as already mentioned before, maximal surfaces in $\mathbb{R}^{2,1}$ lie in the class of isothermic surfaces. Furthermore, here we define a semi-discrete version of Christoffel duals of isothermic surfaces.

Definition 3 Lat x be a semi-discrete isothermic surface in $\mathbb{R}^{2,1}$ with tangent cross ratio $-\tau/\sigma$. Then x^* is called a *Christoffel dual* (or, dual surface) of x if x^* solves

$$\partial x^* = -\tau \frac{\partial x}{\|\partial x\|^2}, \quad \partial x^* = \sigma \frac{\Delta x}{\|\Delta x\|^2}.$$

Remark 2 There always exists a Christoffel dual x^* of x, and x^* is also semi-discrete isothermic with the same tangent cross ratio as for x.

In the smooth and discrete cases, smooth or discrete isothermic maximal surfaces in $\mathbb{R}^{2,1}$ can be characterized as having Christoffel duals that are inscribed in $\mathbb{H}^2_+ \cup \mathbb{H}^2_-$ (see [20] and Proposition 1). Motivated by this, we define semi-discrete isothermic maximal surfaces as follows[1]:

Definition 4 A semi-discrete isothermic surface x is *semi-discrete isothermic maximal* if its dual x^* can be inscribed in $\mathbb{H}^2_+ \cup \mathbb{H}^2_-$.

3 Proof of Theorem 1

We start by proving the first half of Theorem 1. Let g be a semi-discrete holomorphic function with tangent cross ratio $-\dfrac{\tau}{\sigma}$. First we temporarily assume that $|g| \neq 1$, and then

$$x^* := \frac{1}{|g|^2 - 1} \left(-2\mathrm{Re}(g), -2\mathrm{Im}(g), |g|^2 + 1\right)^t \in \mathbb{H}^2_+ \cup \mathbb{H}^2_- \subset \mathbb{R}^{2,1} \cong \mathbb{C} \times \mathbb{R}$$

is semi-discrete isothermic, because x^* is the image of g under the inverse of stereographic projection. (Here, the symbol "$\cong$" is used because we are identifying $\mathbb{R}^{2,1}$ and $\mathbb{C} \times \mathbb{R}$.) Then

[1] We will define mean curvatures of general semi-discrete surfaces later. More details can be found in Sect. 5.

$$\partial x^* \cong \frac{2}{(|g|^2-1)^2}\begin{pmatrix} g' + \overline{g'} g^2 \\ -(g'\bar{g} + \overline{g'} g) \end{pmatrix},$$
$$\Delta x^* \cong \frac{2}{(|g|^2-1)(|g_1|^2-1)}\begin{pmatrix} \Delta g + \overline{\Delta g} g g_1 \\ -(\Delta g \bar{g_1} + \overline{\Delta g} g) \end{pmatrix}.$$

It follows that $\|x^*\|^2 = \dfrac{4|g'|^2}{(|g|^2-1)^2}$, $\|\Delta x^*\|^2 = \dfrac{4|\Delta g|^2}{(|g|^2-1)(|g_1|^2-1)}$, and

$$-\tau\frac{\partial x^*}{\|\partial x^*\|^2} \cong -\frac{\tau}{2|g'|^2}\begin{pmatrix} g' - \bar{g}' g^2 \\ g'\bar{g} + \bar{g}' g \end{pmatrix} = -\frac{\tau}{2}\begin{pmatrix} \mathrm{Re}\left(\frac{1+g^2}{g'}\right) + \sqrt{-1}\mathrm{Re}\left(\frac{\sqrt{-1}(1-g^2)}{g'}\right) \\ -\mathrm{Re}\left(\frac{2g}{g'}\right) \end{pmatrix}$$
$$\cong -\frac{\tau}{2}\mathrm{Re}\left(\frac{1}{g'}\begin{pmatrix} 1+g^2 \\ \sqrt{-1}(1-g^2) \\ -2g \end{pmatrix}\right) = \partial x.$$

Similarly, we can compute

$$\frac{\sigma}{\|\Delta x^*\|^2}\Delta x^* = \frac{\sigma}{2}\mathrm{Re}\left(\frac{1}{\Delta g}\begin{pmatrix} 1+gg_1 \\ \sqrt{-1}(1-gg_1) \\ -(g+g_1) \end{pmatrix}\right) = \Delta x.$$

Thus if x solving (1) exists, x and x^* are dual to each other, and x will be semi-discrete isothermic. Since x^* is inscribed in $\mathbb{H}^2_+ \cup \mathbb{H}^2_-$, x is semi-discrete isothermic maximal if x exists. To show existence of x, we need to show compatibility of the two equations in (1), and this amounts to showing that the two operators Δ and ∂ in (1) commute, that is,

$$\partial\left(\mathrm{Re}(\psi)\right) = \Delta\left(\mathrm{Re}(\phi)\right), \tag{4}$$

where

$$\psi := \frac{\sigma}{2\Delta g}(1+gg_1, \sqrt{-1}(1-gg_1), -(g+g_1))^t,$$
$$\phi := -\frac{\tau}{2\,g'}(1+g^2, \sqrt{-1}(1-g^2), -2g)^t. \tag{5}$$

For our later convenience, we show the following lemma.

Lemma 1 *Let ϕ and ψ be vector-valued functions in $\mathbb{C}^3$ described by g as in Eq. (5). Then*

$$\partial\psi = \Delta\phi \tag{6}$$

holds for any given semi-discrete holomorphic function g.

Proof Using Eq. (3), one can compute:

$$\text{Left-hand side of (6)} = \frac{\sigma}{2(\Delta g)^2}\begin{pmatrix} g'(1+g_1^2) - g_1'(1+g^2) \\ \sqrt{-1}g'(1-g_1^2) - \sqrt{-1}g_1'(1-g^2) \\ -2g'g_1 + 2gg_1' \end{pmatrix}$$

$$= \frac{\sigma}{2}\left(-\frac{\tau}{\sigma g' g_1'}\right)\begin{pmatrix} g'(1+g_1^2) - g_1'(1+g^2) \\ \sqrt{-1}g'(1-g_1^2) - \sqrt{-1}g_1'(1-g^2) \\ -2g'g_1 + 2gg_1' \end{pmatrix} = \text{right-hand side of (6)},$$

proving the lemma.

Now we prove the final sentence of Theorem 1. Let x be a semi-discrete maximal surface, and ψ be stereographic projection $\psi : \mathbb{H}^2_+ \cup \mathbb{H}^2_- \to \mathbb{C}$. Then by definition, there exists a dual x^* that is semi-discrete isothermic and inscribed in $\mathbb{H}^2_+ \cup \mathbb{H}^2_-$. Taking

$$g := \psi \circ x^*,$$

g is a semi-discrete holomorphic function. Then g produces x via Eq. (1), which completes the proof.

4 Examples of Semi-discrete Isothermic Maximal Surfaces

Example 1 We can construct a semi-discrete isothermic maximal Enneper surface by taking (Fig. 2)

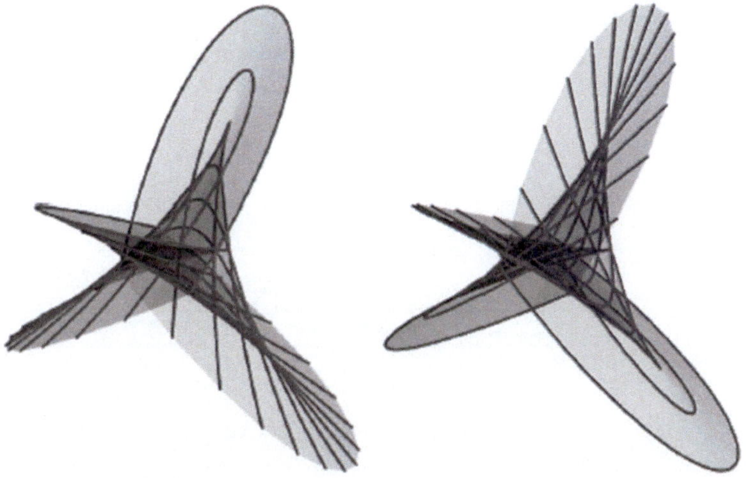

Fig. 2 Two semi-discrete maximal Enneper surfaces

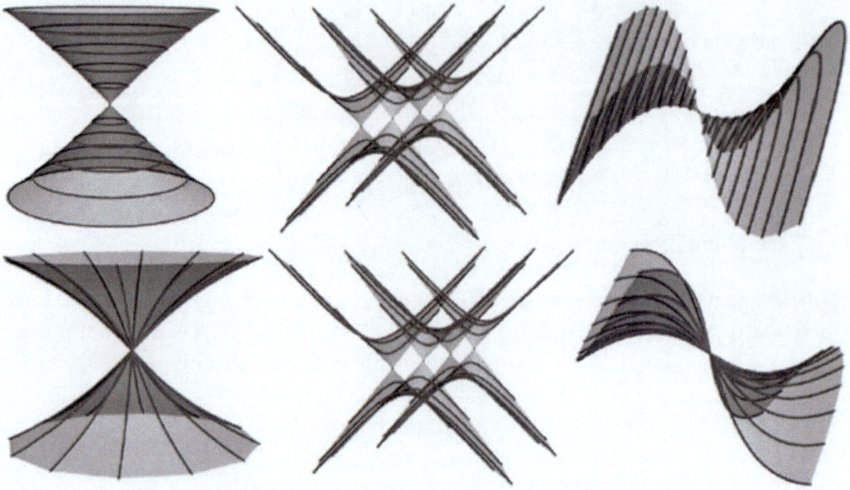

Fig. 3 At upper column: From left to right, semi-discrete isothermic maximal surfaces of revolution with discrete profile curve and timelike, spacelike, and lightlike axis respectively. At lower column: From left to right, semi-discrete isothermic maximal surfaces of revolution with smooth profile curve and timelike, spacelike, and lightlike axis respectively

$$g(k,t) = c(k + \sqrt{-1}t) \text{ (resp. } g(k,t) = c(k + \sqrt{-1}t)) \; (c \in \mathbb{R} \setminus \{0\})$$

in Theorem 1.

Example 2 We can construct semi-discrete isothermic maximal surfaces of revolution with timelike, spacelike, or lightlike axes. Pictures can be found in Fig. 3. In the semi-discrete case, we can consider two kinds of semi-discrete maximal surfaces of revolution as follows:

Table 1 Semi-discrete holomorphic functions of semi-discrete isothermic maximal surfaces of revolution

Rotational axis	Profile curve	Rotational direction	Semi-discrete holomorphic function
Timelike	Discrete	Smooth	$e^{-hk+\sqrt{-1}t}$
Timelike	Smooth	Discrete	$e^{-t+\sqrt{-1}hk}$
Spacelike	Discrete	Smooth	$\dfrac{-\sinh(hk)+\sqrt{-1}\sin(t)}{\cos(t)-\cosh(hk)}$
Spacelike	Smooth	Discrete	$\dfrac{-\sinh(hk)+\sqrt{-1}\sin(t)}{\cos(t)-\cosh(hk)}$
Lightlike	Discrete	Smooth	$1-\dfrac{4h}{2h(1+\sqrt{-1}t)+\sqrt{\beta}k}$
Lightlike	Smooth	Discrete	$\dfrac{(\sqrt{-1}+hk)^2+3\beta t^2}{1+h^2k^2-2\sqrt{3\beta}t+3\beta t^2}$

1. Semi-discrete maximal surfaces of revolution with discrete profile curve,
2. Semi-discrete maximal surfaces of revolution with smooth profile curve.

The corresponding semi-discrete holomorphic functions are listed in Table 1.

5 Associated Families of Semi-discrete Maximal Surfaces

In this section we discuss associated families of semi-discrete isothermic maximal surfaces in $\mathbb{R}^{2,1}$. The associated families of semi-discrete maximal surfaces in $\mathbb{R}^{2,1}$ are defined as follows:

Definition 5 Let x be a semi-discrete isothermic maximal surface obtained by solving Eq. (1). Then a one-parameter family x^θ of x defined by

$$\partial x^\theta := -\frac{\tau}{2}\mathrm{Re}(\lambda\phi), \quad \Delta x^\theta := \frac{\sigma}{2}\mathrm{Re}(\lambda\psi), \quad \lambda = e^{\sqrt{-1}\theta} \ (\theta \in [0, 2\pi)) \tag{7}$$

is called an *associated family* of x for $\theta \in [0, 2\pi)$.

Note that the compatibility condition $\partial(\Delta x^\theta) = \Delta(\partial x^\theta)$ holds. In fact, using Lemma 1, we have

$$\begin{aligned}
\partial(\Delta x^\theta) &= \partial\left(\mathrm{Re}\left((\cos\theta + \sqrt{-1}\sin\theta)\cdot(\mathrm{Re}(\psi) + \sqrt{-1}\mathrm{Im}(\psi))\right)\right)\\
&= \cos\theta\cdot\partial(\mathrm{Re}(\psi)) - \sin\theta\cdot\partial(\mathrm{Im}(\psi)) = \cos\theta\cdot\Delta(\mathrm{Re}(\phi)) - \sin\theta\cdot\Delta(\mathrm{Im}(\phi))\\
&= \Delta(\partial x^\theta).
\end{aligned}$$

Furthermore, by the same argument as in [4], Eq. (7) can be generically rewritten as

$$\begin{aligned}
\partial x^\theta &= -\frac{\tau}{\|\partial n\|^2}\left(\cos\theta\cdot\partial n + \sin\theta\cdot n\times\partial n\right),\\
\Delta x^\theta &= \frac{\sigma}{\|\Delta n\|^2}\left(\cos\theta\cdot\Delta n + \sin\theta\cdot n\times\Delta n\right).
\end{aligned} \tag{8}$$

Here we define curvatures of semi-discrete surfaces in $\mathbb{R}^{2,1}$. Curvatures of semi-discrete circular surfaces in $\mathbb{R}^3$ were first introduced in [9]. After that, Carl [4] introduced a curvature theory for more general classes of semi-discrete surfaces. Motivated by these works, here we formulate curvatures of semi-discrete surfaces in $\mathbb{R}^{2,1}$. We consider the following class of semi-discrete surfaces, and we define Gaussian and mean curvatures of such semi-discrete surfaces. The main idea comes from the previous works [4, 8].

Definition 6 Let $f : \mathbb{Z}\times\mathbb{R} \to \mathbb{R}^{2,1}$ be a semi-discrete surface and let $n : \mathbb{Z}\times\mathbb{R} \to \mathbb{H}^2_+ \cup \mathbb{H}^2_-$ be another semi-discrete surface with $(n + n_1)\times\partial n$ timelike. Then the pair (f, n) of f and n is called *coupled* if

$$\Delta f \perp (n + n_1) \quad \text{and} \quad \partial f \perp n$$

holds. For a given coupled pair (f, n), we define *partial derivatives* by

$$\partial_1 f := \Delta f,\ \partial_1 n := \Delta n,\ \partial_2 f := \frac{\partial f + \partial f_1}{2},\ \partial_2 n := \frac{\partial n + \partial n_1}{2},$$

and the first, second, and third fundamental forms of f are defined by

$$I := \begin{pmatrix} \langle \pi(\partial_1 f), \pi(\partial_1 f)\rangle & \langle \pi(\partial_1 f), \pi(\partial_2 f)\rangle \\ \langle \pi(\partial_1 f), \pi(\partial_2 f)\rangle & \langle \pi(\partial_2 f), \pi(\partial_2 f)\rangle \end{pmatrix}, \quad II := \begin{pmatrix} \langle \partial_1 f, \partial_1 n\rangle & \langle \partial_1 f, \partial_2 n\rangle \\ \langle \partial_2 f, \partial_1 n\rangle & \langle \partial_2 f, \partial_2 n\rangle \end{pmatrix},$$

$$III := \begin{pmatrix} \langle \partial_1 n, \partial_1 n\rangle & \langle \partial_1 n, \partial_2 n\rangle \\ \langle \partial_1 n, \partial_2 n\rangle & \langle \partial_2 n, \partial_2 n\rangle \end{pmatrix},$$

where $\pi(X) := X + \langle X, N\rangle N$ and $N := \dfrac{\partial_1 n \times \partial_2 n}{\sqrt{-\|\partial_1 n \times \partial_2 n\|^2}}$. Then

$$K := \det I^{-1} II, \quad H = \frac{1}{2}\mathrm{tr} I^{-1} II$$

are called the *Gaussian and mean curvatures* of f, respectively.

Remark 3 If (f, n) is coupled, the second fundamental form II of f is symmetric. The proof is similar to the one in [4].

Like in the case of semi-discrete minimal surfaces in $\mathbb{R}^3$, we show the following property. Note that $\partial_1 n \times \partial_2 n$ is not necessarily timelike. When considering curvatures of semi-discrete surfaces, we assume that $\partial_1 n \times \partial_2 n$ is timelike.

Theorem 3 *The x^θ described in Definition 5 has vanishing mean curvature.*

In order to show Theorem 3, we first introduce the wedge product of two vectors in $\mathbb{R}^{2,1}$ and its basic property.

Definition 7 *(and Lemma)* For $x = (x_1, x_2, x_0)^t$, $y = (y_1, y_2, y_0)^t \in \mathbb{R}^{2,1}$ and $\mathbf{e}_1 = (1, 0, 0)^t$, $\mathbf{e}_2 = (0, 1, 0)^t$, $\mathbf{e}_0 = (0, 0, 1)^t$,

$$x \times y := \begin{vmatrix} \mathbf{e}_1 & \mathbf{e}_2 & -\mathbf{e}_0 \\ x_1 & x_2 & x_0 \\ y_1 & y_2 & y_0 \end{vmatrix}$$

denotes the wedge product of x and y in $\mathbb{R}^{2,1}$, and $x \times y \perp x$ and y. And

$$(\alpha \times \beta) \times \gamma = -(\langle \alpha, \gamma\rangle \beta - \langle \beta, \gamma\rangle \gamma), \quad \|\alpha \times \beta\|^2 = -\|\alpha\|^2\|\beta\|^2 + \langle \alpha, \beta\rangle^2 \tag{9}$$

hold for all $\alpha, \beta, \gamma \in \mathbb{R}^{2,1}$.

In [4], Carl computed the mean curvature of semi-discrete isothermic minimal surfaces and their associated families by observing a geometric property. Here we prove Theorem 3 directly. The computation is tedious, but the advantage of this proof is that we can extend it to more general cases. In Definition 6, set

$$(f, n) = \left(x^\theta, n\right) \quad \left(n = \frac{1}{|g|^2 - 1}(-2\mathrm{Re}(g), -2\mathrm{Im}(g), |g|^2 + 1)^t\right).$$

By Eq. (8), we can easily check that (x^θ, n) is generically coupled. So we can compute the mean curvature of x^θ. By definition, the mean curvature H of x^θ is expressed by

$$H = \frac{\xi(k,t)}{\|\pi(\partial_1 x^\theta)\|^2\|\pi(\partial_2 x^\theta)\|^2 - \langle\pi(\partial_1 x^\theta), \pi(\partial_2 x^\theta)\rangle^2}, \text{ where}$$
$$\xi(k,t) = \|\pi(\partial_1 x^\theta)\|^2\langle\partial_2 x^\theta, \partial_2 n\rangle + \|\pi(\partial_2 x^\theta)\|^2\langle\partial_1 x^\theta, \partial_1 n\rangle$$
$$-2\langle\pi(\partial_1 x^\theta), \pi(\partial_2 x^\theta)\rangle\langle\partial_1 x^\theta, \partial_2 n\rangle.$$

Here we show Theorem 3. Our goal is to show that x^θ has vanishing mean curvature, so we only have to show that $\xi(k,t) = 0$ identically. First, we can calculate that

$$\|\pi(\partial_1 x^\theta)\|^2 = -\frac{\|\Delta x^\theta \times \{\Delta n \times (\partial n + \partial n_1)\}\|^2}{\|\Delta n \times (\partial n + \partial_1)\|^2},$$
$$\langle\pi(\partial_1 x^\theta), \pi(\partial_2 x^\theta)\rangle = \frac{1}{2\|\Delta n \times (\partial n + \partial_1)\|^2}\Big(\langle\Delta x^\theta, \partial x^\theta + \partial x_1^\theta\rangle\|\Delta n \times (\partial n + \partial_1)\|^2$$
$$-\langle\Delta x^\theta, \Delta n \times (\partial n + \partial n_1)\rangle\langle\partial x^\theta + \partial x_1^\theta, \Delta n \times (\partial n + \partial n_1)\rangle\Big),$$
$$\|\pi(\partial_2 x^\theta)\|^2 = -\frac{\|(\partial x^\theta + \partial x_1^\theta) \times \{\Delta n \times (\partial n + \partial n_1)\}\|^2}{4\|\Delta n \times (\partial n + \partial_1)\|^2},$$
$$\langle\partial_1 x^\theta, \partial_1 n\rangle = -\frac{\sigma\cos\theta}{2}, \ \langle\partial_2 x^\theta, \partial_2 n\rangle = -\frac{\tau}{4}\left\{2\cos\theta + \cos\theta\left(\frac{1}{\|\partial n\|^2} + \frac{1}{\|\partial n_1\|^2}\right)\right.$$
$$\left.+\sin\theta\left(\frac{\langle n \times \partial n, \partial n_1\rangle}{\|\partial n\|^2} + \frac{\langle n_1 \times \partial n_1, \partial n\rangle}{\|\partial n_1\|^2}\right)\right\},$$
$$\langle\partial_1 x^\theta, \partial_2 n\rangle = \frac{\sigma}{2\|\Delta n\|^2}\{\cos\theta\langle\Delta n, \partial n + \partial n_1\rangle + \sin\theta\langle n \times \Delta n, \partial n + \partial n_1\rangle\}.$$

Substituting $n = \dfrac{1}{|g|^2 - 1}\left(-2\mathrm{Re}(g), -2\mathrm{Im}(g), |g|^2 + 1\right)^t$ and Eqs. (3), (8) into $\xi(k,t)$, we can directly show that $\xi(k,t) = 0$, proving the theorem.

6 Singularities of Semi-discrete Maximal Surfaces

In this section we discuss singularities of semi-discrete maximal surfaces in $\mathbb{R}^{2,1}$. An interesting point is that singularities of associated families of semi-discrete maximal surfaces can be described by the direction of special vectors. More analysis on singularities of semi-discrete circular nets with Weierstrass-type representations can be found in [16, 21].

6.1 Singularities of Semi-discrete Isothermic Maximal Surfaces

In this subsection we discuss singularities of semi-discrete isothermic maximal surfaces. Note that, as in Fig. 4, even when we consider semi-discrete circular surfaces in $\mathbb{R}^3$, they may have configurations that appear to have "singularities". Unlike the case of discrete surfaces, semi-discrete surfaces can be considered as a discrete one-parameter family of smooth curves, and each curve may have singularities. We can also consider semi-discrete surfaces as a discrete one-parameter family of strips $\{(1-s)x_k + s \cdot x_{k+1} | s \in [0, 1]\}$, and we can imagine finding singularities of each strip in terms of the two smooth parameters t, s. In this sense, Müller, Wallner [15] showed that, when a semi-discrete surface in $\mathbb{R}^3$ is isothermic, these strips do not have singularities.

Returning to the case of $\mathbb{R}^{2,1}$, semi-discrete surfaces actually depend on only a discrete parameter k and a smooth parameter t, so we would like to define "singularities" of semi-discrete surfaces in a way that depends on only those two parameters. From this viewpoint, it is suitable to define singularities of semi-discrete surfaces as we do in Definition 8 below. At present, we do not have a general definition of singularities on semi-discrete surfaces, but in the case of semi-discrete maximal surfaces, we can define their singular edges as below. In the smooth case, except at singularities, the tangent planes at all points are spacelike, which leads to this natural description of singularities of semi-discrete maximal surfaces:

Definition 8 Let x be a semi-discrete maximal surface. Then the edge $[x, x_1]$ for some k, t is a *singular edge* if the plane $\mathcal{P}(x, x_1)$ spanned by $\{\partial x, \Delta x, \partial\Delta x\}$ is not spacelike.

Fig. 4 Parallel surfaces of a semi-discrete minimal Enneper surface in $\mathbb{R}^3$, with mesh shown and without mesh shown. This surface appears to have singularities

Here we prove Theorem 2 for semi-discrete isothermic maximal surfaces. Let g be a semi-discrete holomorphic function, and let x be the semi-discrete maximal surface given by g. Let $[x, x_1]$ be an edge for which the corresponding tangent circle $\mathscr{C}$ of g may or may not intersect $\mathbb{S}^1$. The dual surface x^* has tangent planes along edges that are parallel to those of x, so it suffices to consider whether the conic section passing through x^* and x_1^* lies in a non-spacelike plane.

Let $\mathscr{C} \subset \mathbb{C}$ be a circle tangent to g', g_1' at g, g_1 with center $a + \sqrt{-1}b$ and radius r, and let $[x^*, x_1^*]$ be the edge given by the image of the inverse of stereographic projection. Taking $p \in \mathscr{C}$, the image of the map

$$\mathbb{C} \ni p := x + \sqrt{-1}y \mapsto \left(-\frac{2x}{|p|^2 - 1}, -\frac{2y}{|p|^2 - 1}, \frac{|p|^2 + 1}{|p|^2 - 1}\right) \in \mathbb{H}^2_+ \cup \mathbb{H}^2_-$$

lies in the plane

$$\mathcal{P} := \{(x_1, x_2, x_0)^t \in \mathbb{R}^{2,1} | \langle (x_1, x_2, x_0)^t, N\rangle = -(a^2 + b^2 - r^2 - 1)\},$$

where $N = (-2a, -2b, a^2 + b^2 - r^2 + 1)^t$. $\mathcal{P}$ is not spacelike if and only if n is not timelike. By using the circularity condition of g, one can check that x^*, x_1^*, ∂x^* and ∂x_1^* lie in a plane parallel to

$$\mathcal{P} := \{(x_1, x_2, x_0)^t \in \mathbb{R}^{2,1} | \langle (x_1, x_2, x_0)^t, N\rangle = 0\}.$$

One can compute that

$$\langle N, N\rangle = -\{(a^2 + b^2) - (r - 1)^2\}\{(a^2 + b^2) - (r + 1)^2\},$$

so the condition that n is not timelike gives

$$|r - 1| \le \sqrt{a^2 + b^2} \le r + 1. \tag{10}$$

On the other hand, considering the distance between the origin and $a + ib$, and the radii of $\mathbb{S}^1$ and $\mathscr{C}$, one can check that $\mathscr{C}$ intersects $\mathbb{S}^1$ if and only if (10) holds, proving the theorem.

6.2 *Singularities of Associated Families of Semi-discrete Isothermic Maximal Surfaces*

In this subsection we further discuss singularities of associated families of semi-discrete maximal surfaces. Note that the associated family of a semi-discrete isothermic maximal surface is no longer semi-discrete conjugate nets. So the same argument does not work for the associated families.

In order to characterize singularities of semi-discrete maximal surfaces, we define a new type of vectors for semi-discrete maximal surfaces. These vectors can be regarded as normal vector fields of frontals (for the notion of frontals, see [18] for example).

Definition 9 Let x^θ be a semi-discrete maximal surface obtained by Eq. (8). A vector N defined on each edge is called an *edge normal* of $[x, x_1]$ if

$$\left|\left\langle \frac{\partial x^\theta \times \Delta x^\theta}{\sqrt{|\|\partial x^\theta \times \Delta x^\theta\|^2|}}, N \right\rangle\right| = \left|\left\langle \frac{\partial x_1^\theta \times \Delta x^\theta}{\sqrt{|\|\partial x_1^\theta \times \Delta x^\theta\|^2|}}, N \right\rangle\right|.$$

Furthermore, if N is not spacelike, the edge $[x, x_1]$ is called a *singular edge*.

Here we show Theorem 2. One can start from

$$\begin{aligned}
&\partial x^\theta \times \Delta x^\theta \parallel \alpha\beta(\cos\theta \cdot \partial n + \sin\theta \cdot (n \times \partial n)) \times (\cos\theta \cdot \Delta n + \sin\theta \cdot (n \times \Delta n)) \\
&= \cos^2\theta \cdot \partial n \times \Delta n + \sin\theta\cos\theta \cdot \partial n \times (n \times \Delta n) + \sin\theta\cos\theta \cdot (n \times \partial n) \times \Delta n \\
&+ \sin^2\theta \cdot (n \times \partial n) \times (n \times \Delta n).
\end{aligned}$$

Using Eq. (9), we have (Fig. 5)

$$\begin{aligned}
&\partial n \times (n \times \Delta n) = -(n \times \Delta n) \times \partial n = \langle n, \partial n\rangle \Delta n - \langle \Delta n, \partial n\rangle n = -\langle \Delta n, \partial n\rangle n, \\
&(n \times \partial n) \times \Delta = \langle \partial n, n \times \Delta n\rangle n - \langle n, n \times \Delta n\rangle = \langle \partial n, n \times \Delta n\rangle n \\
&(n \times \partial n) \times (n \times \Delta n) = \langle \partial n, n \times \Delta n\rangle n - \langle n, n \times \Delta n\rangle \partial n = \langle \partial n, n \times \Delta n\rangle n, \\
&(\partial n \times \Delta n) \times n = \langle \Delta n, n\rangle \partial n - \langle \partial n, n\rangle \Delta n = \langle \Delta n, n\rangle \partial n.
\end{aligned}$$

Furthermore, differentiating the identity $\langle n, n \times \Delta n\rangle = 0$ implies

$$\langle n, \partial n \times \Delta n\rangle = -\langle \partial n, n \times \Delta n\rangle.$$

Combining them, we have

$$\partial x^\theta \times \Delta x^\theta \parallel \cos^2\theta \cdot N - d\sin^2\theta \cdot n + \sin\theta\cos\theta \cdot (n \times N) \quad (\text{setting } \partial n \times \Delta n = c_1 N).$$

Therefore, we have

$$\left|\left\langle \frac{\partial x^\theta \times \Delta x^\theta}{\sqrt{|\|\partial x^\theta \times \Delta x^\theta\|^2|}}, N \right\rangle\right| = \sqrt{|\cos^2\theta \|N\|^2 - d^2\sin^2\theta|},$$

where $d := \langle n, N\rangle = \langle n_1, N\rangle$. Similarly,

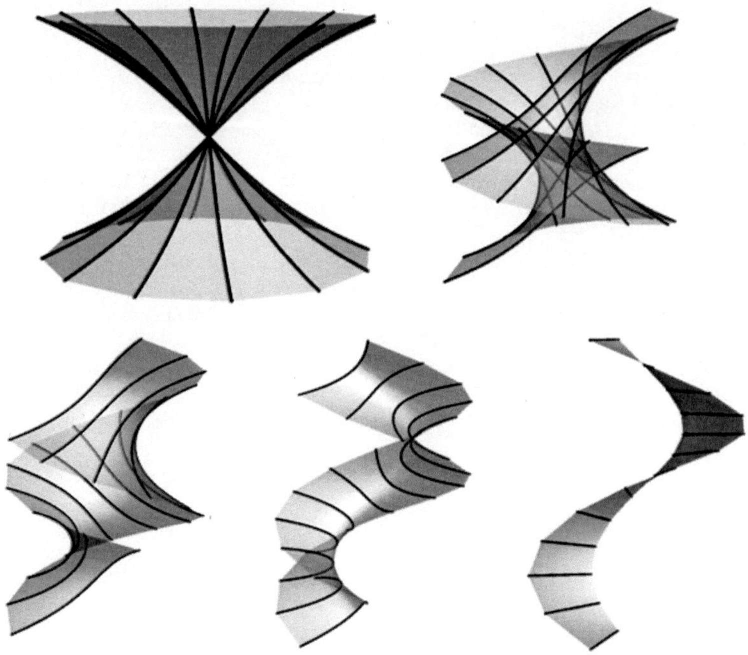

Fig. 5 The associated family of a semi-discrete maximal surface of revolution with timelike axis and smooth profile curve

$$\left|\left\langle \frac{\partial x_1^\theta \times \Delta x^\theta}{\sqrt{|\|\partial x_1^\theta \times \Delta x^\theta\|^2|}}, N \right\rangle\right| = \sqrt{|\cos^2\theta\|N\|^2 - d^2\sin^2\theta|}\,,$$

proving the theorem.

Acknowledgements The author thanks Wayne Rossman for valuable comments and helpful discussions, and the author was partly supported by the Grant-in-Aid for JSPS Fellows: Grant Number 26-3154, 19J02034, JSPS Grant-in-Aid for Scientific Research on Innovative Areas "Discrete Geometric Analysis for Materials Design": Grant Number 18H04489, JSPS/FWF bilateral joint project "Transformations and Singularities" between Austria and Japan, and Osaka City University Advanced Mathematical Institute (MEXT Joint Usage/Research Center on Mathematics and Theoretical Physics).

References

1. Bobenko, A.I., Pinkall, U.: Discrete isothermic surfaces. J. Reine Angew. Math. **475**, 187–208 (1996)
2. Burstall, F., Hertrich-Jeromin, U., Müller, C., Rossman, W.: Semi-discrete isothermic surfaces. Geom. Dedicata **183**, 43–58 (2016)

3. Burstall, F., Hertrich-Jeromin, U., Rossman, W.: Discrete linear Weingarten surfaces. Nagoya Math. J. **231**, 55–88 (2018)
4. Carl, W.: On semidiscrete constant mean curvature surfaces and their associated families. Monatsh. Math. **182**(3), 537–563 (2017)
5. Cho, J., Ogata, Y.: Deformation and singularities of maximal surfaces with planar curvature lines. Beitr. Algebra Geom. **59**(3), 465–489 (2018)
6. Fujimori, S., Kawakami, Y., Kokubu, M., Rossman, W., Umehara, M., Yamada, K.: Entire zero-mean curvature graphs of mixed type in Lorentz-Minkowski 3-space. Q. J. Math. **67**(4), 801–837 (2016)
7. Fujimori, S., Saji, K., Umehara, M., Yamada, K.: Singularities of maximal surfaces. Math. Z. **259**, 827–848 (2008)
8. Hoffmann, T., Sageman-Furnas, A.O., Wardetzky, M.: A discrete parametrized surface theory in $\mathbb{R}^3$. Int. Math. Res. Not. IMRN **14**, 4217–4258 (2017)
9. Karpenkov, O., Wallner, J.: On offsets and curvatures for discrete and semidiscrete surfaces. Beitr. Algebra Geom. **55**, 207–228 (2014)
10. Kobayashi, O.: Maximal surfaces in the 3-dimensional Minkowski space L^3. Tokyo J. Math. **6**(2), 297–309 (1983)
11. Kobayashi, O.: Maximal surfaces with conelike singularities. J. Math. Soc. Japan **36**(4), 609–617 (1984)
12. Lam, W.Y.: Discrete minimal surfaces: critical points of the area functional from integrable systems. Int. Math. Res. Not. IMRN **6**, 1808–1845 (2018)
13. Lam, W.Y., Yasumoto, M.: Trivalent maximal surfaces in Minkowski space. Springer Proc. Math. Stat. **211**, 169–184 (2018)
14. Müller, C.: Semi-discrete constant mean curvature surfaces. Math. Z. **279**(1–2), 459–478 (2015)
15. Müller, C., Wallner, J.: Semi-discrete isothermic surfaces. Results Math. **63**(3–4), 1395–1407 (2013)
16. Rossman, W., Yasumoto, M.: Semi-discrete linear Weingarten surfaces and their singularities in Riemannian and Lorentzian spaceforms, to appear in Osaka J. Math
17. Rossman, W., Yasumoto, M.: Weierstrass representation for semi-discrete minimal surfaces, and comparison of various catenoids. J. Math-for-Ind. **4B**, 109–118 (2012)
18. Saji, K., Umehara, M., Yamada, K.: The geometry of fronts. Ann. Math. **169–2**, 491–529 (2009)
19. Umehara, M., Yamada, K.: Maximal surfaces with singularities in Minkowski space. Hokkaido Math. J. **35**, 13–40 (2006)
20. Yasumoto, M.: Discrete maximal surfaces with singularities in Minkowski space. Differ. Geom. Appl. **43**, 130–154 (2015)
21. Yasumoto, M.: Semi-discrete surfaces of revolution. Kobe J. Math. **34**(1–2), 13–26 (2017)